Molecular Research in Obesity

Molecular Research in Obesity

Guest Editors

Zaida Abad-Jiménez
Teresa Vezza

MDPI

Basel • Beijing • Wuhan • Barcelona • Belgrade • Novi Sad • Cluj • Manchester

Guest Editors

Zaida Abad-Jiménez
Department of Endocrinology
and Nutrition
University Hospital Doctor
Peset
Valencia
Spain

Teresa Vezza
Servicio de Digestivo
Hospital Universitario
Virgen de las Nieves
Granada
Spain

Editorial Office
MDPI AG
Grosspeteranlage 5
4052 Basel, Switzerland

This is a reprint of the Special Issue, published open access by the journal *Biomedicines* (ISSN 2227-9059), freely accessible at: https://www.mdpi.com/journal/biomedicines/special_issues/GL45HVXB22.

For citation purposes, cite each article independently as indicated on the article page online and as indicated below:

Lastname, A.A.; Lastname, B.B. Article Title. *Journal Name* **Year**, *Volume Number*, Page Range.

ISBN 978-3-7258-3787-8 (Hbk)
ISBN 978-3-7258-3788-5 (PDF)
https://doi.org/10.3390/books978-3-7258-3788-5

Cover image courtesy of Makysm

Contents

Preface

Obesity is a major global health concern, contributing to the rising incidence of metabolic disorders such as type 2 diabetes, cardiovascular diseases, and chronic inflammation. Despite significant advancements in obesity research, many underlying molecular mechanisms still need to be fully elucidated. This Special Issue compiles original research articles and reviews that provide novel insights into the molecular pathways regulating adipose tissue function, energy metabolism, and systemic inflammation.

This Reprint will highlight the complex interplay between genetic predisposition, metabolic regulation, and environmental factors in the development of obesity. By presenting cutting-edge findings, we will advance understanding of obesity-related pathophysiology and pave the way for innovative therapeutic approaches.

We extend our sincere gratitude to all the authors for their valuable contributions, as well as to the reviewers for their insightful feedback. We also acknowledge the support of the editorial team in bringing this Reprint to completion. We hope that this collection will serve as a valuable resource for researchers, clinicians, and professionals dedicated to addressing the challenges of obesity and its related comorbidities.

Zaida Abad-Jiménez and Teresa Vezza
Guest Editors

biomedicines

MDPI

Editorial

Obesity: A Global Health Challenge Demanding Urgent Action

Zaida Abad-Jiménez [1,*] and Teresa Vezza [2,3,*]

1 Department of Endocrinology and Nutrition, Doctor Peset University Hospital—Foundation for the Promotion of Health and Biomedical Research in the Valencian Region (FISABIO), 46017 Valencia, Spain
2 Digestive System Service, Virgen de las Nieves University Hospital, 18014 Granada, Spain
3 Instituto de Investigación Biosanitaria (ibs.GRANADA), 18012 Granada, Spain
* Correspondence: zaida.abad@fisabio.es (Z.A.-J.); tvezza@ibsgranada.es (T.V.)

Obesity has become one of the most critical health crises of the modern era, affecting millions of individuals worldwide. In 2021, the World Health Organization (WHO) estimated that over 1 billion individuals were living with obesity, including approximately 650 million adults, 340 million adolescents, and 39 million children [1]. This complex, chronic disease has reached epidemic proportions, significantly contributing to the growing burden of non-communicable diseases. Far from being merely a condition of excess weight, obesity is a central driver of numerous severe health complications, including cardiovascular diseases, type 2 diabetes (T2D), hyperlipidemia, and an elevated risk of premature mortality [2].

The etiology of obesity is multifactorial, involving a dynamic interplay of genetic predispositions, environmental influences, and behavioral factors. Sedentary lifestyles, high-calorie diets, socioeconomic disparities, and urbanization have further exacerbated its prevalence. Equally pivotal are genetic factors, which determine individual susceptibility to obesity and its associated metabolic consequences [3]. These genetic variations help explain why some individuals develop obesity in obesogenic environments, while others remain unaffected [4]. Thus, understanding these factors is critical for advancing personalized preventive and therapeutic strategies. Moreover, disparities in obesity prevalence and its associated risks across demographic groups underscore the need for targeted interventions. For example, non-Hispanic White individuals and Black individuals display varying susceptibilities to obesity and related complications, such as T2D, highlighting the intricate interplay between genetic predispositions and social determinants of health [5,6].

Received: 6 February 2025
Accepted: 15 February 2025
Published: 18 February 2025

Citation: Abad-Jiménez, Z.; Vezza, T. Obesity: A Global Health Challenge Demanding Urgent Action. *Biomedicines* **2025**, *13*, 502. https://doi.org/10.3390/biomedicines13020502

Recent findings published in the Special Issue entitled "Molecular Research in Obesity" of *Biomedicines* provide crucial insights into the multifaceted nature of obesity. Highlights include the identification of specific genetic variants associated with increased risk, as well as novel biomarkers that could serve as targets for early intervention. Moreover, new research underscores the significant role of adipose tissue dysfunction in driving chronic inflammation and metabolic disturbances, offering potential therapeutic pathways for restoring energy homeostasis and reducing the burden of obesity-related comorbidities.

In addition, recent studies in this Special Issue also underscore the intricate relationship between the gut microbiota and obesity. A particularly intriguing study published shows that obesity can be linked to decreased levels of vitamin B12 [7], a micronutrient primarily synthesized by gut microbes. The researchers found that obese mice had significantly fewer enzymes involved in the biosynthesis of vitamin B12 compared to non-obese control. This enzyme reduction was even more pronounced in genetically obese mice, suggesting that the composition of the microbiome plays a key role in regulating vitamin B12 synthesis. As the researchers concluded, "the degree of obesity and the composition of the microbiota are the main factors influencing the expression of genes and pathways for vitamin B12 biosynthesis in the gut", emphasizing the potential for microbiome-based approaches to address nutrient deficiencies associated with obesity [7].

In a complementary approach, other studies in this Special Issue have examined the effects of postbiotics derived from *Lacticaseibacillus paracasei* strains on adipocyte metabolism [8]. These strains, isolated from *Formica rufa anthills*, demonstrated promising results in improving glucose uptake and enhancing lipid turnover in mature 3T3-L1 adipocytes without promoting excessive lipid accumulation. Among them, the P4 strain proved particularly effective in increasing the expression of key enzymes involved in beta-oxidation and adiponectin production, such as adipose triglyceride lipase and peroxisomal beta-oxidation enzymes, suggesting its potential to improve metabolic function in individuals with obesity.

Additionally, the role of inflammation in obesity-related metabolic disorders has been extensively studied and is widely recognized as a central factor driving the progression of these conditions. Chronic low-grade inflammation is a hallmark of obesity, contributing to the development of metabolic diseases such as type 2 diabetes, cardiovascular disease, and insulin resistance [9]. The increased fat mass in obesity triggers an inflammatory response that involves immune cells infiltrating adipose tissue, which in turn leads to the release of pro-inflammatory cytokines and adipokines. These molecules disrupt normal metabolic processes, further exacerbating obesity and its associated complications [10].

Several articles in this Special Issue provide novel insights into the molecular mechanisms underlying this inflammatory response, advancing our understanding of how obesity leads to systemic inflammation and metabolic dysfunction. One study in particular sheds light on the predictive role of certain fatty acids in modulating the leptin/adiponectin ratio, a key marker of adipose tissue dysfunction and metabolic imbalance [11]. The researchers identified that the 22:6n3 fatty acid, also known as docosahexaenoic acid (DHA), was strongly correlated with an improved adiponectin/leptin ratio in individuals with severe obesity. This finding is of particular interest because the leptin/adiponectin ratio is an important biomarker for evaluating adipose tissue function and metabolic health. Leptin, a hormone primarily produced by adipocytes, plays a crucial role in regulating energy balance [12], while adiponectin, another adipocyte-derived hormone, has anti-inflammatory and insulin-sensitizing effects [13]. In obese individuals, the elevated levels of leptin and reduced levels of adiponectin contribute to inflammation and insulin resistance [11].

The strong correlation between DHA and the adiponectin/leptin ratio suggests that DHA could serve as a potential predictive marker for inflammation in severe obesity. As an omega-3 polyunsaturated fatty acid, DHA has well-documented anti-inflammatory properties, and its role in modulating the inflammatory response in obesity may help prevent or reduce the metabolic disturbances associated with excessive adiposity [14]. Furthermore, this research paves the way for the development of targeted nutritional interventions that could enhance the intake of DHA or similar fatty acids to improve adipose tissue function and reduce inflammation, ultimately leading to better metabolic outcomes in obese individuals. Supporting this concept, previous studies have highlighted DHA as a potential biomarker for chronic inflammation in obesity. One such study compared eutrophic individuals with those suffering from obesity, revealing that DHA levels in PBMCs were inversely associated with markers of inflammation in obese individuals [15]. These findings reinforce the idea that DHA not only plays a critical role in modulating inflammation but may also serve as a key factor in improving metabolic health in obese individuals.

Another study underscores the intricate connection between obesity-induced inflammation and hypothalamic dysfunction, with a particular focus on the neuropeptide neurosecretory protein GM (NPGM) [16]. This research provides valuable insights into how

NPGM-expressing neurons in the hypothalamus play a critical role in regulating lipid metabolism and inflammatory responses, which are key processes in maintaining glucose homeostasis. The hypothalamus, a central brain region that controls energy balance, is increasingly recognized as an important mediator of the systemic metabolic disruptions observed in obesity [17]. In this context, the study suggests that NPGM neurons not only contribute to the regulation of lipid storage but also influence inflammation within the central nervous system, linking metabolic dysfunction with immune responses. The findings indicate that NPGM may act as a key mediator in the communication between the brain and peripheral tissues, modulating both lipid metabolism and the inflammatory state characteristic of obesity. By influencing these processes, NPGM has the potential to play a significant role in the dysregulation of energy homeostasis, contributing to the development of obesity-related metabolic disorders such as insulin resistance and type 2 diabetes [16]. This emerging understanding of NPGM's involvement in central regulation presents an exciting opportunity for developing new therapeutic strategies aimed at targeting NPGM pathways to control lipid storage and inflammation in the hypothalamus.

In addition to the well-documented roles of hypothalamic dysfunction, alterations in microbiota composition, and systemic inflammation, microRNAs have emerged as key regulators of metabolic health in obesity. MicroRNAs (miRNAs) are small non-coding RNA molecules that modulate gene expression and play a crucial role in various cellular processes, including lipid metabolism, insulin sensitivity, and inflammation [18]. One study in this Special Issue examines the circulating levels of miR-33a [19], a microRNA that is strongly associated with lipid metabolism and is thought to play a role in regulating cholesterol homeostasis and fat storage. The study found that miR-33a was significantly elevated in obese adolescents compared to their non-obese counterparts, suggesting that this microRNA may serve as a biomarker for obesity-related metabolic dysfunction. Furthermore, the study revealed that miR-33a levels correlated with key metabolic parameters, such as insulin resistance and cholesterol levels, further supporting the idea of miR-33a as a predictor of metabolic disturbances in obesity. Interestingly, although exosomal miR-33a expression was lower in obese individuals, it did not show any significant correlation with metabolic measures. This contrasts with the findings for circulating miR-33a, which appear to have stronger links to metabolic dysfunction.

These results highlight the potential of circulating miR-33a as an early indicator of metabolic dysfunction in obese adolescents, potentially identifying those at higher cardiometabolic risk. Thus, these insights enable possibilities for miRNA-based therapeutic strategies that could target specific miRNAs to correct metabolic imbalances in obesity and improve insulin sensitivity while reducing the risk of associated comorbidities.

In a related area of research, recent studies in this Special Issue examine the effects of bariatric surgery on mitochondrial DNA (mtDNA) levels and insulin sensitivity [20]. Bariatric surgery, a procedure aimed at promoting weight loss, has been shown to significantly improve various metabolic parameters, including insulin sensitivity, weight loss, and reductions in comorbidities like T2D and hypertension. However, the precise molecular mechanisms driving these improvements are still not fully understood. The study found that while no significant changes in circulating mitochondrial DNA (cf-mtDNA) levels were observed post-surgery, there were notable improvements in insulin sensitivity and other key metabolic parameters. This suggests that the benefits of bariatric surgery may not be directly related to changes in mitochondrial DNA, but rather to other factors such as alterations in adipose tissue, gut hormones, or the composition of the gut microbiome. In this sense, the gut microbiome plays a crucial role in digestion and metabolic processes, and shifts in its composition after bariatric surgery may influence metabolic health. Interestingly, these findings align with a recent study of healthy adolescents, which found that circulating cf-DNA was more strongly associated with the risk of metabolic syndrome than obesity itself [21].

The authors propose that bariatric surgery may improve metabolic health through mechanisms that do not necessarily involve significant changes in mitochondrial function or mtDNA levels. This perspective challenges the traditional view that mitochondrial dysfunction plays a central role in the pathophysiology of obesity-related metabolic disorders. Instead, the study encourages further exploration into how other molecular pathways, such as those involved in energy regulation, inflammation, and hormonal signaling, contribute to the metabolic improvements observed after bariatric procedures. These findings provide valuable insights into the complex interactions between mitochondria, obesity, and metabolic health, and offer new avenues for research into novel therapeutic approaches for obesity and its associated metabolic disorders.

Lastly, obesity is a significant risk factor for cardiovascular diseases (CVDs), and identifying early biomarkers to predict CVD risk is crucial for improving clinical outcomes and preventing complications. A study featured in this Special Issue suggests that HOMA-IR, a widely used index for assessing insulin resistance, could serve as an effective predictor of plasminogen activator inhibitor-1 (PAI-1) levels, a cytokine strongly associated with both obesity and CVDs [22]. In the study, women with severe obesity displayed altered glycemic parameters, including elevated insulin resistance, which correlated with higher PAI-1 levels. These findings indicate that HOMA-IR could be a valuable marker for identifying individuals at higher cardiovascular risk, especially in the context of obesity. By monitoring insulin resistance and its relationship with PAI-1, clinicians may gain critical insights into metabolic dysfunction and improve early intervention strategies for CVD prevention. Moreover, data from this study revealed a positive correlation between the Atherogenic Index of Plasma (AIP) and Alanine Aminotransferase (ALT), which is consistent with findings from a large-scale study in China involving 7838 participants [23]. The study demonstrated that AIP could serve as an independent predictor for the presence of fatty liver disease. AIP is a ratio calculated from the levels of triglycerides and high-density lipoprotein cholesterol (HDL-C) in the plasma, reflecting the balance between "good" and "bad" cholesterol, which is an important indicator of cardiovascular risk. A higher AIP is often associated with a greater risk of developing heart disease and metabolic disorders. Alanine Aminotransferase (ALT) is an enzyme primarily found in the liver, and elevated levels of ALT typically indicate liver injury or dysfunction. Since ALT is a sensitive marker of liver damage, its levels are often used to assess liver health, including the presence of liver diseases like non-alcoholic fatty liver disease (NAFLD). The positive correlation between AIP and ALT in this study suggests that AIP may not only be a marker for cardiovascular risk but also for liver health, particularly in individuals with obesity. This connection points to AIP as a potentially valuable marker for assessing the risk of NAFLD, which is commonly seen in obese individuals.

The research presented in this Special Issue emphasizes the multifaceted nature of obesity and its complex molecular mechanisms. The studies highlight the promising potential of microbial, inflammatory, and metabolic interventions to address the obesity epidemic. Specifically, they focus on modulating the gut microbiome, targeting therapies for lipid metabolism, and identifying early biomarkers for metabolic dysfunction. However, much remains to be understood about the precise molecular pathways involved in obesity and its related comorbidities.

Looking forward, future research should continue to explore personalized treatment strategies, integrating microbiome analysis, inflammatory markers, and metabolic indicators to create more effective interventions for obesity. The potential of emerging therapeutic approaches, such as targeting neuropeptides like NPGM or enhancing adipose tissue activity, offers exciting possibilities for the management of obesity and its associated disorders. Continued collaboration across disciplines—ranging from microbiology and endocrinology to molecular biology and clinical medicine—will be crucial in unraveling the complexities of obesity and developing novel therapies for this growing public health challenge.

As the Guest Editors of this Special Issue, we are deeply grateful to all the contributors for providing their views of the current developments and future directions in the field of obesity research at the molecular level. These original research articles and reviews span a broad spectrum of topics, from the genetic architecture of obesity to the intricacies of metabolic dysregulation. These contributions not only enhance our understanding of the disease but also inform evidence-based interventions tailored to reduce the global burden of obesity and its related health challenges.

Funding: T. Vezza is the recipient of a Miguel Servet contract (CP22/00153) from the Carlos III Health Institute.

Data Availability Statement: Data sharing is not applicable to this article because no data sets were generated or analyzed during the present study.

Conflicts of Interest: The authors declare no conflicts of interest.

Abbreviations

DHA, docosahexaenoic acid; HOMA-IR, Homeostasis Model Assessment of Insulin Resistance; mtDNA, mitochondrial DNA; NAFLD, non-alcoholic fatty liver disease; T2D, type 2 diabetes.

References

1. World Health Organization. Obesity and Overweight [Internet]. 2021. Available online: https://www.who.int/news-room/fact-sheets/detail/obesity-and-overweight (accessed on 27 December 2024).
2. Yuen, M.M.A. Health Complications of Obesity. *Gastroenterol. Clin. N. Am.* **2023**, *52*, 363–380. [CrossRef] [PubMed]
3. Khanna, D.; Welch, B.S.; Rehman, A. *Pathophysiology of Obesity*; StatPearls Publishing: Treasure Island, FL, USA, 2025.
4. Loos, R.J.F.; Yeo, G.S.H. The genetics of obesity: From discovery to biology. *Nat. Rev. Genet.* **2022**, *23*, 120–133. [CrossRef] [PubMed]
5. Zare, H.; Aazami, A.; Shalby, N.; Gilmore, D.R.; Thorpe, R.J. Measuring Racial Differences in Obesity Risk Factors in Non-Hispanic Black and White Men Aged 20 Years or Older. *Am. J. Mens. Health* **2023**, *17*, 15579883231205845. [CrossRef] [PubMed]
6. Vezza, T.; Víctor, V.M. The Mitochondrial Puzzle: Why Obesity Affects Insulin Sensitivity Differently in Black and White Women. *J. Clin. Endocrinol. Metab.* **2024**. [CrossRef]
7. Zabolotneva, A.A.; Kolesnikova, I.M.; Vasiliev, I.Y.; Grigoryeva, T.V.; Roumiantsev, S.A.; Shestopalov, A.V. The Obesogenic Gut Microbiota as a Crucial Factor Defining the Depletion of Predicted Enzyme Abundance for Vitamin B12 Synthesis in the Mouse Intestine. *Biomedicines* **2024**, *12*, 1280. [CrossRef]
8. Grigorova, N.; Ivanova, Z.; Petrova, V.; Vachkova, E.; Beev, G. Supernatants from Newly Isolated Lacticaseibacillus paracasei P4 Ameliorate Adipocyte Metabolism in Differentiated 3T3-L1 Cells. *Biomedicines* **2024**, *12*, 2785. [CrossRef]
9. Rohm, T.V.; Meier, D.T.; Olefsky, J.M.; Donath, M.Y. Inflammation in obesity, diabetes, and related disorders. *Immunity* **2022**, *55*, 31–55. [CrossRef]
10. Kawai, T.; Autieri, M.V.; Scalia, R. Adipose tissue inflammation and metabolic dysfunction in obesity. *Am. J. Physiol.-Cell Physiol.* **2021**, *320*, C375–C391. [CrossRef]
11. Lima, G.B.; Figueiredo, N.; Kattah, F.M.; Oliveira, E.S.; Horst, M.A.; Dâmaso, A.R.; Oyama, L.M.; Whitton, R.G.M.; de Souza, G.I.M.H.; Lima, G.C.; et al. Serum Fatty Acids and Inflammatory Patterns in Severe Obesity: A Preliminary Investigation in Women. *Biomedicines* **2024**, *12*, 2248. [CrossRef]
12. Picó, C.; Palou, M.; Pomar, C.A.; Rodríguez, A.M.; Palou, A. Leptin as a key regulator of the adipose organ. *Rev. Endocr. Metab. Disord.* **2022**, *23*, 13–30. [CrossRef]
13. Achari, A.; Jain, S. Adiponectin, a Therapeutic Target for Obesity, Diabetes, and Endothelial Dysfunction. *Int. J. Mol. Sci.* **2017**, *18*, 1321. [CrossRef] [PubMed]
14. Ferguson, J.F.; Roberts-Lee, K.; Borcea, C.; Smith, H.M.; Midgette, Y.; Shah, R. Omega-3 polyunsaturated fatty acids attenuate inflammatory activation and alter differentiation in human adipocytes. *J. Nutr. Biochem.* **2019**, *64*, 45–49. [CrossRef] [PubMed]

15. Liquian, S.; Fang-Hui, L.; Minghui, Q.; Yanan, Y.; Haichun, C. Free fatty acids and peripheral blood mononuclear cells (PBMC) are correlated with chronic inflammation in obesity. *Lipids Health Dis.* **2023**, *22*, 93. [CrossRef]
16. Narimatsu, Y.; Kato, M.; Iwakoshi-Ukena, E.; Moriwaki, S.; Ogasawara, A.; Furumitsu, M.; Ukena, K. Neurosecretory Protein GM–Expressing Neurons Participate in Lipid Storage and Inflammation in Newly Developed Cre Driver Male Mice. *Biomedicines* **2023**, *11*, 3230. [CrossRef]
17. Tran, L.T.; Park, S.; Kim, S.K.; Lee, J.S.; Kim, K.W.; Kwon, O. Hypothalamic control of energy expenditure and thermogenesis. *Exp. Mol. Med.* **2022**, *54*, 358–369. [CrossRef]
18. Saliminejad, K.; Khorram Khorshid, H.R.; Soleymani Fard, S.; Ghaffari, S.H. An overview of microRNAs: Biology, functions, therapeutics, and analysis methods. *J. Cell. Physiol.* **2019**, *234*, 5451–5465. [CrossRef]
19. Cabiati, M.; Guiducci, L.; Randazzo, E.; Casieri, V.; Federico, G.; Del Ry, S. Circulating and Exosomal microRNA-33 in Childhood Obesity. *Biomedicines* **2023**, *11*, 2295. [CrossRef] [PubMed]
20. Yuzefovych, L.V.; Pastukh, V.M.; Mulekar, M.S.; Ledbetter, K.; Richards, W.O.; Rachek, L.I. Effect of Bariatric Surgery on Plasma Cell-Free Mitochondrial DNA, Insulin Sensitivity and Metabolic Changes in Obese Patients. *Biomedicines* **2023**, *11*, 2514. [CrossRef]
21. Celec, P.; Janovičová, Ĺ.; Gurecká, R.; Koborová, I.; Gardlík, R.; Šebeková, K. Circulating extracellular DNA is in association with continuous metabolic syndrome score in healthy adolescents. *Physiol. Genom.* **2021**, *53*, 309–318. [CrossRef]
22. Martins Kattah, F.; Janjusevic, M.; Figueiredo, N.; Santos Oliveira, E.; Carielo Lima, G.; Dâmaso, A.R.; Oyama, L.M.; Fluca, A.L.; de Melo, P.R.E.; Horst, M.A.; et al. HOMA-IR as a Predictor of PAI-1 Levels in Women with Severe Obesity. *Biomedicines* **2024**, *12*, 1222. [CrossRef]
23. Xie, F.; Zhou, H.; Wang, Y. Atherogenic index of plasma is a novel and strong predictor associated with fatty liver: A cross-sectional study in the Chinese Han population. *Lipids Health Dis.* **2019**, *18*, 170. [CrossRef]

biomedicines

MDPI

Article

The Obesogenic Gut Microbiota as a Crucial Factor Defining the Depletion of Predicted Enzyme Abundance for Vitamin B12 Synthesis in the Mouse Intestine

Anastasia A. Zabolotneva [1,2,*], Irina M. Kolesnikova [1,2], Ilya Yu. Vasiliev [2,3], Tatiana V. Grigoryeva [2,3], Sergei A. Roumiantsev [1,2] and Aleksandr V. Shestopalov [1,2]

1 Department of Biochemistry and Molecular Biology, Faculty of Medicine, N. I. Pirogov Russian National Research Medical University, 1 Ostrovitianov Str., Moscow 117997, Russia; sel_irka@list.ru (I.M.K.); rumyantsev.sergey@endocrincentr.ru (S.A.R.); al-shest@yandex.ru (A.V.S.)
2 Laboratory of Biochemistry of Signaling Pathways, Endocrinology Research Center, 11 Dm. Ulyanova Str., Moscow 117036, Russia; mepk_m6@mail.ru (I.Y.V.); tatabio@inbox.ru (T.V.G.)
3 Institute of Fundamental Medicine and Biology, Kazan Federal University, 18 Kremlyovskaya Str., Kazan 420008, Russia
* Correspondence: a.zabolotneva@gmail.com; Tel.: +7-9067039402

Abstract: Currently, obesity is a critical global public health burden. Numerous studies have demonstrated the regulation of the pathogenesis of obesity and metabolic abnormalities by the gut microbiota and microbial factors; however, their involvement in the various degrees of obesity is not yet well understood. Previously, obesity has been shown to be associated with decreased levels of vitamin B12. Considering exclusive microbial production of vitamin B12, we hypothesized that a decrease in cobalamin levels in obese individuals may be at least partially caused by its depleted production in the intestinal tract by the commensal microbiota. In the present study, our aim was to estimate the abundance of enzymes and metabolic pathways for vitamin B12 synthesis in the gut microbiota of mouse models of alimentary and genetically determined obesity, to evaluate the contribution of the obesogenic microbiome to vitamin B12 synthesis in the gut. We have defined a significantly lower predicted abundance of enzymes and metabolic pathways for vitamin B12 biosynthesis in obese mice compared to non-obese mice, wherein enzyme depletion was more pronounced in *lepr*(−/−) (*db*/*db*) mice, which developed severe obesity. The predicted abundance of enzymes involved in cobalamin synthesis is strongly correlated with the representation of several microbes in high-fat diet-fed mice, while there were almost no correlations in *db*/*db* mice. Therefore, the degree of obesity and the composition of the correspondent microbiota are the main contributors to the representation of genes and pathways for cobalamin biosynthesis in the mouse gut.

Keywords: gut microbiota; obesity; cobalamin; vitamin b12; *db*/*db* mice

Citation: Zabolotneva, A.A.; Kolesnikova, I.M.; Vasiliev, I.Y.; Grigoryeva, T.V.; Roumiantsev, S.A.; Shestopalov, A.V. The Obesogenic Gut Microbiota as a Crucial Factor Defining the Depletion of Predicted Enzyme Abundance for Vitamin B12 Synthesis in the Mouse Intestine. *Biomedicines* **2024**, *12*, 1280. https://doi.org/10.3390/biomedicines12061280

Academic Editors: Zaida Abad-Jiménez, Teresa Vezza and Yoshinori Nagai

Received: 29 March 2024
Revised: 25 May 2024
Accepted: 7 June 2024
Published: 9 June 2024

1. Introduction

The last few decades have been marked by extensive research on the metabolic activity of the intestinal microbiota and its impact on host health [1]. The concept of an obesogenic microbiome, which is characterized by compositional and metabolic changes in the gut microbiota that are detrimental to the host, has been formed [2]. Although the characteristics of such an obesogenic microbiome may vary significantly, which is confirmed by contradictory results of microbiota analysis in different studies, some common trends have been described. Enrichment by enzymes for improved energy extraction from nutrients, decreased production of short chain fatty acids, and increased production of lipopolysaccharides stimulating proinflammatory response by the host immune system are some common signs of obesogenic microbiota [3]. Furthermore, it has been observed that the obese state of the host is associated with a decrease in the level of vitamin B12 in the blood, although the underlying mechanisms are unclear [4–6]. As shown in previous studies, vitamin B12 is a critical micronutrient associated with the foetal programming of obesity. Researchers from India and UK have revealed that low vitamin B12 status in the gestation period is associated with increased obesity and insulin resistance of the mother and birth weight and risk of insulin resistance and obesity in the offspring. In the study of obese adolescents, it was found that obesity has a 1.6-fold decreasing effect on vitamin B12 levels [7].

In preclinical studies, it was demonstrated that low levels of vitamin B12 could increase lipid accumulation in adipocytes and trigger dyslipidaemia in mice, suggesting that low levels of B12 and dyslipidaemia could be causally related [8]. A low serum B12 level has also been shown in people with nonalcoholic fatty liver disease, especially with grade 2 and 3 steatosis [9]. In the study of Boachie et al. [10], it was shown that low B12 in HepG2 cells increased the gene expression of lipogenesis and decreased lipid oxidation, resulting in increased intracellular triglycerides and the accumulation of lipid droplets.

It has been suggested that the primary cause of B12 malabsorption in certain human diseases is the gut microbiota's depletion of vitamin B12 [11]. Cobalamin (vitamin B12) production is restricted to microorganisms, specifically anaerobes, and cannot be achieved by plants, animals, or fungi [12]. Although it is still not known whether the host uses microbially produced B12 in the gut [13], it was shown that human commensal gut microbes can produce extracellular vesicles containing multiple high-affinity vitamin B12 binding proteins, suggesting that the vesicles play a role in the elimination of micronutrients and their transfer to bacterial and host cells [14]. Many commensal gut bacteria have been indicated as vitamin B12 producers [15]. Vitamin B12 synthesis is a complex multistep process and requires a lot of enzymes (Figure 1).

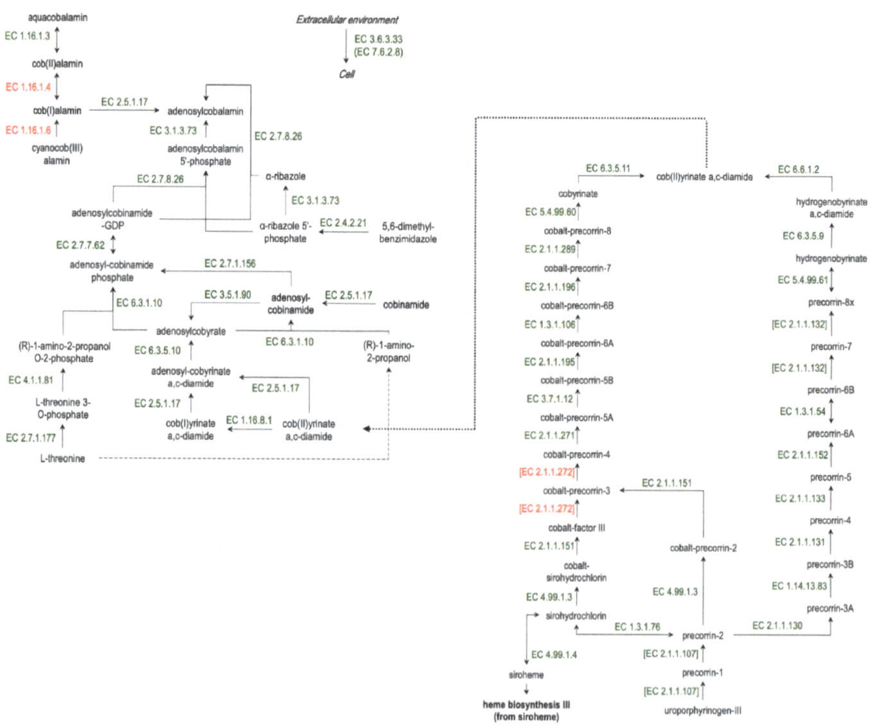

Figure 1. The scheme of vitamin B12 synthesis in bacteria. The predicted abundance of all enzymes, except EC 1.16.1.4, 1.16.1.6, [2.1.1.272] (red font), was investigated in the present study.

We hypothesized that the decrease in cobalamin level in obese individuals may be at least partially caused by its depleted production by the commensal microbiota in the intestine. In our study, our aim was to estimate the abundance of enzymes and metabolic pathways for vitamin B12 synthesis in the gut microbiota of mouse models of alimentary and genetically determined obesity, to evaluate the contribution of the obesogenic microbiome to vitamin B12 synthesis in the gut. To this end, we performed a high-throughput metagenome sequencing analysis followed by a reconstruction of gut microbiota metabolic activity for high-fat diet-fed C57BL/6SPF mice (which are used as a model of alimentary obesity), standard diet-fed *db/db* mice (leptin receptor-deficient mice) (which are used as a model of genetically determined obesity), and standard diet-fed C57BL/6SPF mice (which are used as a control group).

2. Materials and Methods

2.1. Experimental Animals

C57BL/6SPF mice (n = 20, males) (bred at the Nursery of Laboratory Animals in Puschino, Puschino, Russia) and *db/db* (n = 10, males) mice (bred at JAX-East and JAX-West Nurseries of Laboratory Animals, Sacramento, CA, USA) were acclimated to housing conditions (22 °C, 55% humidity, 12 h:12 h light: dark cycle) in an SPF level animal center of I.M. Sechenov First Moscow State Medical University (Moscow, Russia) with ad libitum access to sterile food (Altromin 1324, Lage, Germany. The food includes 4.1% crude fat, 19.2% crude protein, 6.1% crude fiber, 5.9% crude ash, 53.4% nitrogen-free extractives, and 11.3% moisture) and water for 1 week prior to formal study. Following the adaption phase, the mice were split into groups based on genotype, with each group consisting of ten individual animals. The mice were 8 weeks old and weighed 19 $\pm$ 2 g on average at the start of the experiment. Laboratory animals were fed a high-fat diet enriched with triglycerides derived from animals, up to 30% of total calories (Altromin C 1090-30, Lage, Germany. The food includes 13.3% crude fat, 21.1% crude protein, 5.1% crude fiber, 3.9% crude ash, 50.8% nitrogen-free extractives, and 5.8% moisture), beginning at 8 weeks of age and continuing for 90 days until the end of the experiment. This allowed for the replication of the alimentary obesity model.

For the duration of the experiment, the mice in the *db/db* group and the animals in the control group (C57BL/6SPF) were fed a standard chow diet (Altromin 1324 FORTI, Lage, Germany). Mice were anesthetized with isofurane and euthanized after 90 days of feeding (RWD Life Science, Chenzhen, Guangdong, China). Sterile colon tissues were obtained. The tissue samples were immediately frozen in liquid nitrogen and stored at −80 °C until analysis. The colon samples were sent for high-throughput sequencing analysis after being divided into 1 cm sections under sterile conditions, deposited into different sterile Eppendorf tubes, and maintained on dry ice (10 samples for each group) (Figure 2).

All animal experiments were approved by the Ethics Committee for Animal Research, I.M. Sechenov First Moscow State Medical University, Russia (protocol number 96 from 2 September 2021).

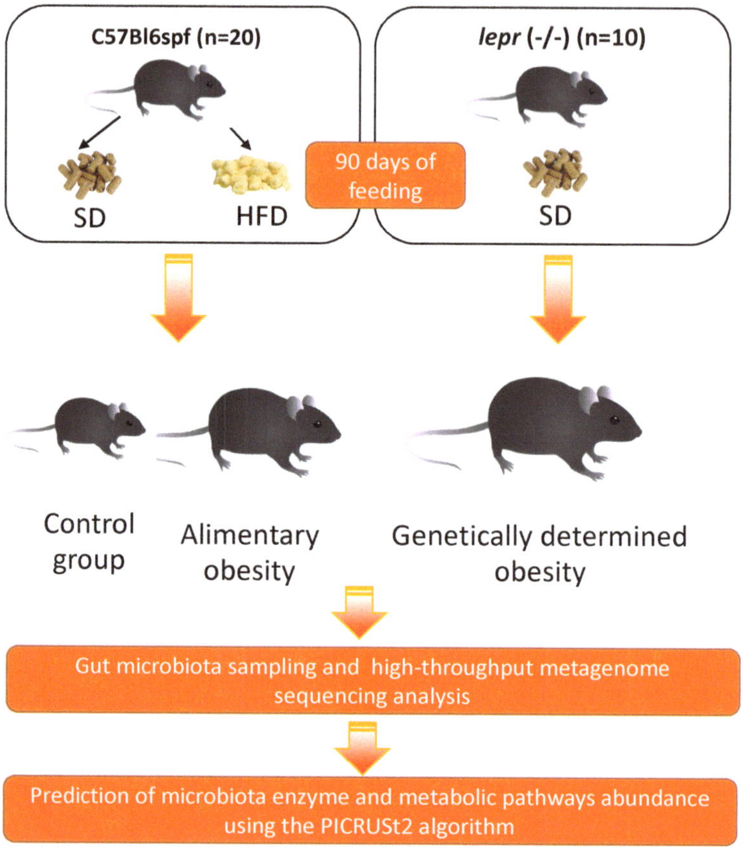

Figure 2. Experimental design showing the distribution of mice into study groups. SD, standard diet; HFD, high-fat diet.

2.2. Measurement of Serum Insulin, Leptin, and Adiponectin Levels

The mice fasted for 8 h. Blood samples were collected after fasting from anaesthetized mice using the eyeball enucleation method. Blood was collected in sterile collection tubes immediately. Serum was spun down at 8000× *g* for 8 min at 4 °C to remove remaining cellular debris. Insulin concentration was measured using the Mouse Insulin ELISA Kit (Abcam, ab277390), leptin concentration was measured using Duo Set ELISA Development system Mouse Leptin (R&DSystem, DY495-05), and adiponectin concentration was measured using Adiponectin/Acrp30 DuoSet ELISA (R&DSystem, DY1119) according to the manufacturer's instructions.

2.3. Measurement of Fasting Blood Glucose Level and HOMA-IR

Fasting levels of glucose were measured every 7 days for every group of mice; mouse tail snip blood samples were used for the analysis. Blood glucose was measured using an ACCU-CHEK Aviva glucometer (Roche, Switzerland). The animals fasted for 5 h with free access to water before a fasting blood glucose test. The HOMA-IR (homeostasis model assessment of insulin resistance) index was calculated as [fasting serum glucose × fasting serum insulin/22.5] to assess insulin resistance.

2.4. Histopathology of Mouse Liver, Pancreas, Skeletal Muscle, and Adipose Tissues

Organ fixation after collection was carried out for at least 72 h. After fixation, the organs were dehydrated, soaked in paraffin, and cut into 4–5-micron-thick sections. Sections were stained with hematoxylin-eosin and examined by light microscopy (magnification ×10, ×40). The research was carried out using Leica (Germany) histological equipment.

2.5. High-Throughput Sequencing Analysis and Reconstruction of Intestinal Microbiota Metabolic Activity

The scientific research laboratory «Multiomics technologies of living systems» (Kazan, Russia) performed the microbiota analysis. Mouse stool samples were treated with the FastDNATM Spin Kit for Faeces (MP Biomedicals, Santa Ana, CA, USA) to extract genomic DNA. The bacterial 16S rRNA gene's V3–V4 region was amplified using specific primers (see Supplementary Table S1). The PCR product purification process using AMPure XP Beads (Beckman Coulter, Brea, CA, USA, CB55766755) to barcode each sample was followed by secondary-round PCR amplification utilizing index primers. Using a Qubit 2.0 Fluorometer and the Qubit dsDNA High Sensitivity Assay Kit (Invitrogen, Carlsbad, CA, USA), the concentration of amplicons was determined. Prior to sequencing, the samples were mixed in an identical mole ratio to complete the sample preparation.

The libraries were then high-throughput sequenced (2 × 300 bp reads) (Illumina Miseq, Illumina, CA, USA). The raw reads were processed using QIIME2 v2023.7.0 [16] and PICRUSt2 v2.5.2 software [17] (accessed on 12 September 2023). According to the results of data sequencing using PICRUSt2 v2.5.2 software, the microbial metabolic pathways encoded by the detected bacterial genomes were scored, and the most abundant pathways were detected by multiple *t*-test analysis.

2.6. Statistical Data Analysis

Statistical processing of the data was carried out using the method of nonparametric statistics using the GraphPad Prism 10 v10.0.2 (171) statistical software package. The mean and standard deviation were used to present all data. All in vivo experimental data were analyzed using Welch's one-way analysis of variance (ANOVA) or multiple Mann–Whitney tests using the two-stage step-up method (Benjamini, Krieger, and Yekutieli) (false discovery rate Q = 5%). *p*-values less than 0.05 were considered statistically significant (* $p < 0.05$, ** $p < 0.01$, *** $p < 0.001$). A correlation analysis was performed according to Spearman with an assessment of the statistical significance of the correlation coefficient.

3. Results

3.1. Body Weight and Hormonal Status

Three groups of mice (males, n = 10 for each group) were fed a standard or high-fat diet for 90 days. The dynamics of weight gain is shown in Figure 3a. At the end of the experiment, the final weights of the mice were compared (Figure 3b). Mice in both obesity models showed significantly higher body weight compared to the control group. The mean weight of the mice in control group was 28.5 ± 2.5 g; in the high-fat diet-fed mice, the mean weight was 34.7 ± 1.4 g (weight gain is 17.9% compared to the control group); and in *db*/*db* mice, the mean weight was 59.2 ± 4.1 g (weight gain is 51.9% compared to the control group).

Although there are no strict criteria for the development of alimentary obesity in mice, it is generally accepted that 20–30% weight gain points to reaching the obesity state [18]. In our experiment, weight gain in a high-fat diet mouse model group was 17.9% higher than in the control group, which was not enough to develop complications of severe obesity. However, such weight gain corresponds to stage 1 of obesity and, most importantly, these mice demonstrated dysmetabolic characteristics associated with obesity, as shown below. On the contrary, *db/db* mice gained 51.9% more weight compared to the control group, which indicates the development of severe obesity.

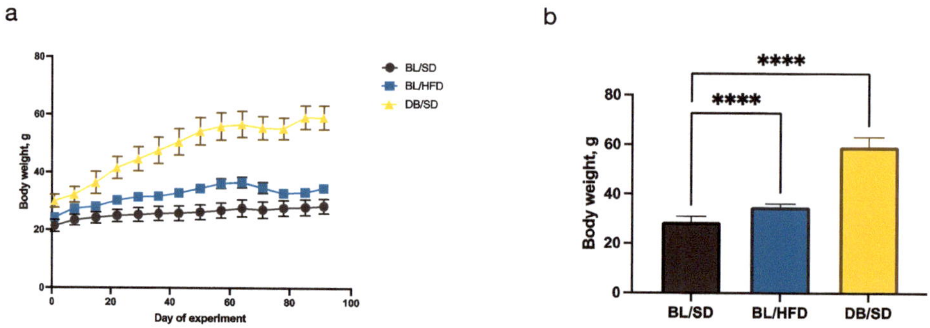

Figure 3. Changes in mouse body weight during the course of the experiment: (**a**)—mouse BW dynamics during the 90-day feeding experiment in different study groups; (**b**)—ANOVA of mouse BW in different experimental groups did not reveal any differences after 90 days of feeding (**** $p < 0.0001$). BL/SD, C57Bl6/spf mice fed a standard diet; BL/HFD, C57Bl6/spf mice fed a high-fat diet; DB/SD, *db/db* mice fed a standard diet.

To confirm the genetically determined character of obesity, we compared insulin, leptin, and adiponectin levels in the blood serum of mice in all experimental groups. We established that insulin (Figure 4a) and leptin (Figure 4b) levels were significantly higher in *db/db* mice compared to C57Bl6 mice (15-times higher for insulin and 21-times higher for leptin), while the level of adiponectin decreased in mice fed a high-fat diet but not in *db/db* mice (Figure 4c). Furthermore, the level of leptin in high-fat diet-fed mice also increased compared to the control group, but this was not as crucial as in *db/db* mice. Significantly increased levels of insulin and leptin in *db/db* mice confirmed strong insulin resistance, which was developed due to leptin receptor deficiency. The increased concentration of leptin and decreased concentration of adiponectin in high-fat diet-fed mice also evidences the dysmetabolic state of the animals.

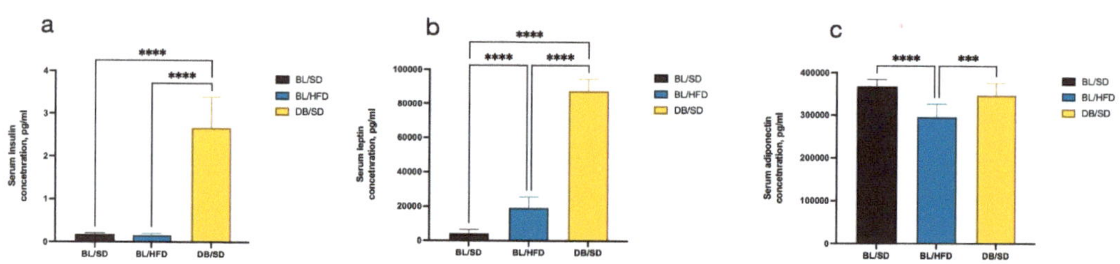

Figure 4. Results of serum insulin (**a**), leptin (**b**) and adiponectin (**c**) levels in different groups of mice at the end of the experiment. Comparisons were carried out using one-way analysis of variance followed by Tukey's multiple comparison test (*** $p < 0.001$, **** $p < 0.0001$). BL/SD, C57Bl6/spf mice fed a standard diet; BL/HFD, C57Bl6/spf mice fed a high-fat diet; DB/SD, *db/db* mice fed a standard diet.

To investigate insulin resistance in mice, we conducted glucose dynamic observation and evaluated HOMA-IR indexes in all experimental groups. Glucose levels and HOMA-IR were significantly higher in *db/db* mice compared to the control group but did not change in HFD-fed mice (Figure 5a,b).

We also investigated the morphological state of the livers of all experimental mice. All mice in the HFD-fed group (Figure 6e–h) and *db/db* group developed liver steatosis (fatty liver) (Figure 6i–l), in contrast to SD-fed C57Bl6 mice, which demonstrated normal liver morphology without fat and inflammatory infiltration (Figure 6a–d). As expected, fat infiltration of the liver in *db/db* mice was significantly more profound than in HFD-fed mice.

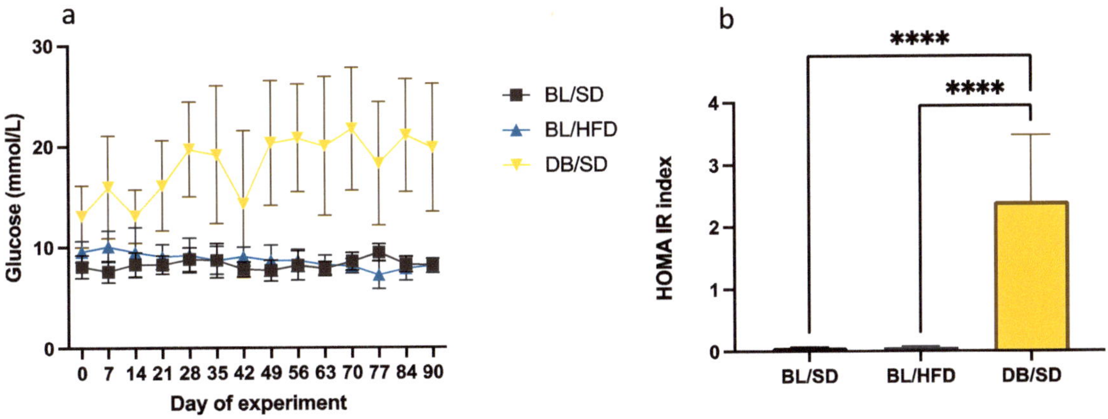

Figure 5. Serum glucose dynamics (**a**) and comparative analysis of HOMA-IR indexes (**b**) observed in different groups of experimental mice (**** $p < 0.0001$). BL/SD, C57Bl6/spf mice fed a standard diet; BL/HFD, C57Bl6/spf mice fed a high-fat diet; DB/SD, *db/db* mice fed a standard diet.

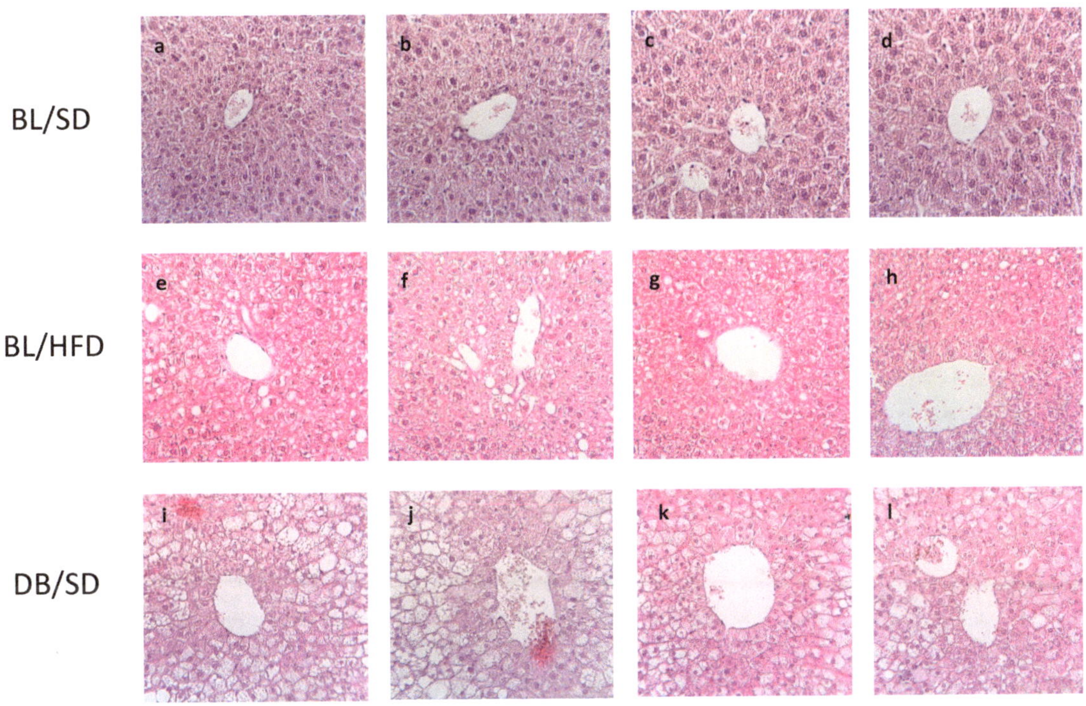

Figure 6. Liver sections of the control group of mice (**a–d**), C57Bl6/spf mice fed a high-fat diet (**e–h**), *db/db* mice fed a standard diet (**i–l**). The portal vein surrounded by hepatocytes in liver lobe is presented in every microphotograph. Fat infiltration is detected as empty vacuoles situated between hepatocytes. BL/SD, C57Bl6/spf mice fed a standard diet; BL/HFD, C57Bl6/spf mice fed a high-fat diet; DB/SD, *db/db* mice fed a standard diet.

Therefore, significant differences in body weight, leptin and adiponectin levels and hepatosteatosis developed in the HFD-fed and *db/db* groups of mice can approve the presence of metabolic changes associated with diet or genetic background, although insulin resistance, as expected, was developed in the group of *db/db* mice only. Thus, we can consider the HFD-fed group of mice as an overweight group with slight metabolic complications, while *db/db* mice simulated severe obesity with strong metabolic disturbances.

3.2. Microbiota Analysis

3.2.1. Taxonomy Composition and Alpha Diversity

To characterize the composition of the mouse gut microbiota, we performed a high-throughput metagenome sequencing analysis of 30 microbiota samples taken from all experimental animals. After OTU annotation, we compared the differences in alpha diversity and microbial community structures. We established that the alpha diversity indexes (observed OTUs, PD whole tree, Chao1, Simpson, and Shannon) were significantly lower ($p < 0.05$) in *db/db* mice compared to the control group (Figure 7a–e), indicating a decrease in the richness and diversity of microbial species in severely obese mice. The microbiota of mice fed a high-fat diet also showed a decrease in alpha diversity; however, only PD whole-tree and Shannon indexes were lower in the BL/HFD group compared to the control group (Figure 7b,e).

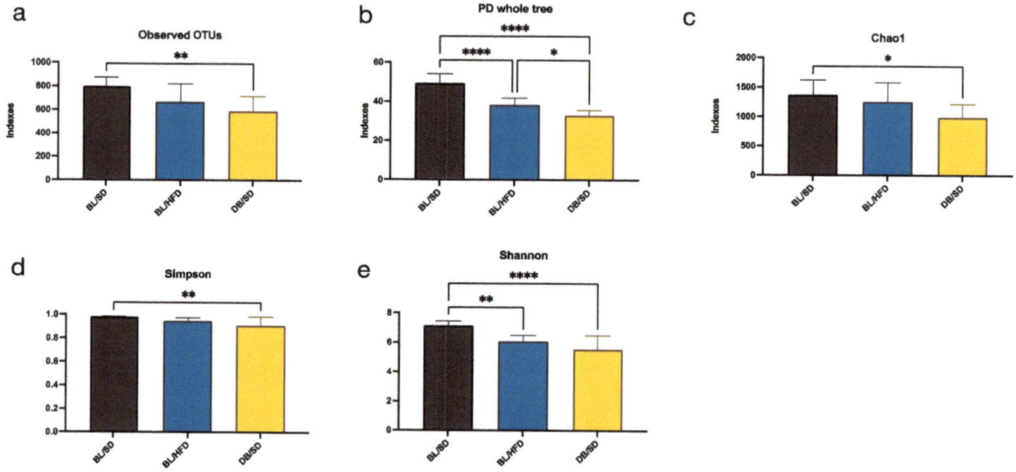

Figure 7. Results of one-way ANOVA followed by Tukey's multiple comparison test for alpha diversity indices ((**a**)—for observed OTUs, (**b**)—for PD whole tree, (**c**)—for Chao1, (**d**)—for Simpson, (**e**)—for Shannon), in different groups of mice. * $p < 0.05$, ** $p < 0.01$, **** $p < 0.0001$. BL/SD, C57Bl6/spf mice fed a standard diet; BL/HFD, C57Bl6/spf mice fed a high-fat diet; DB/SD, *db/db* mice fed a standard diet.

We also investigated the taxonomic structure of bacterial communities (Figures 8a and 9a–c). At the phylum level, we observed a significant decrease in the representation of Verrucomicrobia and Actinobacteria in *db/db* mice compared with HFD- and SD-fed mice (Figure 8b,c). Furthermore, the representation of Bacillota in *db/db* mice was higher than in the HFD feeding group (Figure 8b).

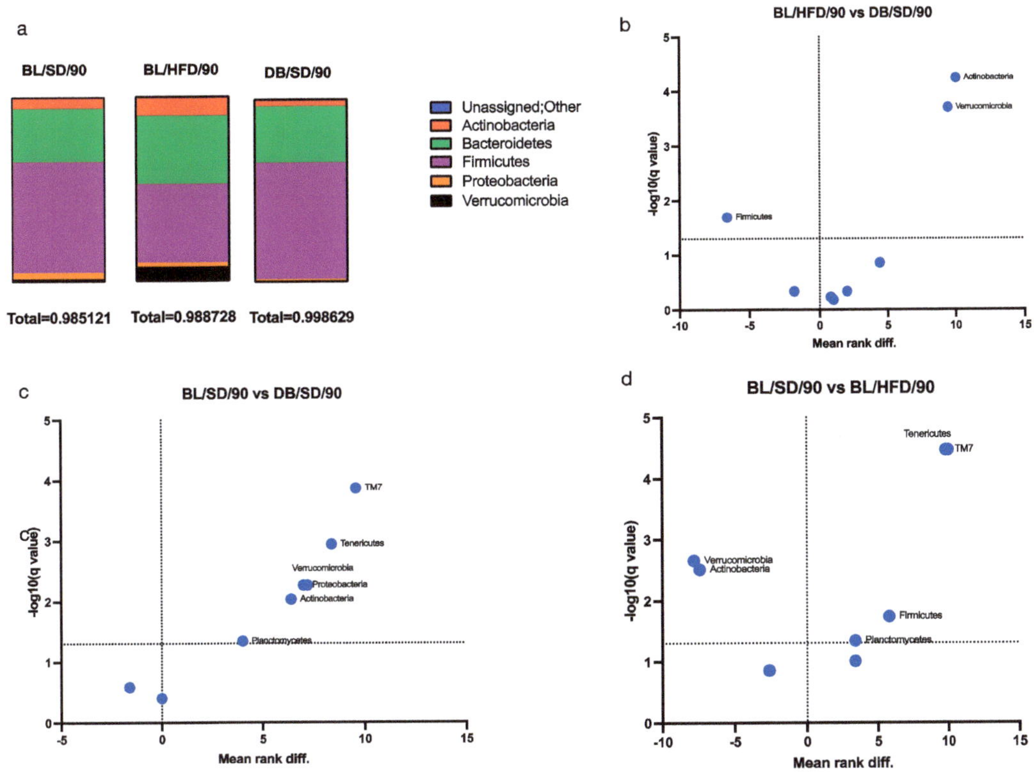

Figure 8. Taxonomy structure of mouse gut microbial communities at the phylum level: (**a**)—diagrams of comparative phyla representation in three groups of mice; (**b**)—volcano plot demonstrating differences in the representation of bacterial phyla in C57Bl6 mice fed a standard or a high-fat diet; (**c**)—volcano plot demonstrating differences in the representation of bacterial phyla in C57Bl6 and *db/db* mice fed a standard diet; (**d**)—volcano plot demonstrating differences in the representation of bacterial phyla in C57Bl6 mice fed a high-fat diet and in *db/db* mice fed a standard diet. The Q value reflects a false discovery rate of 5%. The mean rank difference values reflect the direction of changes in the abundance of bacterial phyla (values less than zero indicate an increased representation of phyla, while values greater than zero indicate a decreased representation of phyla in the microbiota of *db/db* mice (**b**,**c**) or HFD-fed mice (**d**). Statistically significant values ($p < 0.01$) are indicated. BL/SD, C57Bl6/spf mice fed a standard diet; BL/HFD, C57Bl6/spf mice fed a high-fat diet; DB/SD, *db/db* mice fed a standard diet.

At the family level, we observed a significant increase in the representation of Prevotellaceae with a simultaneous decrease in the representation of Verrucomicrobiaceae in the microbiota of *db/db* mice compared to HFD-fed mice (Figure 9a–c). These changes are known to be associated with an obese phenotype [19], although the representation of Verrucomicrobiaceae increased in HFD-fed mice compared to SD-fed mice.

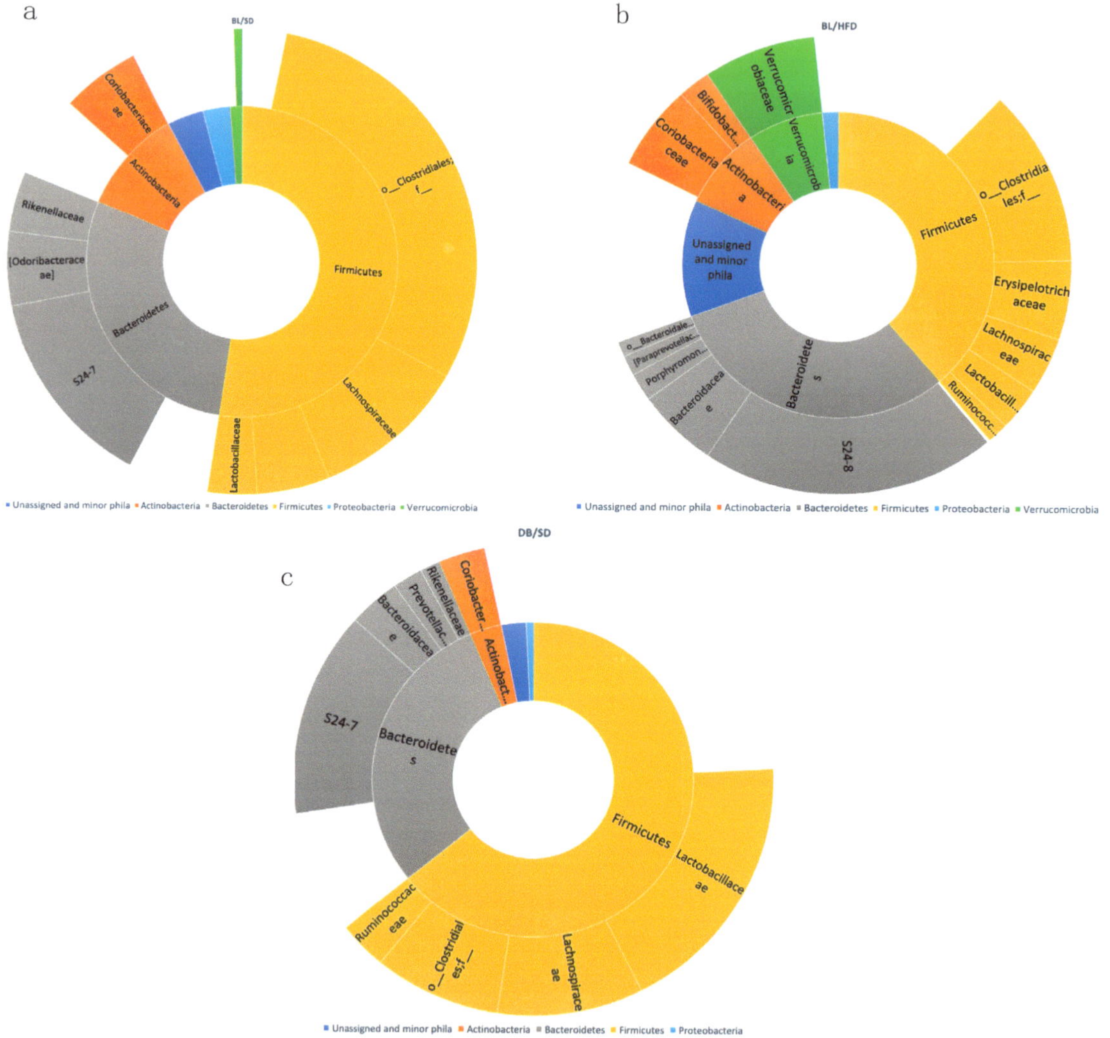

Figure 9. Taxonomy structure of mouse gut microbial communities at family level: (**a**)—diagram of families' representation in C57Bl6 mice fed a standard diet; (**b**)—diagram of families' representation in C57Bl6 mice fed a high-fat diet; (**c**)—diagram of families' representation in *db/db* mice fed a standard diet. BL/SD, C57Bl6/spf mice fed a standard diet; BL/HFD, C57Bl6/spf mice fed a high-fat diet; DB/SD, *db/db* mice fed a standard diet.

At the species level, we found an increase in *Bacteroides acidifaciens* in *db/db* mice compared to BL/SD and BL/HFD mice, while the representation of bacteria beneficial for host health (*Bifidobacterium pseudolongum, Akkermansia muciniphila, Adlercreutzia sp.*) decreased in *db/db* mice compared only to HFD-fed mice, not SD-fed mice (Table 1).

Table 1. Differentially represented bacterial species in different groups of mice according to multiple Mann–Whitney tests with the two-stage step-up method (Benjamini, Krieger, and Yekutieli) (false discovery rate Q = 5%). *p* values less than 0.05 were considered to indicate statistical significance. a. Comparison in BL/HFD and DB/SD groups; arrows show increased or decreased microbe representation in *db/db* mice; b. Comparison in BL/SD and DB/SD groups; arrows show increased or decreased microbe representation in *db/db* mice; c. Comparison in BL/SD and BL/HFD groups; arrows show increased or decreased microbe representation in HFD-fed mice. BL/SD, C57Bl6/spf mice fed a standard diet; BL/HFD, C57Bl6/spf mice fed a high-fat diet; DB/SD, *db/db* mice fed a standard diet.

a. BL/HFD vs. DB/SD		**b. BL/SD vs. DB/SD**		**c. BL/SD vs. BL/HFD**	
Bifidobacterium pseudolongum	↓			*Bifidobacterium pseudolongum*	↑
Bacteroides sp.	↓			*Bacteroides sp.*	↑
Parabacteroides distasonis	↓	*Parabacteroides distasonis*	↓	*Parabacteroides distasonis*	↑
Allobaculum sp.	↓			*Allobaculum sp.*	↑
Akkermansia muciniphila	↓			*Akkermansia muciniphila*	↑
Streptococcus sp.	↓	*Streptococcus sp.*	↓		
Adlercreutzia sp.	↓				
Clostridium sp.	↓	*Clostridium sp.*	↓	*Clostridium sp.*	↓
Ureaplasma sp.	↓				
Anaeroplasma sp.	↑			*Anaeroplasma sp.*	↓
Bacteroides acidifaciens	↑	*Bacteroides acidifaciens*	↑		
Anaerotruncus sp.	↑				
Ruminococcus sp.	↑			*Ruminococcus sp.*	↓
Prevotella sp.	↑				

Table 1. *Cont.*

a. BL/HFD vs. DB/SD		**b. BL/SD vs. DB/SD**		**c. BL/SD vs. BL/HFD**	
Rikenella sp.	↑			*Rikenella sp.*	↓
		AF12 sp.	↓	*AF12 sp.*	↓
		Coprobacillus sp.	↓	*Coprobacillus sp.*	↓
		Dorea sp.	↓	*Dorea sp.*	↓
		Sutterella sp.	↓	*Sutterella sp.*	↓
		Odoribacter sp.	↓	*Odoribacter sp.*	↓
		[Ruminococcus] gnavus	↓		
				Enterococcus sp.	↓
				Coprococcus sp.	↓
				Parabacteroides sp.	↑

All these findings confirm the existence of specific obesogenic changes in the gut microbiota communities of mice, which is more pronounced in *db/db* mice than in HFD-fed mice, which is in concordance with the higher body weight and hormonal disturbances of *db/db* mice.

3.2.2. Reconstruction of the Metabolic Activity of the Mouse Gut Microbiota and Representation of the Pathways Responsible for Vitamin B12 Biosynthesis

To establish whether enzymes and metabolic pathways for B12 synthesis are less abundant in obese mice compared to mice fed a standard diet, we carried out a reconstruction of microbiota metabolic activity by using a PICRUSt2 analytic tool based on metagenome sequencing data analysis, which allowed us to estimate the predicted abundance of bacterial genes in mouse gut microbial communities.

According to the results of PICRUSt2, among the 423 metabolic pathways analyzed, we revealed some differences in the abundance of the cobalamin synthesis pathways of mice fed *db/db* and an HFD compared to C57Bl6 mice fed an SD. Namely, the representation of six pathways for vitamin B12 synthesis in *db/db* mice and four pathways in C57Bl6 mice fed an HFD was decreased (Table 2).

Table 2. Differentially represented metabolic pathways involved in vitamin B12 synthesis according to the results of multiple Mann–Whitney tests with the two-stage step-up method (Benjamini, Krieger, and Yekutieli) (false discovery rate Q = 5%). *p*-values less than 0.05 were considered to indicate statistical significance.

Pathways with Decreased Representation in *db/db* Mice Fed a Standard Chow Diet Compared to a Control Group ($p < 0.01$, $q < 0.01$)	Pathways with Decreased Representation in C57Bl6/spf Mice Fed a High-Fat Diet Compared to a Control Group ($p < 0.001$, $q < 0.001$)
• cob(II)yrinate a,c-diamide biosynthesis I (early cobalt insertion) • adenosylcobalamin biosynthesis I (early cobalt insertion) • adenosylcobalamin biosynthesis II (late cobalt incorporation) • adenosylcobalamin salvage from cobinamide I • adenosylcobalamin biosynthesis from cobyrinate a,c-diamide I • adenosylcobalamin salvage from cobinamide II	• adenosylcobalamin biosynthesis II (late cobalt incorporation) • adenosylcobalamin salvage from cobinamide I • adenosylcobalamin biosynthesis from cobyrinate a,c-diamide I • adenosylcobalamin salvage from cobinamide II

3.2.3. Abundance of Enzymes for Vitamin B12 Synthesis in Gut Bacteria According to the Results of the Metagenome Sequencing

PICRUSt2 analysis was also used to estimate the abundance of enzymes involved in cobalamin synthesis. To this aim, we chose 37 enzymes required for cobalamine synthesis (Figure 1) among more than 8000 enzymes represented in investigated microbiomes. Only three enzymes (EC 1.16.1.4, EC 1.16.1.6, [EC 2.1.2.272]) were not presented in our data and were not investigated.

We found that the representation of almost all enzymes (except for adenosylcobalamine hydrolase and vitamin B12-transporting ATPase) needed for vitamin B12 synthesis was decreased in *db*/*db* mice compared to C57Bl6 mice fed an SD (Table 3), while there was no difference in enzyme representation in other comparisons.

Table 3. Differentially represented enzymes involved in vitamin B12 synthesis in different groups of mice according to the results of multiple Mann–Whitney tests with the two-stage step-up method (Benjamini, Krieger and Yekutieli) (false discovery rate Q = 5%). p values less than 0.05 were considered to indicate statistical significance. a. Comparison in BL/SD and BL/HFD groups; red cells indicate decreased enzyme representation in HFD-fed mice; b. Comparison in BL/SD and DB/SD groups; red cells indicate decreased enzyme representation in *db/db* mice; c. Comparison in BL/HFD and DB/SD groups; red cells indicate decreased enzyme representation in *db/db* mice. BL/SD, C57Bl6/spf mice fed a standard diet; BL/HFD, C57Bl6/spf mice fed a high-fat diet; DB/SD, *db/db* mice fed a standard diet.

Enzymes	**a. SD vs. HFD**	**b. SD vs. DB**	**c. HFD vs. DB**
Precorrin-3B synthase			
Aquacobalamin reductase			
Cob(II)yrinic acid a,c-diamide reductase			
Cobalt-precorrin-6A reductase			
Precorrin-6A reductase			
Precorr in-2 dehydrogenase			
Uroporphyrinogen-III C-methyltransferase			
Precorrin-2 C(20)-methyltransferase			
Precorrin-3B C(17)-methyltransferase			
Precorrin-6B C(5,15)-methyltransferase (decarboxylating)			
Precorrin-4 C(11)-methyltransferase			
Cobalt-factor II C(20)-methyltransferase			

Table 3. *Cont.*

Enzymes	a. SD vs. HFD	b. SD vs. DB	c. HFD vs. DB
Precorrin-6A synthase (deacetylating)			
Cobalt-precorrin-5B (C(1))-methyltransferase			
Cobalt-precorrin-6B (C(15))-methyltransferase (decarboxylating)			
Cobalt-precorrin-4 methyltransferase			
Cobalt-precorrin-7 (C(5))-methyltransferase			
Nicotinate-nucleotide--dimethylbenzimidazole phosphoribosyltransferase			
Cob(I)yrinic acid a,c-diamide adenosyltransferase			
Adenosylcobinamide kinase			
L-threonine kinase			
Adenosylcobinamide-phosphate guanylyltransferase			
Adenosylcobinamide-GDP ribazoletransferase			
Adenosylcobalamin/alpha-ribazole phosphatase			
Adenosylcobinamide hydrolase			
Vitamin B12-transporting ATPase			
Cobalt-precorrin 5A hydrolase			
Threonine-phosphate decarboxylase			
Sirohydrochlorin cobaltochelatase			
Sirohydrochlorin ferrochelatase			
Cobalt-precorrin-8 methylmutase			
Precorrin-8X methylmutase			
Adenosylcobinamide-phosphate synthase			
Adenosylcobyric acid synthase (glutamine-hydrolyzing)			
Cobyrinate a,c-diamide synthase (glutamine-hydrolyzing)			
Hydrogenobyrinic acid a,c-diamide synthase (glutamine-hydrolyzing)			
Cobaltochelatase			

These observations may point to depletion of vitamin B12 synthesis in the gut of obese mice, mainly in *db/db* mice.

3.3. Correlation Analysis

To establish the role of individual species in cobalamin synthesis, we conducted Spearman correlation analysis, which revealed some important trends. We noticed that the abundance of enzymes involved in cobalamin biosynthesis is strongly correlated in a positive and negative way with the representation of several species in the microbiota of C57Bl6 mice fed a high-fat diet (Table 4). For example, we observed strong negative correlations between the representation of *Akkermansia sp.* and strong positive correlations between the representation of Lachnospiraceae bacteria and most of the enzymes issued. According to our results, bacteria of Verrucomicrobia phyla were represented significantly more in the microbiota of HFD-fed mice but not in SD-fed mice. Furthermore, it has previously been shown that in obese people, dietary vitamin B12 intake was inversely correlated with *Akkermansia muciniphila* species and species of the Verrucomicrobia phylum, while it was positively correlated with Bacteroides species [20].

Table 4. Spearman correlations evaluated for the representation of bacteria species in microbiota of HFD-fed mice and enzymes for cobalamin biosynthesis according to the data of metagenome sequencing. Green font shows positive correlations ($p < 0.05$), red font shows negative correlations ($p < 0.05$), black font shows non-significant correlations ($p > 0.05$). All bacteria species were uncultured bacteria, except of Enterococcus sp. and Lactobacillus plantarum.

	g__Bacteroides	g__Muribaculaceae	g__Muribaculaceae	g__Faecalibaculum	s__Enterococcus_sp.	s__Lactobacillus_plantarum	f__Lachnospiraceae	f__Lachnospiraceae	g__Butyricicoccus	g__Colidextribacter	g__Colidextribacter	g__Colidextribacter	f__Oscillospiraceae	g__Akkermansia	g__Akkermansia
Adenosylcobalamin/alpha-ribazole phosphatase	−0.92	0.97	0.55	−0.87	0.48	0.67	0.83	−0.82	0.70	−0.90	−0.63	−0.83	−0.90	−0.89	−0.90
Adenosylcobinamide kinase	−0.78	0.97	0.55	−0.83	0.63	0.66	0.83	−0.68	0.81	−0.82	−0.77	−0.72	−0.86	−0.85	−0.86
Adenosylcobinamide-GDP ribazoletransferase	−0.77	0.97	0.60	−0.87	0.65	0.70	0.85	−0.70	0.70	−0.82	−0.82	−0.75	−0.86	−0.85	−0.86
Adenosylcobinamide-phosphate guanylyltransferase	−0.78	0.97	0.55	−0.83	0.63	0.66	0.83	−0.68	0.81	−0.82	−0.77	−0.72	−0.86	−0.85	−0.86
Adenosylcobinamide-phosphate synthase	−0.80	0.97	0.67	−0.85	0.70	0.68	0.90	−0.73	0.70	−0.82	−0.78	−0.80	−0.83	−0.83	−0.83
Adenosylcobyric acid synthase (glutamine-hydrolyzing)	−0.87	0.97	0.63	−0.77	0.70	0.57	0.90	−0.73	0.81	−0.82	−0.65	−0.78	−0.83	−0.83	−0.83
Cob(I)yrinic acid a,c-diamide adenosyltransferase	−0.82	0.97	0.25	−0.62	0.34	0.32	0.61	−0.62	0.59	−0.90	−0.28	−0.63	−0.90	−0.89	−0.90
Cobalt-factor II C(20)-methyltransferase	−0.68	0.97	0.77	−0.73	0.85	0.69	0.90	−0.65	0.67	−0.79	−0.88	−0.72	−0.71	−0.74	−0.71
Cobalt-precorrin 5A hydrolase	−0.58	0.82	0.72	−0.68	0.82	0.72	0.78	−0.55	0.74	−0.75	−0.93	−0.58	−0.71	−0.74	−0.71
Cobalt-precorrin-4 methyltransferase	−0.58	0.82	0.72	−0.68	0.82	0.72	0.78	−0.55	0.74	−0.75	−0.93	−0.58	−0.71	−0.74	−0.71

Table 4. *Cont.*

	g__Bacteroides	g__Muribaculaceae	g__Muribaculaceae	g__Faecalibaculum	s__Enterococcus_sp.	s__Lactobacillus_plantarum	f__Lachnospiraceae	f__Lachnospiraceae	g__Butyricicoccus	g__Colidextribacter	g__Colidextribacter	g__Colidextribacter	f__Oscillospiraceae	g__Akkermansia	g__Akkermansia
Cobalt-precorrin-5B (C(1))-methyltransferase	−0.72	0.97	0.68	−0.82	0.79	0.72	0.92	−0.68	0.67	−0.82	−0.90	−0.75	−0.76	−0.79	−0.76
Cobalt-precorrin-6A reductase	−0.68	0.82	0.48	−0.93	0.21	0.80	0.63	−0.75	0.34	−0.79	−0.72	−0.77	−0.83	−0.80	−0.83
Cobalt-precorrin-6B (C(15))-methyltransferase (decarboxylating)	0.60	−0.05	−0.17	0.67	−0.12	−0.51	−0.57	0.50	−0.81	0.60	0.55	0.45	0.64	0.67	0.64
Cobalt-precorrin-8 methylmutase	−0.58	0.82	0.72	−0.68	0.82	0.72	0.78	−0.55	0.74	−0.75	−0.93	−0.58	−0.71	−0.74	−0.71
Cobyrinate a,c-diamide synthase (glutamine-hydrolyzing)	−0.82	0.97	0.68	−0.67	0.83	0.53	0.91	−0.68	0.81	−0.79	−0.67	−0.73	−0.76	−0.78	−0.76
Cobaltochelatase	0.52	−0.87	−0.82	0.62	−0.76	−0.70	−0.71	0.62	−0.29	0.74	0.82	0.67	0.69	0.72	0.69
Hydrogenobyrinic acid a,c-diamide synthase (glutamine-hydrolyzing)	−0.82	0.97	0.68	−0.67	0.83	0.53	0.91	−0.68	0.81	−0.79	−0.67	−0.73	−0.76	−0.78	−0.76
Nicotinate-nucleotide--dimethylbenzimidazole phosphoribosyltransferase	−0.85	0.97	0.60	−0.75	0.76	0.55	0.93	−0.72	0.81	−0.82	−0.68	−0.77	−0.81	−0.83	−0.81
Precorrin-2 C(20)-methyltransferase	−0.68	0.97	0.77	−0.73	0.85	0.69	0.90	−0.65	0.67	−0.79	−0.88	−0.72	−0.71	−0.74	−0.71
Precorrin-2 dehydrogenase	−0.73	0.67	0.67	−0.58	0.85	0.48	0.87	−0.57	0.95	−0.65	−0.68	−0.62	−0.60	−0.62	−0.60
Precorrin-3B C(17)-methyltransferase	−0.58	0.82	0.72	−0.68	0.82	0.72	0.78	−0.55	0.74	−0.75	−0.93	−0.58	−0.71	−0.74	−0.71
Precorrin-3B synthase	−0.27	0.36	−0.20	−0.17	−0.75	0.06	−0.12	−0.30	−0.02	−0.09	0.42	−0.23	−0.07	−0.01	−0.07

Table 4. *Cont.*

	g__Bacteroides	g__Muribaculaceae	g__Muribaculaceae	g__Faecalibaculum	s__Enterococcus_sp.	s__Lactobacillus_plantarum	f__Lachnospiraceae	f__Lachnospiraceae	g__Butyricicoccus	g__Colidextribacter	g__Colidextribacter	g__Colidextribacter	f__Oscillospiraceae	g__Akkermansia	g__Akkermansia
Precorrin-4 C(11)-methyltransferase	−0.58	0.82	0.72	−0.68	0.82	0.72	0.78	−0.55	0.74	−0.75	−0.93	−0.58	−0.71	−0.74	−0.71
Precorrin-6A reductase	−0.68	0.82	0.48	−0.93	0.21	0.80	0.63	−0.75	0.34	−0.79	−0.72	−0.77	−0.83	−0.80	−0.83
Precorrin-6B C(5,15)-methyltransferase (decarboxylating)	−0.68	0.97	0.67	−0.83	0.69	0.74	0.80	−0.65	0.67	−0.82	−0.88	−0.70	−0.86	−0.85	−0.86
Precorrin-8X methylmutase	−0.58	0.82	0.72	−0.68	0.82	0.72	0.78	−0.55	0.74	−0.75	−0.93	−0.58	−0.71	−0.74	−0.71
Sirohydrochlorin cobaltochelatase	−0.57	0.67	0.78	−0.62	0.90	0.61	0.79	−0.50	0.81	−0.65	−0.87	−0.57	−0.64	−0.66	−0.64
Sirohydrochlorin ferrochelatase	−0.73	0.67	0.67	−0.58	0.85	0.48	0.87	−0.57	0.95	−0.65	−0.68	−0.62	−0.60	−0.62	−0.60
Threonine-phosphate decarboxylase	−0.88	0.97	0.73	−0.68	0.70	0.59	0.90	−0.78	0.81	−0.82	−0.62	−0.82	−0.71	−0.73	−0.71
Uroporphyrinogen-III C-methyltransferase	−0.72	0.97	0.68	−0.82	0.79	0.72	0.92	−0.68	0.67	−0.82	−0.90	−0.75	−0.76	−0.79	−0.76

At the same time, we observed almost a complete loss of correlations in the microbiota of *db/db* mice, except for positive correlations for species of the Lachnospiraceae family (Table 5). This fact is apparently caused by a significant decrease in the alpha diversity of microbiota communities and a loss of microbes involved in cobalamine synthesis.

Table 5. Spearman correlations evaluated for the representation of bacteria species in the microbiota of *db/db* mice and enzymes for cobalamin biosynthesis according to the data from metagenome sequencing. Green font shows positive correlations ($p < 0.05$), red font shows negative correlations ($p < 0.05$), black font shows nonsignificant correlations ($p > 0.05$).

	s__Burkholderia_sp.	s__Muribacter_muris	s__Pseudomonas_sp.	f__Lachnospiraceae; g__uncultured; s__uncultured_bacterium	g__Colidextribacter; s__uncultured_bacterium
Cobalt-precorrin-6A reductase	−0.43	0.65	0.33	0.84	0.75
Precorrin-6A reductase	−0.43	0.65	0.33	0.84	0.75
Precorrin-2 dehydrogenase	−0.77	0.88	0.11	0.23	0.79
Uroporphyrinogen-III C-methyltransferase	−0.77	0.88	0.11	0.23	0.79
Precorrin-2 C(20)-methyltransferase	−0.58	0.74	0.54	0.78	0.79
Precorrin-3B C(17)-methyltransferase	−0.52	0.74	0.71	0.84	0.61
Precorrin-6B C(5,15)-methyltransferase (decarboxylating)	−0.67	0.74	0.71	0.74	0.64
Precorrin-4 C(11)-methyltransferase	−0.52	0.74	0.71	0.84	0.61
Cobalt-factor II C(20)-methyltransferase	−0.58	0.74	0.54	0.78	0.79
Precorrin-6A synthase (deacetylating)	−0.21	0.12	−0.93	−0.35	0.20
Cobalt-precorrin-5B (C(1))-methyltransferase	−0.67	0.74	0.65	0.74	0.61
Cobalt-precorrin-4 methyltransferase	−0.52	0.74	0.71	0.84	0.61
Nicotinate-nucleotide--dimethylbenzimidazole phosphoribosyltransferase	−0.86	0.77	0.37	0.63	0.50
Cob(I)yrinic acid a,c-diamide adenosyltransferase	−0.73	0.88	0.07	0.51	0.64
Adenosylcobinamide kinase	−0.82	0.77	0.31	0.56	0.54
Adenosylcobinamide-phosphate guanylyltransferase	−0.82	0.77	0.31	0.56	0.54
Adenosylcobinamide-GDP ribazoletransferase	−0.86	0.77	0.37	0.63	0.50
Cobalt-precorrin 5A hydrolase	−0.52	0.74	0.77	0.84	0.64
Threonine-phosphate decarboxylase	−0.67	0.74	0.65	0.74	0.61
Sirohydrochlorin cobaltochelatase	−0.52	0.74	0.71	0.84	0.61
Sirohydrochlorin ferrochelatase	−0.77	0.88	0.17	0.23	0.82
Cobalt-precorrin-8 methylmutase	−0.52	0.74	0.71	0.84	0.61
Precorrin-8X methylmutase	−0.52	0.74	0.71	0.84	0.61
Adenosylcobinamide-phosphate synthase	−0.67	0.74	0.65	0.74	0.61
Adenosylcobyric acid synthase (glutamine-hydrolyzing)	−0.67	0.74	0.65	0.74	0.61
Cobyrinate a,c-diamide synthase (glutamine-hydrolyzing)	−0.52	0.74	0.71	0.84	0.61
Hydrogenobyrinic acid a,c-diamide synthase (glutamine-hydrolyzing)	−0.52	0.74	0.71	0.84	0.61
Cobaltochelatase	0.09	0.12	−0.85	−0.20	−0.07

4. Discussion

Vitamin B12 is an essential component for pro- and eucaryotic living organisms [21]. Because it is synthesized exclusively in bacterial cells, it can be accumulated in animals that receive it with food [22]. Many species of commensal bacteria inhabiting the mammalian gut are capable of synthesizing vitamin B12; however, it is not reliably known whether this microbiota-derived vitamin can be effectively absorbed into the host's circulation and contribute to the systemic level of cobalamin [23]. In mammalian cells, vitamin B12 is used for the methionine synthase reaction and for the metabolism of amino and fatty acids [24]. Vitamin B12 deficiency is associated with the development of megaloblastic anemia and neurological complications [23]. Furthermore, an obese state and other dysmetabolic conditions have been found to be associated with decreased levels of vitamin B12 [6]. The fact that obesity is coupled with disturbances in intestinal microbiota composition, and metabolic activity also points to possible disturbances in cobalamin synthesis by commensal microbes in obesity. Interestingly, there may be a reciprocal relationship between the gut microbiota and vitamin B12 levels [15]. Vitamin B12 deficiency can alter the balance and functional interaction of the gut microbial community [25]. Therefore, by changing the microbial composition of the gut, interactions between vitamin B12 and the gut microbiota can prevent the development of obesity [26]. Thus, there may be a relationship between obesity, intestinal flora, and vitamin B12 levels [24].

To our knowledge, there are few studies investigating the ability of the intestinal microbiota to synthesize vitamin B12 in association with obesity.

Mice deficient in the leptin receptor (*db/db*) develop severe obesity with corresponding hormonal changes, such as hyperinsulinemia and hyperleptinemia, and they are widely used as a model of hyperphagia, obesity, and type 2 diabetes mellitus [27]. Furthermore, as we have shown in the present investigation, the gut microbiota of *db/db* mice differ significantly compared to the microbiota of C57Bl6 mice. That is, we observed a dramatic decrease in alpha diversity and the representation of 'obesoprotective' microbes, such as *Akkermansia miciniphila* and *Bifidobacterium pseudolongum*, along with an increase in the abundance of 'obesogenic' *Prevotella* in *db/db* mice, in contrast to C57Bl6 mice fed a high-fat diet. Mice that received an HFD showed an increased representation of Verrucomicrobia and Actinobacteria phyla, which may be associated with increased fuel sources for bacterial growth. However, the alpha diversity of the gut microbiota community of HFD-fed mice decreased as well, as is confirmed by the Shannon and PD whole-tree indexes.

For the first time, we have shown that severe obesity with corresponding metabolic disturbances, which are developed in *db/db* mice, is associated with a decrease in the representation of enzymes and metabolic pathways for the synthesis of cobalamin. However, C57Bl6 mice were fed a high-fat diet and also developed obesity, and although they gained less weight than *db/db* mice, they showed no differences in enzyme abundance and fewer pathways of vitamin B12 synthesis. Furthermore, strong associations between the representation of gut microbes and an abundance of cobalamin biosynthetic enzymes were described for HFD-fed mice. For example, we have seen strong negative correlations for the genera Colidextribacter, Akkermansia, Bacteroides, and Faecalibacterium, while the genera Muribaculaceae, Enterococcus, Butyricoccus and the uncultured genus of the Lachnospiraceae family were negatively associated with enzyme abundance. On the other hand, in *db/bd* mice, only positive correlations were observed with the uncultured genus of the Lachnospiraceae family, whereas other microbes showed no or unitary correlations with enzyme abundance.

Therefore, the degree of obesity and the composition of the correspondent microbiota are the main contributors to the predicted abundance of genes and pathways for cobalamin biosynthesis in the gut of mice. It can be proposed that the 'obesogenic' microbiota community is dwindling with vitamin B12 microbe producers, which can lead to the formation of a more aberrant microbial community and lower levels of vitamin B12 in the host. However, this hypothesis should be confirmed by studying serum levels of cobalamin in association with the degree of obesity and intestinal microbiota composition, as well as the direct contribution of separate microbes in cobalamin synthesis.

5. Conclusions

In this study, we performed a high-throughput metagenome sequencing analysis followed by a reconstruction of the metabolic activity of the gut microbiota for a high-fat diet-fed C57BL/6SPF mice, standard diet-fed *db/db* mice, and standard diet-fed C57BL/6SPF mice. We observed a specific obesogenic shift in mouse gut microbiota communities, which was more pronounced in *db/db* mice than in HFD-fed mice and was consistent with higher body weight and hormonal disturbances of *db/db* mice. We have defined a significantly lower predicted abundance of enzymes and metabolic pathways for vitamin B12 biosynthesis in obese mice compared to non-obese mice, where enzyme depletion was more pronounced in *db/db* mice, which developed severe obesity. The abundance of enzymes involved in cobalamin synthesis is strongly correlated with the representation of several microbes in HFD-fed mice, whereas there were almost no correlations in *db/db* mice. Therefore, the obesogenic gut microbiota can be implicated in decreased vitamin B12 synthesis in the gut, which, in turn, can influence host levels of cobalamin and make obese mice more prone to the development of cobalamin deficiency. However, because these findings are based on bioinformatic analysis, more direct molecular experiments are needed that confirm the involvement of the gut microbiota in vitamin B12 synthesis in connection with host vitamin B12 status and obesity development.

Supplementary Materials: The following supporting information can be downloaded at: https://www.mdpi.com/article/10.3390/biomedicines12061280/s1, Table S1: Primers used to amplify the V3-V4 region of bacterial 16S rRNA gene.

Author Contributions: Conceptualization, A.A.Z. and A.V.S.; methodology, I.M.K.; validation, I.Y.V. and T.V.G.; formal analysis, A.A.Z. and I.Y.V.; investigation, I.M.K.; data curation, T.V.G.; writing—original draft preparation, A.A.Z.; writing—review and editing, A.V.S.; visualization, A.A.Z.; resources, S.A.R.; supervision, A.V.S. and S.A.R.; project administration, A.V.S.; funding acquisition, A.V.S. and S.A.R. All authors have read and agreed to the published version of the manuscript.

Funding: This research was funded by the Ministry of Science and Higher Education of the Russian Federation (grant number 075-15-2022-310).

Institutional Review Board Statement: The animal study protocol was approved by the Institutional Review Board (or Ethics Committee) of the Ethics Committee for Animal Research of I.M. Sechenov First Moscow State Medical University, Russia (protocol number 96 from 2 September 2021).

Informed Consent Statement: Not applicable.

Data Availability Statement: The data presented in this study are available upon request from the corresponding author.

Conflicts of Interest: The authors declare no conflicts of interest.

References

1. Hou, K.; Wu, Z.-X.; Chen, X.-Y.; Wang, J.-Q.; Zhang, D.; Xiao, C.; Zhu, D.; Koya, J.B.; Wei, L.; Li, J.; et al. Microbiota in Health and Diseases. *Signal Transduct. Target. Ther.* **2022**, *7*, 135. [CrossRef]
2. Van Hul, M.; Cani, P.D. The Gut Microbiota in Obesity and Weight Management: Microbes as Friends or Foe? *Nat. Rev. Endocrinol.* **2023**, *19*, 258–271. [CrossRef] [PubMed]
3. Muscogiuri, G.; Cantone, E.; Cassarano, S.; Tuccinardi, D.; Barrea, L.; Savastano, S.; Colao, A. Gut Microbiota: A New Path to Treat Obesity. *Int. J. Obes. Suppl.* **2019**, *9*, 10–19. [CrossRef] [PubMed]
4. Baltaci, D.; Kutlucan, A.; Turker, Y.; Yilmaz, A.; Karacam, S.; Deler, H.; Ucgun, T.; Kara, I.H. Association of Vitamin B12 with Obesity, Overweight, Insulin Resistance and Metabolic Syndrome, and Body Fat Composition; Primary Care-Based Study. *Med. Glas.* **2013**, *10*, 203–210.
5. Sun, Y.; Sun, M.; Liu, B.; Du, Y.; Rong, S.; Xu, G.; Snetselaar, L.G.; Bao, W. Inverse Association Between Serum Vitamin B12 Concentration and Obesity among Adults in the United States. *Front. Endocrinol.* **2019**, *10*, 414. [CrossRef] [PubMed]
6. Boachie, J.; Adaikalakoteswari, A.; Samavat, J.; Saravanan, P. Low Vitamin B_{12} and Lipid Metabolism: Evidence from Pre-Clinical and Clinical Studies. *Nutrients* **2020**, *12*, 1925. [CrossRef] [PubMed]
7. Demirtas, M.S.; Kilicaslan, C.; Erdal, H. Evaluation of Vitamin B_{12} Levels among Severe Obese and Obese Adolescents. *J. Investig. Med.* **2024**, *72*, 319–325. [CrossRef] [PubMed]
8. Ghosh, S.; Sinha, J.K.; Putcha, U.K.; Raghunath, M. Severe but Not Moderate Vitamin B_{12} Deficiency Impairs Lipid Profile, Induces Adiposity, and Leads to Adverse Gestational Outcome in Female C57BL/6 Mice. *Front. Nutr.* **2016**, *3*, 1. [CrossRef] [PubMed]
9. Koplay, M.; Gulcan, E.; Ozkan, F. Association between Serum Vitamin B_{12} Levels and the Degree of Steatosis in Patients with Nonalcoholic Fatty Liver Disease. *J. Investig. Med.* **2011**, *59*, 1137–1140. [CrossRef]

10. Boachie, J.; Adaikalakoteswari, A.; Gázquez, A.; Zammit, V.; Larqué, E.; Saravanan, P. Vitamin B12 Induces Hepatic Fatty Infiltration through Altered Fatty Acid Metabolism. *Cell. Physiol. Biochem.* **2021**, *55*, 241–255. [CrossRef]
11. Polter, D.E.; Boyle, J.D.; Miller, L.G.; Finegold, S.M. Anaerobic Bacteria as Cause of the Blind Loop Syndrome. *Gastroenterology* **1968**, *54*, 1148–1154. [CrossRef] [PubMed]
12. Gille, D.; Schmid, A. Vitamin B_{12} in Meat and Dairy Products. *Nutr. Rev.* **2015**, *73*, 106–115. [CrossRef]
13. Kundra, P.; Geirnaert, A.; Pugin, B.; Morales Martinez, P.; Lacroix, C.; Greppi, A. Healthy Adult Gut Microbiota Sustains Its Own Vitamin B_{12} Requirement in an in Vitro Batch Fermentation Model. *Front. Nutr.* **2022**, *9*, 155. [CrossRef]
14. Juodeikis, R.; Jones, E.; Deery, E.; Beal, D.M.; Stentz, R.; Kräutler, B.; Carding, S.R.; Warren, M.J. Nutrient Smuggling: Commensal Gut Bacteria-derived Extracellular Vesicles Scavenge Vitamin B_{12} and Related Cobamides for Microbe and Host Acquisition. *J. Extracell. Biol.* **2022**, *1*, e61. [CrossRef]
15. Guetterman, H.M.; Huey, S.L.; Knight, R.; Fox, A.M.; Mehta, S.; Finkelstein, J.L. Vitamin B-12 and the Gastrointestinal Microbiome: A Systematic Review. *Adv. Nutr.* **2022**, *13*, 530–558. [CrossRef] [PubMed]
16. Bolyen, E.; Rideout, J.R.; Dillon, M.R.; Bokulich, N.A.; Abnet, C.C.; Al-Ghalith, G.A.; Alexander, H.; Alm, E.J.; Arumugam, M.; Asnicar, F.; et al. Reproducible, Interactive, Scalable and Extensible Microbiome Data Science Using QIIME 2. *Nat. Biotechnol.* **2019**, *37*, 852–857. [CrossRef] [PubMed]
17. Douglas, G.M.; Maffei, V.J.; Zaneveld, J.R.; Yurgel, S.N.; Brown, J.R.; Taylor, C.M.; Huttenhower, C.; Langille, M.G.I. PICRUSt2 for Prediction of Metagenome Functions. *Nat. Biotechnol.* **2020**, *38*, 685–688. [CrossRef]
18. Wang, C.-Y.; Liao, J.K. A Mouse Model of Diet-Induced Obesity and Insulin Resistance. *Methods Mol. Biol.* **2012**, *821*, 421–433.
19. Dong, T.S.; Guan, M.; Mayer, E.A.; Stains, J.; Liu, C.; Vora, P.; Jacobs, J.P.; Lagishetty, V.; Chang, L.; Barry, R.L.; et al. Obesity Is Associated with a Distinct Brain-Gut Microbiome Signature That Connects Prevotella and Bacteroides to the Brain's Reward Center. *Gut Microbes* **2022**, *14*, 2051999. [CrossRef]
20. Al-Musharaf, S.; Aljuraiban, G.S.; Al-Ajllan, L.; Al-Khaldi, N.; Aljazairy, E.A.; Hussain, S.D.; Alnaami, A.M.; Sabico, S.; Al-Daghri, N. Vitamin B_{12} Status and Gut Microbiota among Saudi Females with Obesity. *Foods* **2022**, *11*, 4007. [CrossRef]
21. Halczuk, K.; Kaźmierczak-Barańska, J.; Karwowski, B.T.; Karmańska, A.; Cieślak, M. Vitamin B_{12}—Multifaceted In Vivo Functions and In Vitro Applications. *Nutrients* **2023**, *15*, 2734. [CrossRef] [PubMed]
22. Watanabe, F.; Bito, T. Vitamin B_{12} Sources and Microbial Interaction. *Exp. Biol. Med.* **2018**, *243*, 148–158. [CrossRef] [PubMed]
23. Green, R.; Allen, L.H.; Bjørke-Monsen, A.-L.; Brito, A.; Guéant, J.-L.; Miller, J.W.; Molloy, A.M.; Nexo, E.; Stabler, S.; Toh, B.-H.; et al. Vitamin B_{12} Deficiency. *Nat. Rev. Dis. Primers* **2017**, *3*, 17040. [CrossRef] [PubMed]
24. Green, R.; Miller, J.W. Vitamin B_{12} Deficiency. *Vitam. Horm.* **2022**, *119*, 405–439. [PubMed]
25. Sharma, V.; Rodionov, D.A.; Leyn, S.A.; Tran, D.; Iablokov, S.N.; Ding, H.; Peterson, D.A.; Osterman, A.L.; Peterson, S.N. B-Vitamin Sharing Promotes Stability of Gut Microbial Communities. *Front. Microbiol.* **2019**, *10*, 1485. [CrossRef] [PubMed]
26. Voland, L.; Le Roy, T.; Debédat, J.; Clément, K. Gut Microbiota and Vitamin Status in Persons with Obesity: A Key Interplay. *Obes. Rev.* **2022**, *23*, e13377. [CrossRef]
27. Wang, B.; Chandrasekera, P.; Pippin, J. Leptin- and Leptin Receptor-Deficient Rodent Models: Relevance for Human Type 2 Diabetes. *Curr. Diabetes Rev.* **2014**, *10*, 131–145. [CrossRef]

biomedicines

MDPI

Article

Supernatants from Newly Isolated *Lacticaseibacillus paracasei* P4 Ameliorate Adipocyte Metabolism in Differentiated 3T3-L1 Cells

Natalia Grigorova [1], Zhenya Ivanova [1], Valeria Petrova [1], Ekaterina Vachkova [1] and Georgi Beev [2,*]

1 Department of Pharmacology, Animal Physiology, Biochemistry and Chemistry, Faculty of Veterinary Medicine, Trakia University, 6000 Stara Zagora, Bulgaria; nataliya.grigorova@trakia-uni.bg (N.G.); zhenya.ivanova.12@trakia-uni.bg (Z.I.); valeriya.petrova@trakia-uni.bg (V.P.); ekaterina.vachkova@trakia-uni.bg (E.V.)

2 Department of Biochemistry, Microbiology and Physics, Faculty of Agriculture, Trakia University, 6000 Stara Zagora, Bulgaria

* Correspondence: georgi.beev@trakia-uni.bg

Abstract: Background: *Lacticaseibacillus paracasei* (*L. paracasei*) strains and their postbiotics show potential for managing metabolic disorders such as diabetes and obesity. Two newly isolated *L. paracasei* strains, M2.1 and P4, were yielded from *Formica rufa* anthills in Sinite Kamani National Park, Bulgaria. Their metabolic effects on mature 3T3-L1 adipocytes were investigated. **Methods:** Mature 3T3-L1 adipocytes were treated for 24 h with 10% (v/v) cell-free supernatants (CFSs) of M2.1 or P4. Two experimental (M2.1, P4) and two control groups (mature, untreated adipocytes and mature adipocytes, treated with 10% (v/v) MRS broth) were analyzed for intracellular lipid accumulation, glucose uptake, and the mRNA expression of lipid metabolism and beta-oxidation-related genes. Fold changes in gene expression were assessed using RT-qPCR. **Results:** Both M2.1 and P4 CFSs enhanced glucose uptake by over 30% compared to the control. P4 demonstrated a more favorable effect by significantly upregulating adipose triglyceride lipase–patatin-like phospholipase domain containing 2, adiponectin, and peroxisomal beta-oxidation enzymes—acyl-coenzyme A oxidase 1, palmitoyl. Intracellular lipid accumulation increased only with M2.1, while P4 supported improved lipid turnover without promoting excessive lipid storage or lipolysis. **Conclusions:** P4 CFS exhibits the potential to improve adipocyte metabolism by enhancing glucose uptake, promoting beta-oxidation, and increasing adiponectin expression, offering a promising strategy for managing metabolic dysfunctions.

Keywords: *Lacticaseibacillus paracasei* P4 and M2.1 strains; postbiotics; adipocyte metabolism; glucose uptake; beta-oxidation; adiponectin

Citation: Grigorova, N.; Ivanova, Z.; Petrova, V.; Vachkova, E.; Beev, G. Supernatants from Newly Isolated *Lacticaseibacillus paracasei* P4 Ameliorate Adipocyte Metabolism in Differentiated 3T3-L1 Cells. *Biomedicines* **2024**, *12*, 2785. https://doi.org/10.3390/biomedicines12122785

Academic Editors: Teresa Vezza and Zaida Abad-Jiménez

Received: 16 October 2024
Revised: 2 December 2024
Accepted: 5 December 2024
Published: 7 December 2024

1. Introduction

Probiotics are unique microorganisms that confer health benefits to the host when administrated in appropriate amounts. *Lactobacilli*, *Bifidobacteria*, and specific yeast species rank among the microbes with the most robust probiotic characteristics [1,2]. The proper selection and regular consumption of probiotics help balance gut microbiota, enhance gut barrier function, stimulate the immune response, reduce inflammation caused by pathogens, improve digestion, and optimize nutrient absorption [1]. In addition to the well-documented local benefits, there is growing attention to their systemic effects on the host [3,4]. Probiotics are known to improve the lipid profile and insulin sensitivity, reducing inflammation, and the risk of cardiovascular and liver diseases [3–8]. These effects are primarily determined by the resorbed metabolic products released during the probiotic's life cycle, such as short-chain fatty acids, peptides, enzymes, vitamins, and others, collectively known as postbiotics [9,10]. Therefore, a contemporary trend in nutrigenomics is the direct intake of postbiotics, specifically targeted at particular physiological pathways rather than probiotic intake per se [11,12]. The advantages of this approach include greater stability of postbiotics during storage and transport compared to live microorganisms, significantly faster impact, easier and more precise dosing, and strictly specific targeting of a certain health issue, with their effectiveness not dependent on the potential for microbial colonization—a limiting factor for probiotics [9,12,13]. Postbiotics are the preferred choice for people with compromised immune systems or those undergoing severe medical interventions due to the existing risk of bacteremia and other infectious complications [14,15].

Research into the direct effects of postbiotics remains limited. However, evidence suggests that certain probiotic strains exhibit significant antiadipogenic and anti-obesity potential, highlighting their role in modulating body weight and metabolic functions [8,16,17]. These systemic health benefits are mainly based on modulating the physiological state of adipose tissue and the overall level of body inflammation. Besides its well-documented function as an energy storage depot, adipose tissue is a highly metabolic endocrine organ that significantly influences systematic metabolism by producing hundreds of adipokines and biologically active substances [18,19]. Adiponectin, for instance, is known for enhancing insulin sensitivity, inflammation, and the lipid profile and has a significant role in cancer prevention [19,20]. Leptin is well known for regulating appetite and energy homeostasis but also profoundly affects immune system modulations. Thus, the expression levels of adiponectin and leptin influence not only the physiological state of adipocytes but also extend to the overall health of the entire body [19,20]. To improve health outcomes for individuals with obesity, scientists are focusing on dietary supplements that increase the production of beneficial adipokines and enhance beta-oxidation, avoiding the stimulation of lipolysis [21–23]. Such targets for developing nutritional strategies extend beyond traditional views of diet and disease, focusing on metabolic modulation at the cellular level for broader systemic health benefits.

It should be noted that different strains, even those closely related, produce distinct, postbiotics and thereby modulate physiological processes in the microorganism in a highly specific manner [24–26]. Exploring new sources and strains and deepening our understanding of postbiotic mediators suggests a path for developing dietary supplements for personalized pharmacological strategies that meet the body's unique needs. There are many beneficial effects described in the literature concerning *Lacticaseibacillus paracasei* (*L. paracasei*) strains [27–29]. They comprise antimicrobial, anti-inflammatory, antioxidant, anti-obesity, and lipid metabolism improvement; hypocholesterolemic and stress modulator effects; immune system stimulation; intestinal bacterial microbiota enhancement; and many others.

This study aimed to evaluate the effects of potential postbiotics derived from newly isolated *L. paracasei* strains M2.1 and P4 on adipocyte differentiation, lipid metabolism, and the expression of key genes involved in beta-oxidation. Additionally, the impact on adiponectin, a major health-promoting adipokine, was investigated.

2. Materials and Methods

2.1. Materials and Chemical Reagents

The current study employed supernatants from newly isolated Bulgarian *L. paracasei* strains M2.1 and P4, as well as 3T3-L1 mouse embryonic fibroblasts (ATCC® CRL-3242™) from the American Type Culture Collection (ATCC, Washington, DC, USA). The reagents used in the current investigation were Dulbecco's Modified Eagle's Medium (DMEM) with high glucose content (4500 mg/L), fetal bovine serum (FBS), L-glutamine, an antibiotic solution (Penicillin G, Streptomycin, Amphotericin B), indomethacin, dexamethasone, phosphate-buffered saline (PBS), 100% isopropanol, sodium chloride (HCL), Oil Red O powder, 3-(4,5-dimethyl-2-thiazolyl)-2,5-diphenyl-2*H*-tetrazolium bromide (MTT) powder, a trypsin solution, dimethyl sulfoxide (DMSO), and adipolysis assay kit MAK313—all suitable for cell cultures and purchased from Sigma-Aldrich, Chemie, GmbH (Merk KGaA, Darmstadt, Germany). Insulin (cell application, San Diego, CA, USA) and 3-isobutyl-1-

methylxanthine IBMX (Cayman Chemical, Ann Arbor, MI, USA) were also used. The investigated microorganisms were pre-cultured in de Man, Rogosa, and Sharpe (MRS) broth, which was supplied by Oxoid, UK. A Glucose GOD-PAD reagent was purchased from Biolabo SAS (Maizy, France). The plates and pipettes used were sterile and single-use, produced by Corning Incorporated, Costar, Washington, DC, USA. For gene expression analyses, we used the RNeasy Mini Lipid Tissue Kit (QIAGEN Sciences, Inc.; Germantown, MD, USA), the RevertAid First Strand cDNA Synthesis Kit (Thermo Scientific, Waltham, MA, USA), and the KAPA SYBR® fast qPCR Master Mix kit (QIAGEN Sciences, Inc.; Germantown, MD, USA).

2.2. Preparation of Microbial Supernatants

L. paracasei strains (M2.1 and P4) were isolated from *Formica rufa* anthills in Sinite Kamani National Park, Bulgaria, and identified and cultivated in MRS at 37 °C for 24 h, as previously described [26]. Supernatants from both microorganisms, M2.1 and P4, were filtered through a sterile syringe filter with a pore size of 20 µm (Corning, New York, NY, USA) after centrifugation at 9000 rpm for ten minutes. The obtained supernatants were adjusted to pH 7 with 0.1 N NaOH and then ex tempore included in the freshly prepared adipocyte maintenance culture media (AMM) at a 10% *v*/*v* concentration.

2.3. Cultivation and Adipogenesis of 3T3-L1 Cells

3T3-L1 preadipocytes were propagated in basal media (BM) consisting of DMEM, 10% (*v*/*v*) FBS, and 1% antibiotic solution in T75 flasks. They were then seeded in 12- and 24-well plates at 10^4 cells/mL concentrations and cultured at 37 °C in a humidified atmosphere of 95% air and 5% CO_2. Upon reaching 100% confluence, the cells underwent a 24 h growth arrest followed by adipogenic differentiation. They were cultured for 48 h in adipogenic induction media (AIM) containing DMEM, 10% (*v*/*v*) FBS, 2% L-glutamine, 0.1 mM IBMX, 0.05 mM indomethacin, 1 µM dexamethasone, 10 µg/mL insulin, and 1% antibiotic solution. In order to achieve full maturation by day 8, the cells were maintained in AMM composed of DMEM, 10% (*v*/*v*) FBS, 2% L-glutamine, 10 µg/mL insulin, and 1% antibiotic solution.

On day 9, the mature adipocytes were divided into two experimental and two control groups. The experimental groups, M2.1 and P4, were treated for 24 h with 10% *v*/*v* *L. paracasei* M2.1 or P4 CFS in AMM. As a positive control group, mature, untreated cells (IC) were cultured, serving as a baseline for successful adipogenesis. As a Supplementary Material Figure S1 comparison of intracellular neutral lipid accumulation between IC and non-differentiated, untreated (NC) groups at day 8 was provided (Supplementary Materials (Figure S1)). However, an additional control group of mature adipocytes supplemented with 10% (*v*/*v*) (MRS) was included to evaluate the possible metabolic influence of the pure MRS for the same duration.

Three parallel replicates were conducted concurrently for each group (n = 6): Replicate 1 (24-well plates) involved cell viability assessment using the MTT assay; Replicate 2 (12-well plates) included the evaluation of neutral lipid accumulation microscopically and spectrophotometrically by adding Oil Red O staining; and Replicate 3 (12-well plates) was used for measuring glucose and glycerol concentrations in cell supernatants and isolating mRNA from the same 3T3-L1 adipocytes for an RT-PCR gene expression analysis.

2.4. Cell Viability Assay

The cell viability of 3T3-L1 upon 24 h treatment with 10% supernatants was determined using the MTT assay. This colorimetric method exploits the reduction in MTT by NAD(P)H-dependent cellular oxidoreductase enzymes, producing an insoluble purple formazan product, as detailed by Yang et al. [30]. In this procedure, cells were incubated with the MTT solution (5 mg/mL) at 37 °C for 80 min. Following incubation, the formazan product was solubilized using an isopropanol solution containing 0.04 N HCl. Absorbance was measured at 570 nm (a reference wavelength of 630 nm) using a Synergy TM Lee Multi-Mode Microplate Reader from BioTek Instruments, Inc., Santa Clara, CA, USA. The results were expressed as a percentage of the control (NC), in line with the approach described by Park et al. [31].

2.5. Oil Red O Staining and Intracellular Lipid Accumulation Assessment

At the end of the experiment, differentiated adipocytes were fixed with 10% formalin, dried with 60% isopropanol, and then colored with Oil Red O for 30 min. To quantify neutral lipid accumulation, the stain was extracted from the lipid droplets with 100% isopropanol, and the dye absorbance was measured at 490 nm [32].

2.6. Glycerol Concentration Measurement and Lipolysis Rate Estimation

We quantified the glycerol released into the supernatants to evaluate lipolysis in mature adipocytes treated with 10% *L. paracasei* CFS. This analysis used the adipolysis assay kit specifically designed for cell culture supernatants. Each sample was measured twice, at a 570 nm wavelength with a reference correction at 630 nm, to ensure accuracy. Glycerol concentrations were then calculated based on a pre-established concentration curve, following the protocol provided by the manufacturer, and expressed relative to the MRS group as a percent.

2.7. Glucose Concentration in Cell Supernatants

After the mature adipocytes were exposed for 24 h to *L. paracasei* M2.1 or P4 CFS, or a 10% solution of MRS broth, we assessed the glucose levels in cell supernatants. This measurement was performed using the Mindrey BS-120 automatic biochemical analyzer manufactured in Guangzhou, China. For the glucose assay, we used the Glucose GOD-PAD reagent. The procedure was carried out according to the manufacturers' guidelines. To determine the glucose level taken from the treated cells in each group, we subtracted each value (experimental glucose concentration (EG)) from the corresponding amount established in the freshly prepared media just before application (initial glucose concentration (IG)) based on the equation provided by Diaz et al. [33].

$$\text{Glucose uptake UG (mg/L)} = \text{IG} - \text{EG}$$

Finally, the data were presented as a percentage of the control group (MRS).

2.8. Real-Time PCR

The RNeasy Mini Lipid Tissue Kit was used for total mRNA isolation from pre-lysed mature adipocytes. The quality and quantity of the obtained mRNA were evaluated spectrophotometrically, ensuring that only high-quality mRNA (with absorbance ratios of approximately 2 at 260/280 nm) was used for subsequent analyses. The reverse transcription of equal amounts of mRNA from each sample was performed using the RevertAid First Strand cDNA Synthesis Kit. RT-qPCR was conducted using the KAPA SYBR Green Master Mix, employing self-designed primers as previously described by Grigorova et al. [26]. The RT-PCR data were analyzed using the modified ΔΔCt method, which incorporates normalization based on multiple housekeeping genes. Six housekeeping genes were analyzed: Hypoxanthine Phosphoribosyltransferase (*Hprt*), 18S ribosomal RNA (*18S*), Beta-actin (*Actb*), Ribosomal Protein L19 (Rpl19, also known as 36B4), Glyceraldehyde 3-Phosphate Dehydrogenase (*Gapdh*), and Hydroxymethylbilane Synthase (*Hmbs*). The identification of the least variable gene or combination of genes was performed by the web-based software RefFinder (http://blooge.cn//RefFinder/) accessed on 15 October 2024 [34]. In the final analysis, a combination of *Hprt* and *Actb* was utilized. The sequences of the employed housekeeping and target genes are presented in Table 1.

Table 1. Primer Sequences for RT-PCR Analysis.

Abbreviation	Full Name	Forward Primer	Reverse Primer	Product Size (bp)
Cpt1 NM_013495.2	Carnitine palmitoyltransferase 1a	AAGAACATCGTGAGTGGCGT	GACCTTGACCATAGCCATCCA	165
Cpt2 NM_009949.2	Carnitine palmitoyltransferase 2	CATCGTACCCACCATGCACT	CTCCTTCCCAATGCCGTTCT	169
Acaca NM_133360.3	Acetyl-CoA carboxylase	TGCTCATGTTCCTTGCCCAA	TGCCACCACCATATTTGAGATT	247
Fasn NM_007988.3	Fatty acid synthase	CTGAAGCCGAACACCTCTGT	GGGAATGTTACACCTTGCTCCT	218
Pnpla2 NM_001163689.1	Patatin-like phospholipase domain containing 2	CCTTCACCATCCGCTTGTTG	CCCAGTGAGAGGTTGTTTCG	250
Plin1 NM_001113471.1	Perilipin 1	ACCCTCCAGAAAAGATCGCC	CTTCCCAGAGCCAGATCAGC	229
Fabp4 NM_024406.4	Fatty acid-binding protein 4	AACTGGGCGTGGAATTCGAT	CCACCAGCTTGTCACCATCT	150
Acox1 NM_015729.4	Acyl-coenzyme A oxidase 1, palmitoyl	ACAGAGATGGGTCATGGAACT	ATGTAACCCGTAGCACTCCC	195
Adipoq NM_028320.4	Adiponectin	TCCCGTATGATGTGCTTCCT	AGCACAAAACCAAGCAGATGT	157
Actb NM_007393.5	β-actin	CCTCTATGCCAACACAGTGC	GTACTCCTGCTTGCTGATCC	211
Hprt NM_013556.2	Hypoxanthine guanine phosphoribosyl transferase	ACAGGCCAGACTTTGTTGGA	ACTTGCGCTCATCTTAGGCT	150

2.9. Statistical Analyses

We utilized Statistica version 10 (StatSoft Inc., 2011, Tulsa, OK, USA) to analyze the obtained data. Initially, descriptive statistics were conducted to calculate the mean and standard error of the mean, which are represented in the figures with values and error bars. The statistical significance of differences was assessed using the non-parametric Mann–Whitney U test.

3. Results

3.1. Results from MTT Cell Viability Assay

On day 8 following adipogenic induction, mature 3T3-L1 cells were treated with 10% cell-free supernatants (CFSs) from *L. paracasei* M2.1 and P4 for 24 h. Subsequently, an MTT assay was performed to assess any cytotoxic effects of the applied supernatant concentration. As illustrated in Figure 1, no cell-damaging impacts were observed. In fact, cell viability was notably increased by 18% in the samples treated with M2.1 supernatants compared to the control (MRS) ($p < 0.001$).

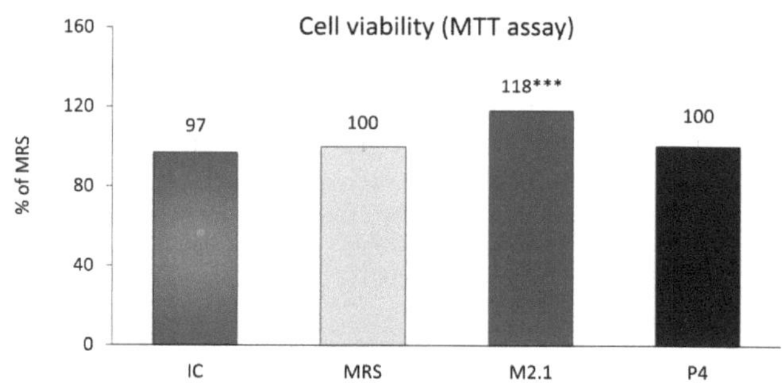

Figure 1. Assessment of cell viability in mature 3T3-L1 cells, treated with 10% cell-free supernatants (CFSs) from *L. paracasei* M2.1 or P4 strains, using the MTT assay: Abbreviations. IC—mature, untreated adipocytes; MRS—mature adipocytes treated with 10% (*v*/*v*) MRS broth (control); M2.1 and P4—experimental groups of mature adipocytes exposed to 10% (*v*/*v*) M2.1 or P4 CFSs. The statistical significance of differences between each experimental group (M2.1 or P4) and the control (MRS) group was evaluated using the non-parametric Mann–Whitney U test. The "asterisk" symbol shows the degree of significance in the figure as follows: *** for $p < 0.001$.

3.2. Intracellular Lipid Accumulation

Neutral lipid deposition in already differentiated 3T3-L1 adipocytes exposed to 10% cell supernatants from *L. paracasei* M2.1 or P4 was observed microscopically (Figure 2a) and then quantified spectrophotometrically following isopropanol extraction (Figure 2b). Data showed a 19% increase in the intracellular lipid accumulation in adipocytes treated with M2.1 supernatants compared to MRS ($p < 0.01$), with no change observed in the P4 group.

3.3. Glucose Uptake and Lipolysis Rate

At the end of the experiment, the glucose concentration and glycerol release in the cell supernatants from all groups were measured. Based on these results, the glucose uptake and lipolysis rate in adipocytes were evaluated (Figure 3). The treated cells demonstrated a significant increase of over 30% in glucose uptake compared to the MRS group ($p < 0.001$). Concurrently, a 28% reduction in the lipolysis rate in the M2.1 group ($p < 0.001$) and no significant change in lipolysis in the P4 group were observed.

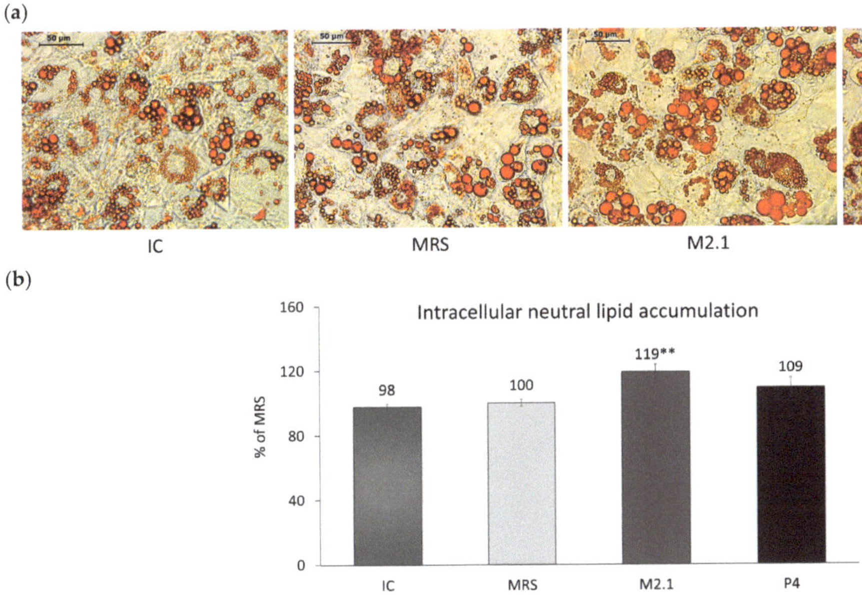

Figure 2. The effect of *L. paracasei* M2.1 and P4 cell-free supernatants (CFSs) on the intracellular neutral lipid accumulation in mature 3T3-L1 cells. (**a**) Microscopic images, stained with Oil Red O (magnification 40×, bars: 50 µm); (**b**) intracellular lipid accumulation after the isopropanol extraction of Oil Red O. Abbreviations: IC—mature, untreated adipocytes; MRS—mature adipocytes treated with 10% (*v*/*v*) MRS broth (control); M2.1 and P4—experimental groups of mature adipocytes exposed to 10% (*v*/*v*) M2.1 or P4 CFSs. The statistical significance of differences between each experimental group (M2.1 or P4) and the control (MRS) group was evaluated using the non-parametric Mann–Whitney U test. The "asterisk" symbol shows the degree of significance in the figure as follows: ** for $p < 0.01$.

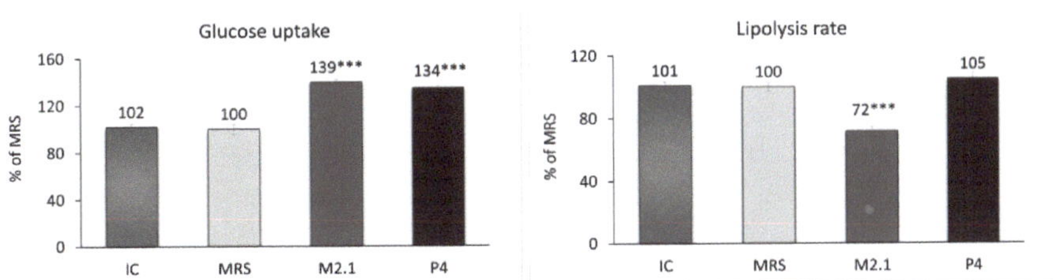

Figure 3. Assessment of the glucose uptake and lipolysis rate after 24 h of treatment of mature 3T3-L1 adipocytes with 10% cell-free supernatants (CFSs) from *L. paracasei* M2.1 or P4 strains. Abbreviations: IC—mature, untreated adipocytes; MRS—mature adipocytes treated with 10% (*v*/*v*) MRS broth (control); M2.1 and P4—experimental groups of mature adipocytes exposed to 10% (*v*/*v*) M2.1 or P4 CFSs. The statistical significance of differences between each experimental group (M2.1 or P4) and the control (MRS) group was evaluated using the non-parametric Mann–Whitney U test. The asterisk symbol shows the degree of significance in the figures as follows: *** for $p < 0.001$.

3.4. Relative Gene Expression

The expression of genes related to mitochondrial and peroxisomal beta-oxidation in adipocytes was analyzed, as illustrated in Figure 4a–c.

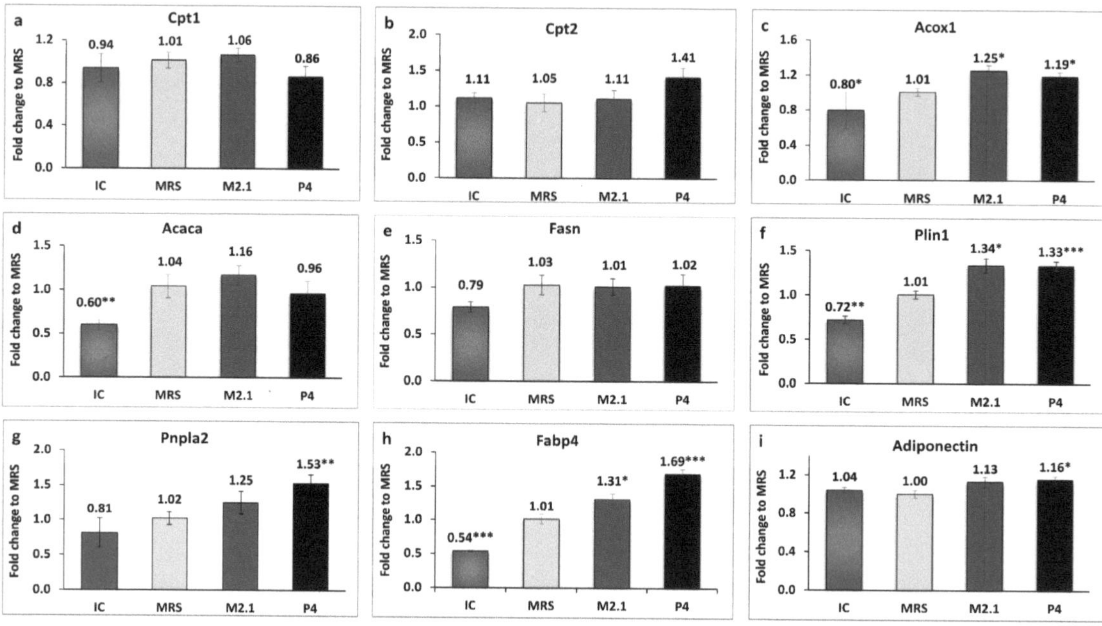

Figure 4. The effect of *L. paracasei* M2.1 and P4 cell-free supernatants (CFSs) on the relative fold change in the gene expression of (**a**) carnitine palmitoyltransferase 1a (*Cpt1*); (**b**) carnitine palmitoyltransferase 2 (*Cpt2*); (**c**) acyl-coenzyme A oxidase 1, palmitoyl (*Acox1*); (**d**) acetyl-CoA carboxylase (*Acaca*); (**e**) fatty acid synthase (*Fasn*); (**f**) perilipin 1 (*Plin1*); (**g**) patatin-like phospholipase domain containing 2 (*Pnpla2*); (**h**) fatty acid-binding protein 4 (*Fabp4*); and (**i**) adiponectin. Abbreviations: IC—mature, untreated adipocytes; MRS—mature adipocytes treated with 10% (*v*/*v*) MRS broth (control); M2.1 and P4—experimental groups of mature adipocytes exposed to 10% (*v*/*v*) M2.1 or P4 cell-free supernatants (CFSs). The statistical significance of differences between each experimental group (M2.1 or P4) and the control (MRS) group was evaluated using the non-parametric Mann–Whitney U test. The "asterisk" symbol shows the degree of significance in the figures as follows: * for $p < 0.05$, ** for $p < 0.01$, and *** for $p < 0.001$.

A statistically significant upregulation, exceeding 10%, was observed exclusively in the acyl-coenzyme A oxidase 1, palmitoyl (*Acox1*) gene in both treated adipocyte groups compared to the control ($p < 0.05$). Perilipin 1 (*Plin1*) and fatty acid-binding protein 4 (*Fabp4*) were also significantly upregulated in the M2.1 ($p < 0.05$) and P4 ($p < 0.001$) compared to the MRS group. Interestingly, the patatin-like phospholipase domain containing 2 (*Pnpla2*) and adiponectin (*Adipoq*) showed significantly higher expression only in the P4 group ($p < 0.01$ and $p < 0.05$, respectively).

4. Discussion

In Bulgaria, several region-specific microorganisms have been identified for their antiadipogenic, anticancer, and longevity-enhancing properties, such as *Lactobacillus bulgaricus*, *Lactobacillus delbrueckii* subsp. *bulgaricus*, *Lactobacillus helveticus*, *Lactobacillus brevis*, *Lactobacillus plantarum*, etc. [35,36]. Therefore, endemic areas in the country have been studied extensively for probiotic microorganisms with unique health benefits [37–41]. One such area is Sinite Kamani National Park, where we previously obtained four indigenous *L. paracasei* strains (M2.1, C8, C15, and P4) exhibiting significant antidiabetic potential. Among these potentially beneficial strains, we highlighted P4 for its ability to enhance insulin sensitivity and M2.1 for promoting the healthy expansion of mature adipocytes by significantly suppressing their lipolysis. In this context, we recommended a further investigation of their beta-oxidation activity to better understand the mechanism of intracellular action [23].

The results of the present study confirm that 10% CFSs of *L. paracasei* M2.1 and P4, applied in mature 3T3-L1 cells for 24 h, increase glucose uptake by over 30% under simulated high-carbohydrate cell feeding without intensifying lipolysis or exhibiting any noticeable antiadipogenic effects.

The analyzed CFSs appear to regulate the metabolism of mature adipocytes finely, thereby preventing "lipid overload"— a condition associated with sharply reduced adipocyte metabolism, disrupted insulin signaling, induced oxidative stress, and increased inflammatory response [42–44]. The data again draw our attention to the possible enhancement of fatty acid beta-oxidation.

Beta-oxidation in mature adipocytes occurs mainly in peroxisomes and mitochondria, with the latter being the primary pathway. This process reduces intracellular lipid accumulation by converting fatty acids into acetyl-CoA and other metabolites. The transport of fatty acids into mitochondria, a critical step in beta-oxidation, is regulated by the enzymes Cpt1 and Cpt2 on the outer and inner mitochondrial membranes, respectively. Cpt1 converts long-chain fatty acyl-CoA into acyl-carnitine, while Cpt2 converts acyl-carnitine back to fatty acyl-CoA in the mitochondrial matrix. The initiation of mitochondrial beta-oxidation is highly dependent on acetyl-CoA carboxylase, which transforms acetyl-CoA to malonyl-CoA, subsequently inhibiting Cpt1 and preventing fatty acid transport into mitochondria [45].

Our study did not establish significant changes in the *Cpt1*, *Cpt2*, and *Acaca* gene expression among the groups, despite the observed trend of increased *Cpt2* gene expression in P4. Therefore, we cannot conclude that the suspected enhanced beta-oxidation occurred in adipocyte mitochondria. Both supernatants, M2.1 and P4, upregulated the expression of the *Acox1* gene, which encodes the first enzyme in the peroxisomal fatty acid beta-oxidation pathway. Its upregulation in adipocytes signifies an increase in the cellular capacity to process fatty acids, thus promoting their utilization for energy, rather than storage [46,47]. Peroxisomal beta-oxidation serves as an additional mechanism for regulating fatty acid balance within the cell, primarily handling medium- and long-chain fatty acids [48,49]. Under certain conditions, such as dietary supplements, fasting, or mitochondrial overload, the peroxisomal beta-oxidation rate can increase significantly, reducing intracellular lipid accumulation in adipocytes. This activation is crucial for maintaining normal metabolic activity and insulin sensitivity in mature adipocytes subject to substantial energy supply [47,50]. The increased breakdown of long-chain fatty acids in peroxisomes carries a risk of intracellular oxidative stress due to the generation of hydrogen peroxide (H_2O_2) as a byproduct. However, catalase within peroxisomes rapidly converts H_2O_2 into water and oxygen, thereby mitigating the risk of oxidative damage [51].

As mentioned above, peroxisomal beta-oxidation primarily handles medium- and long-chain fatty acids. In our experiment, the source of these fatty acids in both treated groups was questionable. All cells were provided with identical nutrient media, and the groups differed only when the MRS broth in the experimental groups was subjected to microbial fermentation by either M2.1 or P4 for 24 h prior to treatment.

The strains under investigation belong to the *L. paracasei* group [52], known for preferentially fermenting environmental sugars. As a result of their metabolic activity, predominantly short-chain fatty acids, especially lactic acid and acetate, are produced [53,54]. These further influence the expression of specific membrane receptors such as G-protein-coupled receptors 43 and 41, enhancing insulin sensitivity and carbohydrate metabolism rather than beta-oxidation [55]. Nevertheless, they are recognized to affect intracellular fat metabolism positively [45].

Digging deeper, the results outlined the notable discrepancy between the increased gene expression of *Pnpla2* in P4 and the absence of a corresponding rise in lipolysis levels measured in the cell supernatants. Adipose triglyceride lipase (ATGL) (encoded by the *Pnpla2* gene) plays a pivotal role in breaking down triglycerides into free fatty acids, and its upregulation is typically associated with a higher rate of lipolysis [56,57]. Depending on the energy needs, the released fatty acids can also be redirected toward energy production pathways [47]. Our results suggest that these fatty acids probably remain in the cell, serving as substrates for peroxisomal beta-oxidation. Moreover, the upregulation of *Pnpla2* was combined with increased *Plin1* and *Fabp4* gene expressions, which indicate enhanced intracellular lipid mobilization instead of increased lipolysis. The expression of *Plin1* directly influences the accessibility of ATGL to lipid droplets. Plin1 protects lipid droplets from premature or uncontrolled lipolysis and regulates the access of lipases, particularly in adipocytes with increased insulin sensitivity [58,59]. The phosphatidylinositol 3-kinase and cAMP pathway activation reduces protein kinase A activity, decreases Plin1 phosphorylation, and inhibits lipolysis by restricting ATGL access to lipid droplets [60]. However, under certain metabolic conditions, including nutritional stimuli, Plin1 undergoes post-translational modifications such as phosphorylation and facilitates controlled lipolysis, enabling ATGL and other lipases to access and hydrolyze triglycerides [56,61]. This fine regulation of Plin1 activity is part of a dynamic mechanism regulating fat storage and mobilization in adipocytes. The dual function of fat protection and release facilitation is not contradictory, but rather componential within the complex regulation of adipocyte metabolic processes, balancing lipid homeostasis. Therefore, in humans with insulin resistance and obesity, an enhanced, as well as highly suppressed, lipolysis has been reported [62]. The cells treated with P4 and M2.1 CFSs established a notable increase in *Fabp4* gene expression. Fabp4 is a protein with a critical role in the intracellular transport of fatty acids, whose protein expression is often upregulated due to lipolytic stimulation [63]. Its elevated expression enhances the movement of fatty acids, particularly long chains, into various cellular compartments [64]. Thus, in our investigation, the simultaneous upregulation of *Fabp4*, *Pnpla2*, and *Acox1* in white adipocytes from the P4 group suggests that the fatty acids released from intracellular lipid droplets are likely redirected to peroxisomes, where they undergo partial oxidation for energy production.

These interactions contribute to a fine adjustment of the metabolic profile, enhancing the management of the entire cellular machinery within the mature adipocyte. Improved adipocyte metabolism features enhance glucose utilization, established herein, and boost the production of so-called "good adipokines" rather than "bad" ones. Adiponectin is of significant importance to the health status of the entire organism, especially in obesity. It plays a critical role in energy homeostasis, promotes healthy weight maintenance, and prevents obesity-related complications. We established that P4 CFS supplementation to mature adipocytes increases adiponectin gene expression. Adiponectin's beneficial effects are strongly linked to enhanced fatty acid oxidation, improved lipid metabolism, and increased glucose uptake by adipocytes, all of which are consistent with the findings of this study. Its increased expression is associated with enhanced insulin sensitivity—a fact established in our previous study, where 3T3-L1 cells were treated with the same dose of P4 CFS [26]. Therefore, the upregulation of adiponectin expression along with enhanced beta-oxidation, established in the P4 group, could offer a protective effect against various obesity-related comorbidities.

As a preliminary study investigating potential improvements in mature adipocyte metabolism following supplementation with *L. paracasei* M2.1 and P4 CFSs, it has several limitations. We used gene expression as a proxy for metabolic changes, which may not always reflect functional activity. This approach provides valuable insights but should be further expanded with protein-level validation. Moreover, the obtained results are based on in vitro experiments. Future studies should involve human or animal models to validate our findings and further explore the mechanisms underlying the metabolic improvements observed. Despite these limitations, the current research establishes a strong foundation for future investigations into the metabolic potential of these unique *L. paracasei* strains.

5. Conclusions

The study demonstrates that both *L. paracasei* M2.1 and P4 strains increased glucose uptake in mature 3T3-L1 adipocytes without affecting lipolysis or showing antiadipogenic effects. Both strains regulate adipocyte metabolism to prevent "lipid overload", a condition that disrupts insulin signaling and increases oxidative stress. Notably, P4 CFS upregulated the gene expression of *Acox1*, encoding a key enzyme in peroxisomal beta-oxidation, suggesting enhanced fatty acid processing. This process could reduce lipid accumulation and improve insulin sensitivity. Despite some changes in gene expression, including the upregulation of *Pnpla2*, *Fabp4*, and *Plin1*, lipolysis was not significantly increased, indicating that fatty acids were likely redirected to peroxisomes for partial oxidation. Additionally, P4 CFS increased adiponectin expression, enhancing insulin sensitivity, glucose uptake, and fatty acid oxidation, which could protect against obesity-related complications. These findings highlight the potential of *L. paracasei* strains in improving metabolic health.

6. Patents

GB Patent Application No.: 2411546.1.

Supplementary Materials: The following supporting information can be downloaded at: https://www.mdpi.com/article/10.3390/biomedicines12122785/s1, Figure S1: Intracellular neutral lipid accumulation in 3T3-L1 cells. (a) Microscopic images, stained with Oil Red O (magnification 40×, bars: 50 μm); (b) intracellular lipid accumulation after isopropanol extraction of Oil Red O.

Author Contributions: Conceptualization, E.V. and G.B.; Methodology, N.G., G.B., E.V. and Z.I.; Software, N.G. and Z.I.; Validation, E.V., V.P. and G.B.; Formal analysis, Z.I., N.G. and V.P.; Investigation, N.G. and Z.I.; Resources, G.B., E.V., N.G. and Z.I.; Data curation, Z.I. and N.G.; Writing—Original draft preparation, N.G. and Z.I.; Writing—Review and editing, G.B., E.V., N.G., Z.I. and V.P.; Visualization, N.G. and Z.I.; Supervision, E.V. and G.B.; Project administration, G.B.; Funding acquisition, G.B. All authors have read and agreed to the published version of the manuscript.

Funding: This work was supported by the Bulgarian Ministry of Education and Science within the framework of the Bulgarian National Recovery and Resilience Plan, Component "Innovative Bulgaria", Project No. BG-RRP-2.004-0006-C02 "Development of research and innovation at Trakia University in the service of health and sustainable well-being".

Institutional Review Board Statement: Not applicable.

Informed Consent Statement: Not applicable.

Data Availability Statement: The datasets generated for this study are available from the corresponding author upon request.

Acknowledgments: The authors acknowledge the Sinite Kamani National Park administration for their technical support. The experimental work described in the current study was conducted in the Laboratory of Experimental Cellular Physiology and Therapeutic Drug Monitoring at the Faculty of Veterinary Medicine, Trakia University, Stara Zagora, Bulgaria.

Conflicts of Interest: The authors declare no conflicts of interest.

References

1. Prajapati, K.; Bisani, K.; Prajapati, H.; Prajapati, S.; Agrawal, D.; Singh, S.; Saraf, M.; Goswami, D. Advances in Probiotics Research: Mechanisms of Action, Health Benefits, and Limitations in Applications. *Syst. Microbiol. Biomanuf.* **2024**, *4*, 386–406. [CrossRef]
2. Williams, N.T. Probiotics. *Am. J. Health-Syst. Pharm.* **2010**, *67*, 449–458. [CrossRef] [PubMed]
3. Solito, A.; Bozzi Cionci, N.; Calgaro, M.; Caputo, M.; Vannini, L.; Hasballa, I.; Archero, F.; Giglione, E.; Ricotti, R.; Walker, G.E.; et al. Supplementation with *Bifidobacterium breve* BR03 and B632 Strains Improved Insulin Sensitivity in Children and Adolescents with Obesity in a Cross-over, Randomized Double-Blind Placebo-Controlled Trial. *Clin. Nutr.* **2021**, *40*, 4585–4594. [CrossRef] [PubMed]
4. Pan, Y.-Q.; Zheng, Q.-X.; Jiang, X.-M.; Chen, X.-Q.; Zhang, X.-Y.; Wu, J.-L. Probiotic Supplements Improve Blood Glucose and Insulin Resistance/Sensitivity among Healthy and GDM Pregnant Women: A Systematic Review and Meta-Analysis of Randomized Controlled Trials. *Evid.-Based Complement. Altern. Med.* **2021**, *2021*, 9830200. [CrossRef]

5. Ahn, S.B.; Jun, D.W.; Kang, B.-K.; Lim, J.H.; Lim, S.; Chung, M.-J. Randomized, Double-Blind, Placebo-Controlled Study of a Multispecies Probiotic Mixture in Nonalcoholic Fatty Liver Disease. *Sci. Rep.* **2019**, *9*, 5688. [CrossRef]
6. Moravejolahkami, A.R.; Hojjati Kermani, M.A.; Balouch Zehi, Z.; Mirenayat, S.M.S.; Mansourian, M. The Effect of Probiotics on Lipid Profile & Anthropometric Indices in Diabetic Nephropathy; a Systematic Review and Meta-Analysis of Clinical Trials. *J. Diabetes. Metab. Disord.* **2021**, *20*, 893–904. [CrossRef]
7. Oniszczuk, A.; Oniszczuk, T.; Gancarz, M.; Szymańska, J. Role of Gut Microbiota, Probiotics and Prebiotics in the Cardiovascular Diseases. *Molecules* **2021**, *26*, 1172. [CrossRef]
8. Le Barz, M.; Daniel, N.; Varin, T.V.; Naimi, S.; Demers-Mathieu, V.; Pilon, G.; Audy, J.; Laurin, É.; Roy, D.; Urdaci, M.C.; et al. In Vivo Screening of Multiple Bacterial Strains Identifies *Lactobacillus rhamnosus* Lb102 and *Bifidobacterium animalis* ssp. lactis Bf141 as Probiotics That Improve Metabolic Disorders in a Mouse Model of Obesity. *FASEB J.* **2019**, *33*, 4921–4935. [CrossRef]
9. Salminen, S.; Collado, M.C.; Endo, A.; Hill, C.; Lebeer, S.; Quigley, E.M.M.; Sanders, M.E.; Shamir, R.; Swann, J.R.; Szajewska, H.; et al. The International Scientific Association of Probiotics and Prebiotics (ISAPP) Consensus Statement on the Definition and Scope of Postbiotics. *Nat. Rev. Gastroenterol. Hepatol.* **2021**, *18*, 649–667. [CrossRef]
10. van der Hee, B.; Wells, J.M. Microbial Regulation of Host Physiology by Short-Chain Fatty Acids. *Trends Microbiol.* **2021**, *29*, 700–712. [CrossRef]
11. Aguilar-Toalá, J.E.; Garcia-Varela, R.; Garcia, H.S.; Mata-Haro, V.; González-Córdova, A.F.; Vallejo-Cordoba, B.; Hernández-Mendoza, A. Postbiotics: An Evolving Term Within the Functional Foods Field. *Trends Food Sci. Technol.* **2018**, *75*, 105–114. [CrossRef]
12. Zhang, T.; Zhang, W.; Feng, C.; Kwok, L.-Y.; He, Q.; Sun, Z. Stronger Gut Microbiome Modulatory Effects by Postbiotics than Probiotics in a Mouse Colitis Model. *NPJ Sci. Food* **2022**, *6*, 53. [CrossRef] [PubMed]
13. Taverniti, V.; Guglielmetti, S. The Immunomodulatory Properties of Probiotic Microorganisms beyond Their Viability (Ghost Probiotics: Proposal of Paraprobiotic Concept). *Genes. Nutr.* **2011**, *6*, 261–274. [CrossRef]
14. Kim, H.; Lim, J.-J.; Shin, H.Y.; Suh, H.J.; Choi, H.-S. *Lactobacillus plantarum* K8-Based Paraprobiotics Suppress Lipid Accumulation during Adipogenesis by the Regulation of JAK/STAT and AMPK Signaling Pathways. *J. Funct. Foods* **2021**, *87*, 104824. [CrossRef]
15. Ayichew, T.; Belete, A.; Alebachew, T.; Tsehaye, H.; Berhanu, H.; Minwuyelet, A. Bacterial Probiotics Their Importances and Limitations: A Review. *J. Nutr. Health Sci.* **2017**, *4*, 1. [CrossRef]
16. Tang, C.; Kong, L.; Shan, M.; Lu, Z.; Lu, Y. Protective and Ameliorating Effects of Probiotics against Diet-Induced Obesity: A Review. *Food Res. Int.* **2021**, *147*, 110490. [CrossRef]
17. Kim, H.; Jeong, Y.; Kim, J.-E.; Kim, Y.; Paek, N.-S.; Kang, C.-H. Anti-Obesity Potential of *Lactobacillus* spp. Isolated from Infant Feces. *Biotechnol. Bioprocess Eng.* **2021**, *26*, 575–585. [CrossRef]
18. Scheja, L.; Heeren, J. The Endocrine Function of Adipose Tissues in Health and Cardiometabolic Disease. *Nat. Rev. Endocrinol.* **2019**, *15*, 507–524. [CrossRef] [PubMed]
19. Fasshauer, M.; Blüher, M. Adipokines in Health and Disease. *Trends Pharmacol. Sci.* **2015**, *36*, 461–470. [CrossRef]
20. Cinti, S. Adipose Organ Development and Remodeling. *Compr. Physiol.* **2018**, *8*, 1357–1431. [CrossRef]
21. Rupasinghe, H.P.V.; Sekhon-Loodu, S.; Mantso, T.; Panayiotidis, M.I. Phytochemicals in Regulating Fatty Acid β-Oxidation: Potential Underlying Mechanisms and Their Involvement in Obesity and Weight Loss. *Pharmacol. Ther.* **2016**, *165*, 153–163. [CrossRef] [PubMed]
22. Subash-Babu, P.; Mohammed Alowaidh, H.; Al-Harbi, L.N.; Shamlan, G.; Aloud, A.A.; AlSedairy, S.A.; Alshatwi, A.A. *Ocimum basilicum* L. Methanol Extract Enhances Mitochondrial Efficiency and Decreases Adipokine Levels in Maturing Adipocytes Which Regulate Macrophage Systemic Inflammation. *Molecules* **2022**, *27*, 1388. [CrossRef] [PubMed]
23. Rondanelli, M.; Gasparri, C.; Perna, S.; Petrangolini, G.; Allegrini, P.; Fazia, T.; Bernardinelli, L.; Cavioni, A.; Mansueto, F.; Oberto, L.; et al. A 60-Day Green Tea Extract Supplementation Counteracts the Dysfunction of Adipose Tissue in Overweight Post-Menopausal and Class I Obese Women. *Nutrients* **2022**, *14*, 5209. [CrossRef]
24. Lin, W.; Kuo, Y.-W.; Chen, C.-W.; Hsu, Y.-C.; Huang, Y.-F.; Hsu, C.-H.; Lin, J.-H.; Lin, C.-H.; Lin, C.-C.; Yi, T.-H.; et al. The Function of Mixed Postbiotic PE0401 in Improving Intestinal Health via Elevating Anti-Inflammation, Anti-Oxidation, Epithelial Tight Junction Gene Expression and Promoting Beneficial Bacteria Growth. *J. Pure Appl. Microbiol.* **2022**, *16*, 1771–1782. [CrossRef]
25. Vachkova, E.; Petrova, V.; Grigorova, N.; Ivanova, Z.; Beev, G. Evaluation of the Anticancer and Probiotic Potential of Autochthonous (Wild) *Lacticaseibacillus paracasei* Strains from New Ecological Niches as a Possible Additive for Functional Dairy Foods. *Foods* **2023**, *12*, 185. [CrossRef]
26. Grigorova, N.; Ivanova, Z.; Vachkova, E.; Petrova, V.; Beev, G. Antidiabetic and Hypolipidemic Properties of Newly Isolated Wild *Lacticaseibacillus paracasei* Strains in Mature Adipocytes. *Appl. Sci.* **2023**, *13*, 6489. [CrossRef]
27. Jones, R.M. The use of Lactobacillus casei and Lactobacillus paracasei in clinical trials for the improvement of human health. In *The Microbiota in Gastrointestinal Pathophysiology*; Academic Press: Cambridge, MA, USA, 2017; pp. 99–108.
28. Hill, D.; Sugrue, I.; Tobin, C.; Hill, C.; Stanton, C.; Ross, R.P. The Lactobacillus casei Group: History and Health Related Applications. *Front. Microbiol.* **2018**, *9*, 2107. [CrossRef]
29. Bengoa, A.A.; Dardis, C.; Garrote, G.L.; Abraham, A.G. Health-promoting properties of *Lacticaseibacillus paracasei*: A focus on kefir isolates and exopolysaccharide-producing strains. *Foods* **2021**, *10*, 2239. [CrossRef]
30. Yang, Z.; Tu, Y.; Xia, H.; Jie, G.; Chen, X.; He, P. Suppression of Free-Radicals and Protection against H_2O_2-Induced Oxidative Damage in HPF-1 Cell by Oxidized Phenolic Compounds Present in Black Tea. *Food Chem.* **2007**, *105*, 1349–1356. [CrossRef]

31. Park, Y.J.; Liang, J.F.; Ko, K.S.; Kim, S.W.; Yang, V.C. Low Molecular Weight Protamine as an Efficient and Nontoxic Gene Carrier: In Vitro Study. *J. Gene Med.* **2003**, *5*, 700–711. [CrossRef]
32. Yang, M.T.; Fu, J.; Wang, Y.-K.; Desai, R.A.; Chen, C.S. Assaying Stem Cell Mechanobiology on Microfabricated Elastomeric Substrates with Geometrically Modulated Rigidity. *Nat. Protoc.* **2011**, *6*, 187–213. [CrossRef] [PubMed]
33. Rivera Diaz, P.A.; Gómez Camargo, D.E.; Ondo-Méndez, A.; Gómez-Alegría, C.J. A Colorimetric Bioassay for Quantitation of Both Basal and Insulin-Induced Glucose Consumption in 3T3-L1 Adipose Cells. *Heliyon* **2020**, *6*, e03422. [CrossRef] [PubMed]
34. Xie, F.; Xiao, P.; Chen, D.; Xu, L.; Zhang, B. miRDeepFinder: A miRNA Analysis Tool for Deep Sequencing of Plant Small RNAs. *Plant Mol. Biol.* **2012**, *80*, 75–84. [CrossRef] [PubMed]
35. Yakovlieva, M.; Tacheva, T.; Mihaylova, S.; Tropcheva, R.; Trifonova, K.; Tolекova, A.; Danova, S.; Vlaykova, T. Influence of *Lactobacillus brevis* 15 and *Lactobacillus plantarum* 13 on Blood Glucose and Body Weight in Rats after High-Fructose Diet. *Benef. Microbes* **2015**, *6*, 505–512. [CrossRef] [PubMed]
36. Danova, S.; Tropcheva, R.; Ivanovska, N.; Georgieva, R.; Dobreva-Yosifova, G.; Petrova, M. Characterisation of *Bulgarian lactobacilli* as Probiotics. *New Trends Microbiol.* **2012**, 13–30.
37. Ivanov, I.; Petrov, K.; Lozanov, V.; Hristov, I.; Wu, Z.; Liu, Z.; Petrova, P. Bioactive Compounds Produced by the Accompanying Microflora in Bulgarian Yoghurt. *Processes* **2021**, *9*, 114. [CrossRef]
38. Krastanov, A.; Georgiev, M.; Slavchev, A.; Blazheva, D.; Goranov, B.; Ibrahim, S.A. Design and Volatile Compound Profiling of Starter Cultures for Yogurt Preparation. *Foods* **2023**, *12*, 379. [CrossRef]
39. Petrova, P.; Arsov, A.; Tsvetanova, F.; Parvanova-Mancheva, T.; Vasileva, E.; Tsigoriyna, L.; Petrov, K. The Complex Role of Lactic Acid Bacteria in Food Detoxification. *Nutrients* **2022**, *14*, 2038. [CrossRef] [PubMed]
40. Song, H.; Zhou, L.; Liu, D.; Ge, L.; Li, Y. Probiotic Effect on Helicobacter pylori Attachment and Inhibition of Inflammation in Human Gastric Epithelial Cells. *Exp. Ther. Med.* **2019**, *18*, 1551–1562. [CrossRef]
41. Raheem, A.; Liang, L.; Zhang, G.; Cui, S. Modulatory Effects of Probiotics During Pathogenic Infections with Emphasis on Immune Regulation. *Front. Immunol.* **2021**, *12*, 616713. [CrossRef]
42. Kim, J.I.; Huh, J.Y.; Sohn, J.H.; Choe, S.S.; Lee, Y.S.; Lim, C.Y.; Jo, A.; Park, S.B.; Han, W.; Kim, J.B. Lipid-Overloaded Enlarged Adipocytes Provoke Insulin Resistance Independent of Inflammation. *Mol. Cell. Biol.* **2015**, *35*, 1686–1699. [CrossRef] [PubMed]
43. Chaurasia, B.; Summers, S.A. Ceramides in Metabolism: Key Lipotoxic Players. *Annu. Rev. Physiol.* **2021**, *83*, 303–330. [CrossRef] [PubMed]
44. Schipper, H.S.; Rakhshandehroo, M.; van de Graaf, S.F.J.; Venken, K.; Koppen, A.; Stienstra, R.; Prop, S.; Meerding, J.; Hamers, N.; Besra, G.; et al. Natural Killer T Cells in Adipose Tissue Prevent Insulin Resistance. *J. Clin. Investig.* **2012**, *122*, 3343–3354. [CrossRef] [PubMed]
45. Schreurs, M.; Kuipers, F.; Van Der Leij, F.R. Regulatory Enzymes of Mitochondrial β-Oxidation as Targets for Treatment of the Metabolic Syndrome. *Obes. Rev.* **2010**, *11*, 380–388. [CrossRef] [PubMed]
46. Tahri-Joutey, M.; Andreoletti, P.; Surapureddi, S.; Nasser, B.; Cherkaoui-Malki, M.; Latruffe, N. Mechanisms Mediating the Regulation of Peroxisomal Fatty Acid Beta-Oxidation by PPARα. *Int. J. Mol. Sci.* **2021**, *22*, 8969. [CrossRef]
47. Ding, L.; Sun, W.; Balaz, M.; He, A.; Klug, M.; Wieland, S.; Caiazzo, R.; Raverdy, V.; Pattou, F.; Lefebvre, P.; et al. Peroxisomal β-Oxidation Acts as a Sensor for Intracellular Fatty Acids and Regulates Lipolysis. *Nat. Metab.* **2021**, *3*, 1648–1661. [CrossRef]
48. Wanders, R.J.A.; Waterham, H.R.; Ferdinandusse, S. Metabolic Interplay between Peroxisomes and Other Subcellular Organelles Including Mitochondria and the Endoplasmic Reticulum. *Front. Cell Dev. Biol.* **2016**, *3*, 83. [CrossRef]
49. Violante, S.; Achetib, N.; van Roermund, C.W.T.; Hagen, J.; Dodatko, T.; Vaz, F.M.; Waterham, H.R.; Chen, H.; Baes, M.; Yu, C.; et al. Peroxisomes Can Oxidize Medium- and Long-Chain Fatty Acids through a Pathway Involving ABCD3 and HSD17B4. *FASEB J.* **2019**, *33*, 4355–4364. [CrossRef]
50. Cherkaoui-Malki, M.; Surapureddi, S.; El Hajj, H.I.; Vamecq, J.; Andreoletti, P. Hepatic Steatosis and Peroxisomal Fatty Acid Beta-Oxidation. *Curr. Drug Metab.* **2012**, *13*, 1412–1421. [CrossRef]
51. Sandalio, L.M.; Rodríguez-Serrano, M.; Romero-Puertas, M.C.; del Río, L.A. Role of Peroxisomes as a Source of Reactive Oxygen Species (ROS) Signaling Molecules. In *Peroxisomes and Their Key Role in Cellular Signaling and Metabolism*; del Río, L.A., Ed.; Springer: Dordrecht, The Netherlands, 2013; pp. 231–255, ISBN 978-94-007-6889-5.
52. Beev, G.; Michaylova, M.; Dinev, T.; Naydenova, N.; Tzanova, M.; Urshev, Z. ARDRA Analysis on Biodiversity of *Lactobacilli* Isolated from Bulgarian Raw Buffalo Milk. *Acta Microbiol. Bulg.* **2021**, *37*, 22–26.
53. Shen, Y.-L.; Zhang, L.-Q.; Yang, Y.; Yin, B.-C.; Ye, B.-C.; Zhou, Y. Advances in the Role and Mechanism of Lactic Acid Bacteria in Treating Obesity. *Food Bioeng.* **2022**, *1*, 101–115. [CrossRef]
54. May, K.S.; den Hartigh, L.J. Modulation of Adipocyte Metabolism by Microbial Short-Chain Fatty Acids. *Nutrients* **2021**, *13*, 3666. [CrossRef] [PubMed]
55. Zhao, Y.; Chen, F.; Wu, W.; Sun, M.; Bilotta, A.J.; Yao, S.; Xiao, Y.; Huang, X.; Eaves-Pyles, T.D.; Golovko, G.; et al. GPR_{43} Mediates Microbiota Metabolite SCFA Regulation of Antimicrobial Peptide Expression in Intestinal Epithelial Cells via Activation of mTOR and $STAT_3$. *Mucosal Immunol.* **2018**, *11*, 752–762. [CrossRef]
56. Lass, A.; Zimmermann, R.; Oberer, M.; Zechner, R. Lipolysis—A Highly Regulated Multi-Enzyme Complex Mediates the Catabolism of Cellular Fat Stores. *Prog. Lipid Res.* **2011**, *50*, 14–27. [CrossRef]

57. Zimmermann, R.; Strauss, J.G.; Haemmerle, G.; Schoiswohl, G.; Birner-Gruenberger, R.; Riederer, M.; Lass, A.; Neuberger, G.; Eisenhaber, F.; Hermetter, A.; et al. Fat Mobilization in Adipose Tissue Is Promoted by Adipose Triglyceride Lipase. *Science* **2004**, *306*, 1383–1386. [CrossRef] [PubMed]
58. Brasaemle, D.L.; Subramanian, V.; Garcia, A.; Marcinkiewicz, A.; Rothenberg, A. Perilipin A and the Control of Triacylglycerol Metabolism. *Mol. Cell. Biochem.* **2009**, *326*, 15–21. [CrossRef]
59. Sztalryd, C.; Brasaemle, D.L. The Perilipin Family of Lipid Droplet Proteins: Gatekeepers of Intracellular Lipolysis. *Biochim. Biophys. Acta Mol. Cell Biol. Lipids* **2017**, *1862*, 1221–1232. [CrossRef]
60. Nielsen, T.S.; Jessen, N.; Jørgensen, J.O.L.; Møller, N.; Lund, S. Dissecting Adipose Tissue Lipolysis: Molecular Regulation and Implications for Metabolic Disease. *J. Mol. Endocrinol.* **2014**, *52*, R199–R222. [CrossRef] [PubMed]
61. Yang, X.; Lu, X.; Lombès, M.; Rha, G.B.; Chi, Y.-I.; Guerin, T.M.; Smart, E.J.; Liu, J. The G0/G1 Switch Gene 2 Regulates Adipose Lipolysis through Association with Adipose Triglyceride Lipase. *Cell Metab.* **2010**, *11*, 194–205. [CrossRef]
62. Karczewska-Kupczewska, M.; Nikołajuk, A.; Majewski, R.; Filarski, R.; Stefanowicz, M.; Matulewicz, N.; Strączkowski, M. Changes in Adipose Tissue Lipolysis Gene Expression and Insulin Sensitivity after Weight Loss. *Endocr. Connect.* **2019**, *9*, 90–100. [CrossRef]
63. Mita, T.; Furuhashi, M.; Hiramitsu, S.; Ishii, J.; Hoshina, K.; Ishimura, S.; Fuseya, T.; Watanabe, Y.; Tanaka, M.; Ohno, K.; et al. FABP4 Is Secreted from Adipocytes by Adenyl Cyclase-PKA- and Guanylyl Cyclase-PKG-Dependent Lipolytic Mechanisms. *Obesity* **2015**, *23*, 359–367. [CrossRef] [PubMed]
64. Yang, A.; Mottillo, E.P. Adipocyte Lipolysis: From Molecular Mechanisms of Regulation to Disease and Therapeutics. *Biochem. J.* **2020**, *477*, 985–1008. [CrossRef] [PubMed]

biomedicines

MDPI

Article

Serum Fatty Acids and Inflammatory Patterns in Severe Obesity: A Preliminary Investigation in Women

Gislene B. Lima [1,*], Nayra Figueiredo [2], Fabiana M. Kattah [1], Emilly S. Oliveira [1], Maria A. Horst [1], Ana R. Dâmaso [3], Lila M. Oyama [3], Renata G. M. Whitton [4], Gabriel I. M. H. de Souza [3], Glaucia C. Lima [1], João F. Mota [1,2], Raquel M. S. Campos [3] and Flávia C. Corgosinho [1,2]

[1] Postgraduate Program in Nutrition and Health, Faculty of Nutrition, Federal University of Goiás (UFG), Goiânia 74690-900, Brazil; fabianakattah@discente.ufg.br (F.M.K.); emillysantos@discente.ufg.br (E.S.O.); aderuza@ufg.br (M.A.H.); glauciacarielo@ufg.br (G.C.L.); jfemota@gmail.com (J.F.M.); flaviacorgosinho@ufg.br (F.C.C.)

[2] Postgraduate Program in Health Sciences, Faculty of Medicine, Federal University of Goiás (UFG), Goiânia 74690-900, Brazil; nayra_figueiredo@discente.ufg.br

[3] Interdisciplinary Postgraduate Program in Health Sciences, Federal University of São Paulo (UNIFESP), São Paulo 04039-032, Brazil; ana.damaso@unifesp.br (A.R.D.); lmoyama@unifesp.br (L.M.O.); gabrinacio@gmail.com (G.I.M.H.d.S.); raquelmunhoz@hotmail.com (R.M.S.C.)

[4] Institute of Biosciences, Federal University of São Paulo (USP), São Paulo 05508-900, Brazil; renata.fish@gmail.com

* Correspondence: gislenelima@discente.ufg.br

Abstract: Background: Inflammation plays a central role in many chronic diseases that characterize modern society. Leptin/adiponectin and adiponectin/leptin ratios have been recognized as notable markers of dysfunctional adipose tissue and, consequently, an inflammatory state. **Methods:** Blood samples were collected from 41 adult volunteers (40.2 ± 8.3 years) diagnosed with severe obesity (BMI 46.99; 42.98–51.91 kg/m^2). The adipokines were quantified using an enzyme-linked immunosorbent assay, while the serum fatty acid analysis was conducted using chromatography. **Results:** The results unveiled a positive correlation between the leptin/adiponectin ratio and the 20:3n6 fatty acid (r = 0.52, p = 0.001), as well as a similar positive correlation between the adiponectin/leptin ratio and the 22:6n3 fatty acid (r = 0.74, p = 0.001). In the regression analysis, the 22:6n3 fatty acid predicted the adiponectin/leptin ratio (β = 0.76, p < 0.001), whereas C20:3 n-6 was a predictor for inflammatory markers (β = 4.84, p < 0.001). **Conclusions:** In conclusion, the 22:6n3 fatty acid was demonstrated to be a predictive factor for the adiponectin/leptin ratio and C20:3 n-6 was a predictor for inflammatory markers. This discovery, novel within this population, can help develop new intervention strategies aimed at controlling the inflammatory status in individuals classified as having severe obesity.

Keywords: leptin; adiponectin; omega-3 fatty acid; omega-6 fatty acid; inflammation

Citation: Lima, G.B.; Figueiredo, N.; Kattah, F.M.; Oliveira, E.S.; Horst, M.A.; Dâmaso, A.R.; Oyama, L.M.; Whitton, R.G.M.; de Souza, G.I.M.H.; Lima, G.C.; et al. Serum Fatty Acids and Inflammatory Patterns in Severe Obesity: A Preliminary Investigation in Women. *Biomedicines* **2024**, *12*, 2248. https://doi.org/10.3390/biomedicines12102248

Academic Editors: Teresa Vezza and Zaida Abad-Jiménez

Received: 15 July 2024
Revised: 29 September 2024
Accepted: 30 September 2024
Published: 3 October 2024

1. Introduction

Obesity is a growing global public health concern, characterized by the excessive expansion of adipose tissue, which contributes to a pro-inflammatory state [1,2]. Individuals with obesity often exhibit high leptin concentrations and adipokines with pro-inflammatory properties. In fact, studies have demonstrated that high leptin levels are related to cardiometabolic dysfunction [3,4] in both the adult and pediatric populations [5–7], highlighting the importance of this marker in obesity and its related diseases. Conversely, adiponectin is reduced in individuals with obesity, playing an anti-inflammatory, antidiabetic, and cardioprotective role, underscoring the clinical relevance of hypoadiponectinemia in this population [2,8–11]. Considering that leptin and adiponectin interact with each other in the modulation of cardiometabolic risk, recent studies have suggested that the leptin/adiponectin (lep/adipo) and adiponectin/leptin (adipo/lep) ratios may serve as more precise markers of pro- and anti-inflammatory states, respectively, compared to individual adipokines [12–16].

It has been demonstrated, in both human and experimental studies, that fatty acids might play a role in the pathogenesis of obesity, inflammation, and cardiometabolic disease.

In fact, omega-3 fatty acids have been described as attenuating low-grade inflammation in the adipose tissue of obese individuals [2]. On the other hand, studies suggest that the excessive consumption of trans-saturated fatty acids and omega-6 fatty acids is associated with a pronounced inflammatory state [17,18]. Despite the recognized link between inflammation and fatty acids [19], no study has explored this correlation, particularly in the population with severe obesity, which poses considerably heightened risks.

Considering the increasing prevalence of obesity, the significance of inflammation in disease outcomes, and the potential role of fatty acids in the inflammatory process, there is a gap in the literature regarding this triad. Given the significantly increased risks of severe obesity for cardiometabolic health, especially in women, a better understanding of the relationship between inflammation markers and fatty acids may be crucial for effective obesity management. Therefore, this study aims to assess the association between the adipo/lep and lep/adipo ratios and serum fatty acids in women with severe obesity.

2. Materials and Methods

2.1. Design and Participants

This is a cross-sectional observational study approved by the Research Ethics Committee of the Federal University of Goiás and the Dr. Alberto Rassi State Hospital—HGG (protocol number 961/19). The sample consisted of women from the public health system awaiting bariatric surgery, who were recruited from July to December 2019 (the vast majority of patients undergoing bariatric surgery at HGG are women). All eligible volunteers provided informed consent and signed a form in duplicate before participating in the study. The inclusion criteria were women with severe obesity (BMI $\geq$ 40 kg/m^2) aged between 18 and 59 years. The exclusion criteria included acute inflammatory diseases, infectious diseases, neoplastic diseases, or genetic syndromes; alcohol consumption (>30 g/day); or the use of illicit drugs.

2.2. Anthropometric Assessment

We conducted the measurements encompassing the height, body mass, and waist and hip circumferences. The body mass was determined using a Tanita-UM 080 digital scale with a maximum capacity of 150 kg. The height was measured using an inextensible metric, and the volunteers were asked to stand in an upright position. The waist circumference was measured at the midpoint between the last rib and the iliac crest, following a deep breath, with the volunteers standing. The hip circumference was measured at its maximum identified diameter, with the volunteers instructed to maintain their feet together during the assessment. All measurements were taken in duplicate by the same researcher staff using an inelastic tape. The BMI was calculated and classified according to the World Health Organization, dividing the body mass weight by the square of the height [20].

2.3. Analysis of Adipokines

The blood collection took place following a 12 h overnight fast. In triplicate, the serum samples were collected into vacuum tubes (Labor Import, Osasco, Brazil) after centrifugation (Eppendorf 5702R centrifuge, Hamburg, Germany) at 4000 RPM for 10 min at 10 °C and stored at −80 °C for subsequent analysis. ELISA kits (R&D Systems, Minneapolis, MN, USA) were employed for the quantification of leptin and adiponectin, following the manufacturer's instructions.

2.4. Analysis of Free Fatty Acid Profile in Serum

The analysis of free fatty acid composition in serum was determined using gas chromatography with a Varian 3900 gas chromatograph (Walnut Creek, CA, USA) coupled with flame ionization detection (FID) and an automatic sampler CP-8410. The methylation of each fraction was performed using acetyl chloride (5% HCl and methanol), and the fatty acid composition was determined with methyl esters. Fatty acids were identified by comparing retention times using a known standard of fatty acid methyl ester (FAME). FAMEs were utilized on a capillary column (CP Wax 52 CB, Varian, Lake Forests, CA, USA) with a thickness of 0.25 µm, an inner diameter of 0.25 mm, and a length of 30 m. Hydrogen was used as the carrier gas at a linear velocity of 22 cm/s. The temperature was programmed to 170 °C for 1 min, followed by increments of 2.5 °C/min until 240 °C, with a final hold time of 5 min. The injector temperature was set at 250 °C, and the FID was set at 260 °C. FAMEs were identified by comparing the retention times of the samples with known standards (Supelco, 37 components; Sigma-Aldrich, St. Louis, MO, USA; Mixture, Me93, Larodan, and Qualmix, PUFA Fish M, Menhaden Oil, Larodan, Solna, Sweden). Percentages of total fatty acids were used to express the values of fatty acid composition.

2.5. Statistical Analysis

The statistical analysis was performed using the R Studio program, version 4.3.3. The normality of the variables was assessed using the Shapiro–Wilk test, and the data are presented as mean and standard deviation or median and quartile range, according to normality. To investigate the association between the leptin/adiponectin and adiponectin/leptin ratios and serum fatty acids, and also between C-reactive proteins (CRPs) and omega-3 and -6 fatty acids, Pearson or Spearman correlation analyses were performed based on data normality. A generalized linear regression model (GLM) was performed using the Stepwise strategy, with age, BMI, waist circumference, hip circumference, and neck circumference in the model. The covariates were chosen according to similar models found in the literature [21–23].

3. Results

Forty-nine patients were enrolled in the study; however, one person withdrew and seven were excluded due to a lack of data, such as the waist and hip circumferences (three volunteers) and adipokines measurements (four volunteers). Thus, the analyses were performed with 41 participants. The volunteers had a mean age of 40.2 ± 8.3 years and a BMI of 46.99 (42.97–51.90) kg/m^2. A total of twenty-one plasma fatty acids, including four saturated (56.2% by area), five monounsaturated (17.2% by area), and twelve polyunsaturated fatty acids (26.6% by area), were identified. Regarding the adipokines, 60% of the sample presented hyperleptinemia and 41% presented hypoadiponectinemia. Data from the descriptive analysis of the sample are presented in Table 1.

Table 1. Descriptive analysis of anthropometric data, serum fatty acids, and adipokines from women with severe obesity (n = 41).

	Mean/Median	SD/Quartile Range
	Characteristics and Anthropometric Data	
Age (y)	40.22	±8.3
Height (m)	1.59	±0.05
Body mass (kg)	118.80	111.95–129.10
BMI (kg/m^2)	46.99	42.97–51.90
Waist circumference (cm)	131.46	±12.52
Hip circumference (cm)	142.50	136.1–152.25
Ratio Waist circumference/Hip	0.90	±0.07
Fatty Acids (% by area)		
Saturated (SFA)		
TOTAL	56.19	±5.63
C14:0	4.26	±1.45
C16:0	15.61	±4.75

Table 1. *Cont.*

	Mean/Median	SD/Quartile Range
C20:0	0.55	±0.14
C22:0	0.20	0–0.25
Monounsaturated (MUFA)		
TOTAL	17.15	±2.99
C14:1C	3.17	±1.01
C16:1n7	0.82	0.72–1.08
C18:1n9	11.23	±3.92
C18:1n7	1.11	±0.35
C20:1n9	0.75	±0.22
TOTAL	17.15	+2.99
Polyunsaturated (PUFA)		
TOTAL	26.64	+4.04
Omega-6		
TOTAL	21.68	+4.41
C18:2n6	14.47	±5.16
C18:3n6	3.45	±1.05
C20:2n6	1.73	±0.69
C20:3n6	0.32	±0.99
C20:4n6	0.23	0–0.30
C22:2n6	1.15	0.87–1.44
Omega-3		
TOTAL	4.93	3.35–6.80
C18:3n3	2.53	1.82–3.46
C18:4n3	0.46	0.36–0.57
C20:3n3	0.48	0.37–0.78
C20:4n3	0.29	0.23–0.35
C20:5n3	0.08	±0.32
C22:6n3	0.45	0.29–0.56
Ratio SFA/PUFA	2.02	1.80–2.41
Ratio SFA/MUFA	3.41	±0.89
Ratio n3/n6	0.18	0.13–0.33
Ratio n6/n3	5.39	3.02–7.38
Inflammatory Markers		
CRP	0.95	0.42–1.44
Adiponectin (μg/dL)	7.67	5.76–11.42
Leptin (ng/dL)	32.23	28–44.71
Ratio Adiponectin/Leptin	0.21	0.14–0.29
Ratio Leptin/Adiponectin	4.78	3.41–6.92

BMI: Body mass index; Waist circumference (<80 cm); Ratio circumference Waist/Hip (≤0.8); CRP: C-reactive protein (<1 mg/dL); Adiponectin (>7 μg/dL); Leptin (31.7 ng/dL).

The correlation analyses showed a positive correlation between the Lep/Adipo ratio and Di-homo-γ-linolenic acid (DGLA 20:3n6, r = 0.52, p = 0.00, Table S1), as well as between the Adipo/Lep ratio and Docosahexaenoic Acid—DHA (22:6n3, r = 0.74, p = 0.00, Table S2). Moreover, a positive correlation was found between CRP and 18:3n6 (p = 0.01), 20:4n3 (p = 0.00), and 18:3n-3 (p = 0.01) (Table S3).

The multiple linear regression between C22:6n3 fatty acid and the Adipo/Lep ratio explains 63% of the variability of the ratio. And the regression between C20:3n6 and the Lep/Adipo ratio was adjusted for neck circumference and explains 30% of the variability of this ratio. The results showed C22:6n3 as a predictor of the Adipo/Lep ratio ($p < 0.000$) (Table 2). In addition, C20:3n6 is a predictor factor for the Lep/Adipo ratio ($p < 0.000$) (Table 3).

Table 2. Multiple linear regression model for Adipo/Lep ratio in women with severe obesity.

		β	*p*-Value	OR (95% IC)
Step 1	Age	0.021	0.010	1.021 (0.02–3.27)
	BMI	−0.010	0.609	0.989 (0.95–1.03)
	WC	0.001	0.837	1 (0.988–1.01)
	HC	0.005	0.569	1 (0.98–1.02)
	NC	−0.005	0.781	0.994 (0.95–1.03)
	C22:6n3	0.755	0.000	2.129 (1.740–2.604)
Step 2	Age	0.022	0.006	1 (1.007–1.037)
	BMI	−0.009	0.633	0.991 (0.954–1.028)
	HC	0.005	0.545	1 (0.987–1.024)
	NC	−0.005	0.793	0.994 (0.956–1.034)
	C22:6n3	0.754	0.000	2.127 (1.744–2.594)
Step 3	Age	0.022	0.004	1 (1.007–1.037)
	BMI	−0.010	0.551	0.989 (0.955–1.024)
	HC	0.006	0.517	1 (0.988–1.024)
	C22:6n3	0.755	0.000	2.129 (1.750–2.589)
Step 4	Age	0.021	0.004	1 (1.007–1.036)
	HC	0.001	0.790	1 (0.992–1.010)
	C22:6n3	0.761	0.000	2.142 (1.766–2.598)
Step 5	Age	0.020	0.003	1.020 (1.007–1.034)
	C22:6n3	0.760	0.000	2.139 (1.768–2.588)

BMI: Body mass index; WC: Waist circumference; HC: Hip circumference; NC: Neck circumference.

Table 3. Multiple linear regression model for Lep/Adipo ratio in women with severe obesity.

		β	*p*-Value	OR (95% IC)
Step 1	Age	−0.166	0.328	0.847 (0.698–1.176)
	BMI	−0.116	0.786	0.890 (0.384–2.058)
	WC	0.043	0.777	1 (0.774–1.408)
	HC	−0.052	0.795	0.949 (0.641–1.405)
	NC	0.700	0.120	2 (0.850–4.774)
	C20:3n6	5.215	0.000	184.053 (113.090–298.966)
Step 2	Age	−0.149	0.335	0.861 (0.630–1.162)
	BMI	−0.201	0.469	0.817 (0.476–1.402)
	WC	0.716	0.104	1 (0.775–1.393)
	NC	0.038	0.796	2 (0.881–4.756)
	C20:3n6	5,232	0.000	187.243 (120.015–292.130)
Step 3	Age	−0.143	0.343	0.866 (0.646–1.160)
	BMI	−0.157	0.465	0.854 (0.563–1.297)
	NC	0.735	0.087	2 (0.918–4.736)
	C20:3n6	5.364	0.000	213.575 (170.462–267.592)
Step 4	Age	−0.122	0.343	0.885 (0.665–1.177)
	NC	0.592	0.087	1.809 (0.877–3.728)
	C20:3n6	5.021	0.000	151.640 (145.542–157.994)
Step 5	NC	0.648	0.081	1.911 (0.941–3.883)
	C20:3n6	4.848	0.000	127.495 (12.791–1270.826)

BMI: Body mass index; WC: Waist circumference; HC: Hip circumference; NC: Neck circumference.

4. Discussion

The pro-inflammatory state in individuals with obesity is one of the main factors associated with the development of comorbidities and can be influenced by the circulation of fatty acids. It is already established in the literature that the analysis of the leptin/adiponectin and adiponectin/leptin ratios is a better indicator of inflammation than adipokines alone [12,14,15]. Considering the important role of inflammation in cardiovascular disorders, a more profound understanding of the relationship between these ratios and fatty acids can be of great value for the management of obesity and its comorbidities. For the first time, the present study demonstrated that the C22:6n3 fatty acid was a predictive factor for the increase in the Adipo/Lep ratio ($\beta = 0.760$, $p < 0.000$) in women with severe obesity. The regression model explains that 63% of the variability in the Adipo/Lep ratio is due to C22:6n3.

A previous study, comparing eutrophic individuals and individuals with obesity, identified DHA as a potential biomarker for chronic inflammation in obesity [24], corroborating our findings. Additionally, in a study with post-pubertal adolescents with obesity, it was observed that polyunsaturated fatty acids (PUFA) and, specifically, the n-3/n-6 ratio were positively correlated with adiponectin levels [19]. Finally, both experimental and clinical studies demonstrated a significant increase in adiponectin concentrations after n-3 fatty acid supplementation, reinforcing the positive influence of this type of fatty acid on the concentration of this adipokine [25].

The proposed mechanism for the influence of DHA on adiponectin concentration is related to its ability to act as a ligand for peroxisome proliferator-activated receptors (PPARs), which induces the expression of several genes involved in the metabolism of lipids, glucose, and anti-inflammatory cytokines. In fact, many studies suggest that PPARγ plays a key role in the ability of n-3 PUFA, specifically DHA, to reduce inflammation [25–27]. Additionally, an experimental study raised the hypothesis that n-3 fatty acids stimulate the secretion of adiponectin in epididymal fat in a PPARγ-dependent and PPARα-independent way [9]. Consequently, it is believed that part of the anti-inflammatory association observed for n-3 fatty acid in our study is mediated by these mechanisms.

Furthermore, DHA can reduce the production of eicosanoids derived from arachidonic acid (AA) by competing with it to be incorporated into the phospholipids of the cell membrane, reducing inflammation [28]. Finally, it can act directly on inflammatory cells through membrane receptors, such as GPR120 (G-protein coupled receptor 120), reducing the expression of nuclear factor kappa B (NF-kB) in macrophages and, consequently, the production of inflammatory cytokines, such as tumor necrosis factor alpha (TNF-α) and Toll-like receptors (TRLs) [2,29]. This molecular mechanism is complementary to the PPARs pathway to explain the association between n-3 and the Adipo/Lep ratio.

The findings of the present study reinforce n-3 fatty acid as an inflammation modulator agent and highlight the importance of its consumption in reducing inflammation, including in severe obesity. Several studies show that the Westernized diet, characteristic of the current diet patterns, represents an increase in the consumption of calories in saturated fat and n-6 fatty acids, with a reduction in the intake of n-3 fatty acids. The general population has a low intake of n-3 fatty acids and is unable to reach the recommended levels of n6/n3 [30,31]. Thus, public policies that favor the consumption of foods that are sources of n-3 [29,32], especially in countries where the consumption of these foods is low, can help control inflammation and minimize unfavorable clinical outcomes resulting from obesity.

On the other hand, the regression analyses demonstrated C20:3 n-6 fatty acid as a predictor factor for the Lep/Adipo ratio, revealing the relationship of this type of fatty acid in the pro-inflammatory process. Linoleic fatty acid is a precursor for arachidonic acid (AA), leading to the formation of eicosanoids. These eicosanoids subsequently give rise to series 2 prostaglandins, series 4 leukotrienes, and related metabolites that collectively regulate pro-inflammatory activities, including the production of pro-inflammatory cytokines [32,33]. Considering that, in high concentrations, leptin is also a pro-inflammatory mediator, DGLA corroborates to a more pronounced state of inflammation in obesity [34].

In fact, our study observed a positive correlation between CRP and 18:3 n-6 ($p = 0.01$), a precursor of AA. Positive correlations between CRP and 20:4 n-3 and 18:3 n-3 were also observed in our study, which are two fatty acids that can be converted in n-6 fatty acids depending on the desaturase's concentration [35].

It appears that the increased intake of precursors, such as AA, increases their content in the cell membrane, which may lead to the increased production of pro-inflammatory eicosanoids. Although dietary intake was not assessed in this study, it is known that a

Westernized diet [36] can contribute to an increase in AA precursor consumption. This can be found in vegetable oils and derived products, such as margarine [37]. Older studies demonstrated that high levels of AA have a strong association with the formation of substances that aggregate platelets, such as thromboxanes [37,38]. In any case, investigating the consumption of foods that can increase arachidonic acid levels is important for the management of obesity, mainly from the perspective of inflammation [33].

In the present study, 21 fatty acids were identified in the volunteers' plasma, including AGS, AGMI, and PUFA (Table 1), similarly to a previous study on the topic. Although we do not have a control group, it has been previously demonstrated by Bermúdez-Cardona and Velásquez-Rodríguez (2016) that only the DHA fatty acid differed between women with obesity and normal weight, and others reinforce that DHA might play an important role in obesity and cardiovascular disease [38,39]. Thus, n-3 supplementation can be an ally in controlling chronic low-grade inflammation in this population, either by promoting the anti-inflammatory or reducing the pro-inflammatory pathways, or both [2]. Albracht-Schulte et al. [2] suggest that omega-3 fatty acid can be an adjuvant in the treatment of obesity and metabolic syndrome, along with lifestyle modifications and pharmacotherapy. However, genetic and epigenetic factors require further studies, as they may interfere with the outcomes of greater n-3 consumption [33].

Even with the limitations of the study, such as the small sample size, the lack of a control group, and the lack of information about the volunteers' food consumption, it was possible to observe the association between unsaturated fatty acids and the inflammatory profile. No saturated fatty acids were correlated with pro- and anti-inflammatory markers, which suggests that the strategy should be focused on unsaturated fatty acids. Furthermore, this is the first study carried out only with women with severe obesity, a condition that is increasing in our society. More studies are needed with a larger sample, evaluating dietary intake, to better understand the influence of food on circulating fatty acids.

5. Conclusions

The fatty acid DHA was shown to be a positive predictor for the anti-inflammatory adipo/lep ratio, whereas C20:3 n-6 was a predictor for inflammatory markers in women with severe obesity. The data suggest an intrinsic relationship between polyunsaturated fatty acids and inflammation in obesity, emphasizing the importance of public strategies to encourage the greater consumption of series n-3 fatty acids and to reduce the consumption of series n-6 fatty acids in individuals with severe obesity.

Supplementary Materials: The following supporting information can be downloaded at https://www.mdpi.com/article/10.3390/biomedicines12102248/s1, Table S1. Correlations between fatty acid profile and Lep/Adipo ratios in women with severe obesity. Table S2. Correlations between fatty acid profile and Adipo/Lep ratios in women with severe obesity. Table S3. Correlations between omega 3 and 6 fatty acids and CRP.

Author Contributions: Conceptualization—N.F., M.A.H., A.R.D., L.M.O., G.C.L., J.F.M. and F.C.C.; Data curation—G.B.L., N.F., F.M.K., G.I.M.H.d.S., R.M.S.C. and F.C.C.; Formal analysis—G.B.L., N.F., E.S.O., L.M.O., G.I.M.H.d.S., J.F.M., R.M.S.C. and F.C.C.; Funding acquisition—N.F., F.M.K., E.S.O., A.R.D., G.C.L., J.F.M. and F.C.C.; Investigation—N.F., F.M.K., E.S.O., L.M.O., G.C.L., J.F.M., R.M.S.C. and F.C.C.; Methodology—M.A.H., A.R.D., L.M.O., R.G.M.W., G.I.M.H.d.S., G.C.L., R.M.S.C. and F.C.C.; Project administration—N.F., F.M.K., M.A.H., G.C.L. and F.C.C.; Resources—M.A.H., A.R.D., L.M.O., R.G.M.W., J.F.M., R.M.S.C. and F.C.C.; Software—G.B.L., N.F., E.S.O., R.G.M.W. and G.I.M.H.d.S.; Supervision—G.B.L., N.F., M.A.H., G.C.L. and F.C.C.; Validation—N.F., A.R.D., L.M.O., R.G.M.W., G.I.M.H.d.S., G.C.L. and F.C.C.; Visualization—G.B.L., N.F., F.M.K., E.S.O. and F.C.C.; Writing-original draft—G.B.L., M.A.H., G.C.L., J.F.M., R.M.S.C. and F.C.C. Writing—review and editing—G.B.L., E.S.O., J.F.M., R.M.S.C. and F.C.C. All authors have read and agreed to the published version of the manuscript.

Funding: This work was supported by the National Council for Scientific and Technological Development—CNPQ, (434159/2018-2); Research Support Foundation—FUNAPE–UFG (January 2022). The authors would like to thank the Dâmaso (n. 305240/2021-8).

Institutional Review Board Statement: The study was conducted in accordance with the Declaration of Helsinki, and approved by the Research Ethics Committee of the Federal University of Goiás (no. 3.251.178, 8 April 2019) and the State Hospital of Goiânia, Dr. Alberto Rassi (no. 961/19, 17 June 2019).

Informed Consent Statement: Informed consent was obtained from all subjects involved in the study.

Data Availability Statement: The original contributions presented in this study are included in this article. Further inquiries can be directed to the corresponding authors.

Acknowledgments: We want to thank all the volunteers who agreed to participate in this study.

Conflicts of Interest: The authors declare no conflicts of interest.

References

1. Arroyo-Johnson, C.; Mincey, K.D. Obesity Epidemiology Trends by Race/Ethnicity, Gender, and Education. *Gastroenterol. Clin. N. Am.* **2016**, *45*, 571–579. [CrossRef] [PubMed]
2. Albracht-Schulte, K.; Kalupahana, N.S.; Ramalingam, L.; Rahman, S.M.; Robert-Mccomb, J.; Moustaid, N. Omega-3 fatty acids in obesity and metabolic syndrome: A mechanistic. *J. Nutr. Biochem.* **2020**, *58*, 1–16. [CrossRef] [PubMed]
3. Martin, S.S.; Blaha, M.J.; Muse, E.D.; Qasim, A.N.; Reilly, M.P.; Blumenthal, R.S.; Nasir, K.; Criqui, M.H.; McClelland, R.L.; Hughes-Austin, J.M.; et al. Leptin and incident cardiovascular disease: The Multi-ethnic Study of Atherosclerosis (MESA). *Atherosclerosis* **2015**, *239*, 67–72. [CrossRef] [PubMed]
4. Kang, K.W.; Ok, M.; Lee, S.K. Leptin as a Key between Obesity and Cardiovascular Disease. *J. Obes. Metab. Syndr.* **2020**, *29*, 248–259. [CrossRef]
5. Netto, B.D.; Bettini, S.C.; Clemente, A.P.; Ferreira, J.P.; Boritza, K.; Souza Sde, F.; Von der Heyde, M.E.; Earthman, C.P.; Dâmaso, A.R. Roux-en-Y gastric bypass decreases pro-inflammatory and thrombotic biomarkers in individuals with extreme obesity. *Obes. Surg.* **2015**, *25*, 1010–1018. [CrossRef]
6. Sanches, P.L.; de Mello, M.T.; Elias, N.; Fonseca, F.A.; Campos, R.M.; Carnier, J.; de Piano, A.; Masquio, D.C.; Silva, P.L.; Oyama, L.M.; et al. Hyperleptinemia: Implications on the inflammatory state and vascular protection in obese adolescents submitted to an interdisciplinary therapy. *Inflammation* **2014**, *37*, 35–43. [CrossRef]
7. Bochar, O.M.; Sklyarova, H.Y.; Abrahamovych, K.Y.; Hromnats'ka, N.M.; Bochar, V.T.; Sklyarov, E.Y. Metabolic syndrome, overweight, hyperleptinemia in children and adults. *Wiad. Lek.* **2021**, *74*, 313–316. [CrossRef]
8. Di Filippo, L.; De Lorenzo, R.; Sciorati, C.; Capobianco, A.; Lorè, N.I.; Giustina, A.; Manfredi, A.A.; Rovere-Querini, P.; Conte, C. Adiponectin to leptin ratio reflects inflammatory burden and survival in COVID-19. *Diabetes Metab.* **2021**, *47*, 101268. [CrossRef]
9. Neschen, S.; Morino, K.; Rossbacher, J.C.; Pongratz, R.L.; Cline, G.W.; Sono, S. Fish oil regulates adiponectin secretion by a peroxisome proliferator-activated receptor-γ-dependent mechanism in mice. *Diabetes* **2006**, *55*, 924–928. [CrossRef]
10. Zhao, S.; Kusminski, C.M.; Scherer, P.E. Adiponectin, Leptin and Cardiovascular Disorders. *Circ. Res.* **2022**, *128*, 136–149. [CrossRef]
11. Zaletel, J.; Pongrac Barlovic, D.; Prezelj, J. Adiponectin-leptin ratio: A useful estimate of insulin resistance in patients with type 2 diabetes. *J. Endocrinol. Investig.* **2010**, *33*, 514–518. [CrossRef] [PubMed]
12. López-Jaramillo, P.; Gómez-Arbeláez, D.; López-López, J.; López-López, C.; Martínez-Ortega, J.; Gómez-Rodríguez, A. The role of leptin/adiponectin ratio in metabolic syndrome and diabetes. *Horm. Mol. Biol. Clin. Investig.* **2014**, *18*, 37–45. [CrossRef] [PubMed]
13. Satoh, N.; Naruse, M.; Usui, T.; Tagami, T.; Suganami, T.; Yamada, K.; Kuzuya, H.; Shimatsu, A.; Ogawa, Y. Leptin-to-adiponectin ratio as a potential atherogenic index in obese type 2 diabetic patients. *Diabetes Care* **2004**, *27*, 2488–2490. [CrossRef] [PubMed]
14. Frühbeck, G.; Catalán, V.; Rodríguez, A.; Ramírez, B.; Becerril, S.; Salvador, J.; Colina, I.; Gómez-Ambrosi, J. Adiponectin-leptin ratio is a functional biomarker of adipose tissue inflammation. *Nutrients* **2019**, *11*, 454. [CrossRef] [PubMed]
15. Ferreira, Y.A.M.; Kravchychyn, A.C.P.; Vicente, S.D.C.F.; da Silveira Campos, R.M.; Tock, L.; Oyama, L.M.; Boldarine, V.T.; Masquio, D.C.L.; Damaso, A.R. Influence of magnitude of weight loss on Adipo/lep ratio in adolescents with obesity undergoing multicomponent therapy. *Cytokine* **2020**, *131*, 155111. [CrossRef]
16. Masquio, D.C.L.; Campos, R.M.D.S.; Netto, B.D.M.; Carvalho-Ferreira, J.P.; Bueno, C.R., Jr.; Alouan, S.; Poletto, G.T.; Ganen, A.P.; Tufik, S.; de Mello, M.T.; et al. Interdisciplinary Therapy Improves the Mediators of Inflammation and Cardio-vascular Risk in Adolescents with Obesity. *Int. J. Environ. Res. Public Health* **2023**, *20*, 7114. [CrossRef]
17. Simopoulos, A. The importance of the ratio of omega-6/omega-3 essential fatty acids. *Biomed. Pharmacother.* **2002**, *8*, 365–379. [CrossRef]
18. Monteiro, J.; Leslie, M.; Moghadasian, M.H.; Arendt, B.M.; Allard, J.P.; Ma, D.W.L. The role of n − 6 and n − 3 polyunsaturated fatty acids in the manifestation of the metabolic syndrome in cardiovascular disease and non-alcoholic fatty liver disease. *Food Funct.* **2014**, *5*, 426–435. [CrossRef]

19. Masquio, D.C.L.; de Piano-Ganen, A.; Oyama, L.M.; da Silveira Campos, R.M.; Santamarina, A.B.; Gomes, A.D.O.; Moreira, R.G.; Corgosinho, F.C.; do Nascimento, C.M.O.; Tock, L.; et al. The role of free fatty acids in the inflammatory and cardiometabolic profile in adolescents with metabolic syndrome engaged in interdisciplinary therapy. *J. Nutr. Biochem.* **2016**, *33*, 136–144. [CrossRef]
20. Allain-Regnault, M.; Bwibo, N.O.; Chigier, E. *Young People's Health-a Challenge for Society: Report of a WHO Study Group on Young People and "Health for All by the Year 2000"*; World Health Organization: Geneva, Switzerland, 2000.
21. Akour, A.; Kasabri, V.; Bulatova, N.; Al Muhaissen, S.; Naffa, R.; Fahmawi, H.; Momani, M.; Zayed, A.; Bustanji, Y. Association of Oxytocin with Glucose Intolerance and Inflammation Biomarkers in Metabolic Syndrome Patients with and without Prediabetes. *Rev. Diabet. Stud.* **2018**, *14*, 364–371. [CrossRef]
22. Moraes Ados, S.; Pisani, L.P.; Corgosinho, F.C.; Carvalho, L.O.; Masquio, D.C.; Jamar, G.; Sanches, R.B.; Oyama, L.M.; Dâmaso, A.R.; Belote, C.; et al. The role of leptinemia state as a mediator of inflammation in obese adults. *Horm. Metab. Res.* **2013**, *45*, 605–610. [PubMed]
23. Lopes, K.L.S.; Figueiredo, N.; Kattah, F.M.; Lima, G.C.; Oliveira, E.S.; Horst, M.A.; Oyama, L.M.; Dâmaso, A.R.; Whitton, R.G.M.; de Souza Abreu, V.; et al. The degree of food processing can influence serum fatty acid and lipid profiles in women with severe obesity. *Front. Nutr.* **2023**, *10*, 1046710. [CrossRef] [PubMed]
24. Liqiang, S.; Fang-Hui, L.; Minghui, Q.; Yanan, Y.; Haichun, C. Free fatty acids and peripheral blood mononuclear cells (PBMC) are correlated with chronic inflammation in obesity. *Lipids Health Dis.* **2023**, *22*, 93. [CrossRef] [PubMed]
25. Chang, H.Y.; Lee, H.N.; Kim, W.; Surh, Y.J. Docosahexaenoic acid induces M2 macrophage polarization through peroxisome proliferator-activated receptor γ activation. *Life Sci.* **2015**, *120*, 39–47. [CrossRef]
26. Yu, K.; Bayona, W.; Kallen, C.B.; Harding, H.P.; Ravera, C.P.; McMahon, G.; Brown, M.; Lazar, M.A. Differential activation of peroxisome proliferator-activated receptors by eicosanoids. *J. Biol. Chem.* **1995**, *270*, 23975–23983. [CrossRef]
27. Marion-Letellier, R.; Savoye, G.; Ghosh, S. Fatty acids, eicosanoids and PPAR gamma. *Eur. J. Pharmacol.* **2016**, *785*, 44–49. [CrossRef]
28. Talukdar, S.; Bae, E.J.; Imamura, T.; Morinaga, H.; Fan, W.; Li, P.; Lu, W.J.; Watkins, S.M.; Olefsky, J.M. GPR120 Is an Omega-3 Fatty Acid Receptor Mediating Potent Anti-inflammatory and Insulin-Sensitizing Effects. *Cell* **2010**, *142*, 687–698.
29. Calder, P.C. Polyunsaturated fatty acids, inflammation, and immunity. *Lipids* **2001**, *36*, 1007–1024. [CrossRef]
30. Candela, C.G.; López, L.M.B.; Kohen, L.V. Importance of a balanced omega 6/omega 3 ratio for the mainte-nance of health. Nutritional recommendations. *Nutr. Hosp.* **2011**, *26*, 323–329.
31. Simopoulos, A.P. Evolutionary aspects of diet, the omega-6/omega-3 ratio and genetic variation: Nutritional implications for chronic diseases. *Biomed. Pharmacother.* **2006**, *60*, 502–507. [CrossRef]
32. Sergeant, S.; Rahbar, E.; Chilton, F.H. Gamma-linolenic acid, Dihommo-gamma linolenic, Eicosanoids and Inflammatory Processes. *Eur. J. Pharmacol.* **2016**, *785*, 77–86. [CrossRef] [PubMed]
33. Innes, J.K.; Calder, P.C. Omega-6 fatty acids and inflammation. *Prostaglandins Leukot. Essent. Fat. Acids.* **2018**, *132*, 41–48. [CrossRef]
34. Moreno-Aliaga, M.J.; Lorente-Cebrián, S.; Martínez, J.A. Regulation of adipokine secretion by n-3 fatty acids. *Proc. Nutr. Soc.* **2010**, *69*, 324–332. [CrossRef] [PubMed]
35. Garófolo, A.; Petrilli, A.S. Omega-3 and 6 fatty acids balance in inflammatory response in patients with cancer and cachexia. *Rev. Nutr. Camp.* **2006**, *19*, 611–621. [CrossRef]
36. Willis, A.L.; Kuhn, D.C.; Vane, F.M.; Scott, C.G.; Petrin, M. An endoperoxide aggregator (LASS), formed in platelets in response to thrombotic stimuli-purification, identification and unique biological significance. *Prostaglandins* **1974**, *8*, 453–507. [CrossRef]
37. Hamberg, M.; Svensson, J.; Samuelsson, B. Thromboxanes: A new group of biologically active compounds derived from prostaglandin endoperoxides. *Proc. Natl. Acad. Sci. USA* **1975**, *72*, 2994–2998. [CrossRef]
38. Bermúdez-Cardona, J.; Velásquez-Rodrígues, C. Profile of Free Fatty Acids and Fractions of Phospholipids, Cholesterol Esters and Triglycerides in Serum of Obese Youth with and without Metabolic Syndrome. *Nutrients* **2016**, *8*, 54. [CrossRef]
39. Kim, O.Y.; Lim, H.H.; Lee, M.J.; Kim, J.Y.; Lee, J.H. Association of fatty acid composition in serum phospholipids with metabolic syndrome and arterial stiffness. *Nutr. Metab. Cardiovasc. Dis.* **2013**, *23*, 366–374. [CrossRef]

biomedicines

MDPI

Article

Neurosecretory Protein GM–Expressing Neurons Participate in Lipid Storage and Inflammation in Newly Developed Cre Driver Male Mice

Yuki Narimatsu *, Masaki Kato, Eiko Iwakoshi-Ukena, Shogo Moriwaki, Ayano Ogasawara, Megumi Furumitsu and Kazuyoshi Ukena *

Laboratory of Neurometabolism, Graduate School of Integrated Sciences for Life, Hiroshima University, Higashi-Hiroshima, Hiroshima 739-8521, Japan; iwakoshi@hiroshima-u.ac.jp (E.I.-U.); d221199@hiroshima-u.ac.jp (S.M.)

* Correspondence: ynarimatsu@hiroshima-u.ac.jp (Y.N.); ukena@hiroshima-u.ac.jp (K.U.)

Abstract: Obesity induces inflammation in the hypothalamus and adipose tissue, resulting in metabolic disorders. A novel hypothalamic neuropeptide, neurosecretory protein GM (NPGM), was previously identified in the hypothalamus of vertebrates. While NPGM plays an important role in lipid metabolism in chicks, its metabolic regulatory effects in mammals remain unclear. In this study, a novel Cre driver line, NPGM-Cre, was generated for cell-specific manipulation. Cre-dependent overexpression of *Npgm* led to fat accumulation without increased food consumption in male NPGM-Cre mice. Chemogenetic activation of NPGM neurons in the hypothalamus acutely promoted feeding behavior and chronically resulted in a transient increase in body mass gain. Furthermore, the ablated NPGM neurons exhibited a tendency to be glucose intolerant, with infiltration of proinflammatory macrophages into the adipose tissue. These results suggest that NPGM neurons may regulate lipid storage and inflammatory responses, thereby maintaining glucose homeostasis.

Keywords: neurosecretory protein GM; hypothalamus; neuropeptide; obesity; chronic inflammation; chemogenetics

Citation: Narimatsu, Y.; Kato, M.; Iwakoshi-Ukena, E.; Moriwaki, S.; Ogasawara, A.; Furumitsu, M.; Ukena, K. Neurosecretory Protein GM–Expressing Neurons Participate in Lipid Storage and Inflammation in Newly Developed Cre Driver Male Mice. *Biomedicines* **2023**, *11*, 3230. https://doi.org/10.3390/biomedicines11123230

Academic Editors: Teresa Vezza and Zaida Abad-Jiménez

Received: 31 October 2023
Revised: 4 December 2023
Accepted: 4 December 2023
Published: 6 December 2023

1. Introduction

Obesity is recognized as one of the risk factors associated with metabolic abnormalities, including type 2 diabetes, hypertension, hyperlipidemia, and cardiovascular disease [1–4]. Recent studies have highlighted the important role of chronic inflammation in adipose tissue in the development of metabolic abnormalities [5–8]. Several reports have demonstrated the involvement of neuropeptides and their producing neurons within the hypothalamus in feeding behavior, obesity, chronic inflammation, and related diseases. There is accumulating evidence regarding the arcuate nucleus (Arc). Neuropeptide Y (NPY) and agouti-related protein (AgRP) exhibit potent orexigenic effects [9]. The intracerebroventricular (i.c.v.) injection of AgRP has been shown to upregulate the mRNA expressions of tumor necrosis factor-α (TNF-α), a proinflammatory cytokine, in epididymal white adipose tissue (eWAT) via the sympathetic nervous system (SNS) [10]. AgRP-expressing neurons are also involved in energy homeostasis [11,12]. Conversely, α-melanocyte-stimulating hormone (α-MSH), derived from proopiomelanocortin (POMC), contributes to anorexigenic behavior in rodents through melanocortin receptor type 4 (MC4R) [13,14]. Chronic inhibition of the POMC- and MC4R-expressing neurons results in profound obesity [15]. In addition to the Arc, steroidogenic factor 1 (SF1) neurons, which are vital in regulating energy homeostasis in the ventromedial hypothalamus (VMH), have been found to inhibit inflammatory responses in the inguinal WAT (iWAT) of mice fed a high-fat diet (HFD) [16]. Furthermore, the activation of corticotropin-releasing hormone (CRH)-expressing neurons in the paraventricular nucleus of the hypothalamus (PVH) shifts the preference from a carbohydrate-rich diet to an HFD, and these neurons are involved in the development of HFD-induced obesity [17,18]. Despite accumulating evidence on the hypothalamic regulation of energy homeostasis as described above, the complete picture remains complex and elusive.

Two novel genes, *Fam237a/Gm39653* and *Fam237b/Gm8773*, encoding precursor proteins have been identified in the hypothalamus of chickens, rats, mice, and humans [19–22]. Owing to their distinct C-terminal amino acids, mature proteins derived from these precursor proteins are termed neurosecretory protein GL (NPGL) and neurosecretory protein GM [19]. To date, the essential role of NPGL in energy metabolism has been revealed. NPGL-expressing cells are localized in the lateroposterior part of the Arc (ArcLP) [21]. An acute i.c.v. injection of NPGL has been shown to stimulate feeding behavior [21]. Hypothalamic overexpression of *Npgl* leads to obesity in mice fed normal chow (NC) and high-calorie diets [23]. As for NPGM, we have reported that NPGM and NPGL are co-localized in the medial mammillary nucleus within the hypothalamic infundibulum of chicken [24]. We also found that NPGM accelerated lipid deposition without overeating in chicks [25]. A recent report showed that NPGM-enriched neurons are GABAergic and represent a subpopulation of AgRP-expressing neurons in mice [26]. Furthermore, recent findings revealed that an acute i.c.v. injection of NPGM stimulated feeding behavior in mice [27]. These data indicate that NPGM and/or NPGM-expressing neurons are key hypothalamic regulators of energy metabolism in addition to NPGL. Nevertheless, little is known about the effects of NPGM and NPGM neurons on energy metabolism, primarily due to the yet-to-be-elucidated intramolecular disulfide bond pattern of mammalian endogenous NPGM.

In this study, new Cre-driver mice were generated that bicistronically express NPGM and Cre recombinase, and neuron-specific analyses were conducted to provide further insight into the hypothalamic regulation of energy metabolism by NPGM. The present study analyzes the Cre-dependent overexpression of the precursor gene, chemogenetic manipulation, and neural ablation of NPGM neurons, with a focus on the effects of NPGM neurons on feeding behavior, fat accumulation, and chronic inflammation in the WAT.

2. Materials and Methods

2.1. Animals

NPGM-Cre mice were generated using the CRISPR-Cas9 system (Cyagen Biosciences, Inc., Santa Clara, CA, USA). The gRNAs designed for the CRISPR target sequence (Forward: 5′-AATGGCAGGACGTGATCTGAAGG-3′, Reverse: 5′-CACGTCCTGCCATTTTCTGTGGG-3′) and the donor vector for integration of the P2A (a self-cleaving peptide)-Cre sequence, along with Cas9 mRNA, were co-injected into fertilized mouse eggs to generate targeted knock-in offspring. Additionally, a synonymous mutation, S124 (TCC to AGT), was introduced to prevent sequence binding and recutting by gRNA after homology-directed repair. The desired mutation in the F0 founder mice was identified through PCR, followed by sequence analysis. Germline transmission and generation of F1 mice were confirmed through breeding with wild-type animals. The tail tips of weanlings (4 weeks old) were cut to obtain genomic DNA using the HotSHOT method [28]. PCR amplifications were performed using Quick Taq HS DyeMix (TOYOBO, Osaka, Japan) with the following conditions: 94 °C for 2 min, followed by 35 cycles of 95 °C for 30 s, 55 °C for 30 s, and 68 °C for 30 s. Heterozygous mice were used in all experiments.

Male NPGM-Cre mice (8 weeks old) were housed individually under standard conditions (25 ± 1 °C under a 12 h light/12 h dark cycle) with ad libitum access to water and either NC (CE-2; CLEA Japan, Tokyo, Japan) or HFD (60% of calories from fat, 7.1% of calories from sucrose; D12492, Research Diets, New Brunswick, NJ, USA).

2.2. Production of Plasmid Enabling Cre-Dependent Overexpression of Npgm

The artificial sequence (*Npgm*-P2A-eGFP) designed for the Cre-dependent overexpression of *Npgm* in NPGM-Cre mice was generated using Eurofins Genomics (Tokyo,

Japan). To generate pAAV-hSyn-DIO-*Npgm*-P2A-eGFP, pAAV-hSyn-DIO-hM3Dq-mCherry (#44361, Addgene, Watertown, MA, USA) was modified, replacing the coding sequence of hM3Dq-mCherry with the artificial sequence.

2.3. Preparation of AAV-Based Vectors

AAV-based vectors were prepared following previous reports [23,29,30] using pAAV-hSyn-DIO-*Npgm*-P2A-eGFP, pAAV-hSyn-DIO-hM3Dq-mCherry, pAAV-EF1α-DIO-mCherry (#50462, addgene), and pAAV-flex-taCasp3-TEVp (#45580, addgene). The AAV-based vectors were stored at −80 °C until use.

2.4. Stereotaxic Surgery

AAV injection was performed as described in previous methods [23,29,30]. NPGM-Cre mice were anesthetized using isoflurane. AAVs were bilaterally delivered into the mediobasal hypothalamic region 2.2 mm caudal to the bregma, 0.25 mm lateral to the midline, and 5.8 mm ventral to the skull surface, using a Neuros Syringe (7001 KH; Hamilton, Reno, NV, USA). Next, 0.5 μL of AAV-EF1α-DIO-mCherry (3.2×10^9 genome copies/μL) was delivered into the target regions of NPGM-Cre mice to confirm the neural-specific gene expression. For Cre-dependent overexpression of *Npgm*, 0.5 μL of AAV-hSyn-DIO-*Npgm*-P2A-eGFP (1.8×10^9 genome copies/μL) was injected into the target regions of NPGM-Cre mice. As a control, the same volume of AAV-hSyn-DIO-P2A-eGFP (1.2×10^9 genome copies/μL) was delivered into the target regions of NPGM-Cre mice. Then, 0.5 μL of AAV-hSyn-DIO-hM3Dq-mCherry (7.2×10^8 genome copies/μL) was used for acute/chronic stimulation of NPGM neurons. For the ablation of NPGM-neurons, 0.5 μL of AAV-flex-taCasp3-TEVp (2.2×10^{10} genome copies/μL) or AAV-EF1α-DIO-mCherry, as a control, was injected into the target regions of NPGM-Cre mice.

2.5. Cre-Dependent Overexpression of Npgm

After AAV injection, NPGM-Cre mice were fed the NC for 28 days. Food intake and body mass were measured at the beginning of the light period (9:00). At the end of this study, mice were decapitated between 13:00 and 15:00. The mediobasal hypothalamus, adipose tissue, and the liver were collected, weighed, and frozen in liquid nitrogen. Blood was simultaneously collected and centrifuged for 15 min at 3000× *g* at 4 °C after incubation for 30 min at 25 ± 1 °C. Plasma was stored at −80 °C. After removing lymph nodes in the iWAT, over 80% of the entire iWAT and eWAT were collected as digestion solutions containing Collagenase I (250 U, Worthington, Lakewood, NJ, USA), DNase I (21.6 U, Worthington), and PBS for flow cytometry.

2.6. Stimulation of NPGM Neurons in the Hypothalamus

One week after viral injection, NPGM-Cre mice were fed an HFD for 1 week. Food deprivation was induced at 17:00 in the measurement day. Immediately before the dark period, saline or clozapine-*N*-oxide (CNO) (1 mg/kg) (Tocris Bioscience, Minneapolis, MN, USA) was injected intraperitoneally into NPGM-Cre mice. Simultaneously, re-feeding of an HFD was initiated. Food intake was measured at 1, 2, 4, 6, and 12 h after intraperitoneal (i.p.) injection. The following day, a crossover study was conducted on these mice. Activation of hM3Dq-expressing neurons in these mice was confirmed using immunohistochemistry after an additional CNO injection.

For chronic stimulation of NPGM neurons, NPGM-Cre mice were allowed to recover for 2 weeks with ad libitum access to the NC after stereotaxic surgery. They were then provided drinking water containing CNO (5 mg/kg) to stimulate NPGM neurons via hM3Dq for 2 weeks. Water without CNO was used as the control. Food intake and body mass were measured during the initial 3 days of CNO consumption.

2.7. Ablation of NPGM Neurons

Before AAV injection, NPGM-Cre mice were fed an HFD for 8 weeks. The mice were fed HFDs for 56 days after stereotaxic surgery. An oral glucose tolerance test (OGTT) and an insulin tolerance test (ITT) were conducted 42 and 49 days later, respectively. Moreover, 56 days after stereotaxic surgery, each tissue sample was collected as the same procedure in Section 2.5.

2.8. Quantitative Reverse Transcriptase PCR (qRT-PCR)

RNA isolation and qRT-PCR were performed as previously reported [23,30,31]. Total RNA was isolated using the TRIzol reagent (Life Technologies, Carlsbad, CA, USA). First-strand cDNA was reverse-transcribed using the PrimeScript RT Reagent Kit with a gDNA Eraser (TAKARA, Shiga, Japan). qRT-PCR was performed with a THUNDERBIRD SYBR qPCR Mix (TOYOBO) under the following conditions: 95 °C for 20 s, followed by 40 cycles of 95 °C for 3 s and 60 °C for 30 s. Data were analyzed using the $2^{-\Delta\Delta Ct}$ method with *β-actin* (*Actb*) [32]. The primer sequences used in this study are listed in Table 1.

Table 1. Sequences of oligonucleotide primers for qRT-PCR.

Gene	Sense Primer (5′ to 3′)	Antisense Primer (5′ to 3′)
Npgm	CTCTCTGACGCTGATAGACC	AGATACTGTAATGCCCAGGA
Actb	GGCACCACACCTTCTACAAT	AGGTCTCAAACATGATCTGG
Genotyping 1	CGTTCTGCTGTTCAGTCTCACTG	GATTCCATTCTTCTATGCAACCCAT
Genotyping 2	GCTGATGATCCGAATAACTACCTG	GATTCCATTCTTCTATGCAACCCAT

2.9. Flow Cytometry

The iWAT and the eWAT were minced and shaken at 37 °C for 30 min in a digestion solution. The digested samples were filtered through 300 (pluriSelect Life Science, Leipzig, Germany), 70 (Funakoshi, Tokyo, Japan), and 40 μm (Funakoshi) cell strainers with cation/phenol red-free Hank's balanced salt solution (HBSS, Thermo Fisher Scientific, Waltham, MA, USA) buffer containing 2% bovine serum albumin (BSA, Sigma-Aldrich, St. Louis, MO, USA), 1 mM EDTA, and 25 mM HEPES (Thermo Fisher Scientific). After centrifugation at 400× *g* for 3 min at 4 °C, red blood cells were lysed with a lysis buffer (Abcam, Cambridge, UK) for 5 min and resuspended in HBSS buffer.

Harvested cells in the stromal vascular fractions were blocked with anti-mouse CD16/CD32 antibodies against Fc receptors (BD Biosciences, San Jose, CA, USA) for 10 min on ice. These cells were incubated for 30 min in an HBSS buffer with BV510 mouse anti-F4/80 (Biolegend, San Diego, CA, USA) or FITC mouse anti-F4/80 (Biolegend), PE mouse anti-CD11c (Biolegend) or FITC mouse anti-CD11c (Biolegend), APC mouse anti-CD206 (Biolegend) or PE-Cy7 mouse anti-CD206 (Biolegend), PerCP/Cy5.5 mouse anti-CD3 (Biolegend) or FITC mouse anti-CD3 (Biolegend), PE-Cy7 mouse anti-CD45 (Biolegend) or PE mouse anti-CD45 (Biolegend), BV421 mouse anti-CD19 (Biolegend) or PE-Cy7 mouse anti-CD19 (Biolegend) on ice. The dead cells were stained with propidium iodide. The stained cells were analyzed using a CytoFLEX S (Beckman Coulter, Brea, CA, USA) or a Cell Sorter MA900 (SONY, Tokyo, Japan). The macrophages were identified as $F4/80^+$ cells. The M1 macrophages were identified as $F4/80^+$, $CD11c^+$, and $CD206^-$. The M2 macrophages were identified as $F4/80^+$, $CD11c^-$, and $CD206^+$. The T cells were identified as $CD45^+$ and $CD3^+$ cells. The B cells were identified as $CD45^+$ and $CD19^+$ cells. To compare with a negative control, unstained cells were also analyzed for comparison.

2.10. Immunohistochemistry

After fixation in 4% paraformaldehyde, cryoprotection, and freezing, the brain tissues were sectioned at a thickness of 20 μm with a cryostat at −20 °C. Immunohistochemistry of free-floating sections was performed as previously described [20,21,23]. The sections were incubated in a blocking buffer (1% BSA, 1% normal donkey serum, and 0.3% Triton

X-100 in 10 mM PBS) for 1 h at 25 ± 1 °C before incubation with primary antibodies, including guinea pig anti-NPGM (1:100 or 1:200 dilution), rat anti-GFP (1:50,000 dilution, GF090R; Nacalai Tesque, Kyoto, Japan), goat anti-mCherry (1:500 dilution, AB0040-200; Origene), and rabbit anti-c-fos (1:1000 dilution, sc-52; Santa Cruz Biotechnology, Dallas, TX, USA) overnight at 4 °C. After undergoing three washes with 10 mM PBS, the floating sections were incubated for 1 h at 25 ± 1 °C with secondary antibodies, including Alexa Fluor 488-labeled donkey antibody to rabbit anti-IgG (1:400 or 1:600 dilution, 711-545-152; Jackson ImmunoResearch Laboratories, West Grove, PA, USA), Alexa Fluor 488-labeled donkey antibody to rat anti-IgG (1:500 dilution, 712-545-150; Jackson ImmunoResearch Laboratories), Cy3-labeled donkey antibody to guinea pig anti-IgG (1:400 dilution, 706-165-148; Jackson ImmunoResearch Laboratories), and Alexa Fluor 568-labeled donkey antibody to goat anti-IgG (1:400 dilution, ab175474; Abcam). The immunoreactive labeling was observed using a microscope (Eclipse E600; Nikon, Tokyo, Japan).

2.11. Plasma and Hepatic Biochemical Analysis

A GLUCOCARD G+ meter (Arkray, Kyoto, Japan) was used to measure the glucose content of the plasma. The NEFA C-test (Wako Pure Chemical Industries, Osaka, Japan) was used to measure the free fatty acid levels. Finally, the Triglyceride E-Test (Wako Pure Chemical Industries) was used to measure triglyceride levels.

To extract triglycerides from the liver, a previously reported protocol [33] was followed. The livers were homogenized in PBS, and a chloroform–methanol solution (2:1) was added to the homogenates. The samples were centrifuged at 18,000× *g* for 5 min at 4 °C, and the lower layer was collected and evaporated. The extracted lipids were dissolved in 100% isopropanol and hepatic triglyceride levels were measured using the Triglyceride E-Test (Wako Pure Chemical Industries).

2.12. OGTT and ITT

The OGTT and the ITT were conducted as previously reported [16,30,34–36], 42 and 49 days after stereotaxic surgery, respectively, to induce the ablation of NPGM neurons. Briefly, NPGM-Cre mice were fasted for 16 h for the OGTT or 4 h for the ITT. Using a GLUCOCARD G+ blood glucose meter, blood glucose levels were measured at 0, 15, 30, 60, and 120 min after oral glucose administration for the OGTT (1 g/kg weight) and i.p. injection of insulin for the ITT (0.75 units/kg). A 35 μL blood sample was collected from the tail vein using a heparinized plastic hematocrit tube (Drummond Scientific Company, Broomall, PA, USA), and the plasma was separated through centrifugation at 3000× *g* for 15 min. After centrifugation, the plasma was stored at −80 °C for insulin measurement. The LBIS Insulin-Mouse-U ELISA kit (Shibayagi, Gunma, Japan) was used to measure the insulin levels. The area under the curve (AUC) and inverse AUC for blood glucose were calculated using the linear trapezoidal method for both OGTT and ITT.

2.13. Statistical Analysis

A Student's *t*-test was performed to assess differences between the 2 groups (control and treated mice). To assess the main effects of groups (between control and treated) and time, a two-way, repeated-measures ANOVA was conducted, followed by Sidak's test for multiple comparisons. Differences at p values < 0.05 were considered statistically significant. All results are presented as the mean ± standard error of the mean (SEM).

3. Results

3.1. Generation of NPGM-Cre Mice

To enable a neuron-specific approach to NPGM-expressing cells, we created Cre driver mice using the CRISPR-Cas9 system. The construction of the targeted allele is shown in Figure 1A. The TGA stop codon in exon 2 of *Npgm* was replaced with a 2A self-cleaving peptide and Cre recombinase. To confirm Cre-dependent gene expression in the created mice, we stereotaxically injected AAV-EF1α-DIO-mCherry into the ArcLP.

Immunohistochemistry following stereotaxic injection of AAV-EF1α-DIO-mCherry into the ArcLP confirmed the co-localization of NPGM-immunoreactive cells and AAV-derived mCherry (Figure 1B). These results indicate Cre-dependent gene expression in the generated mice, which were named "NPGM-Cre."

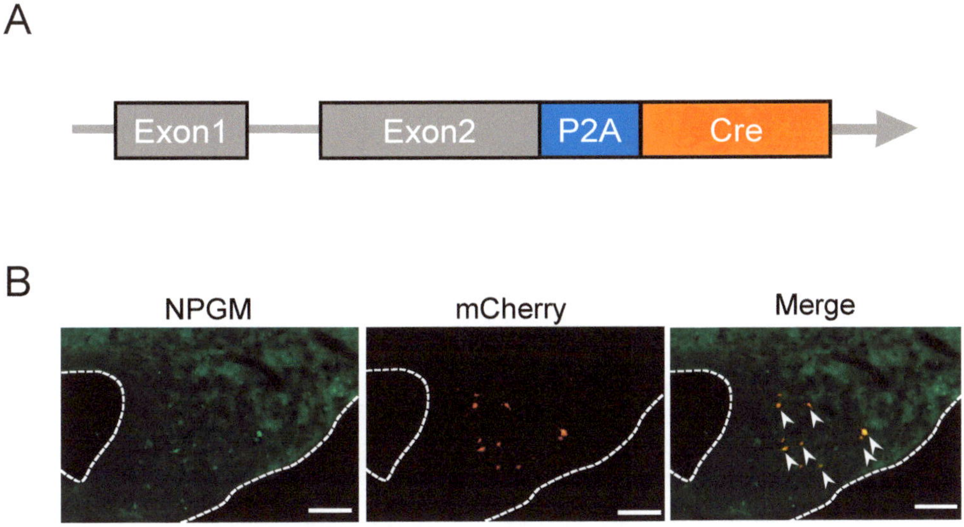

Figure 1. Generation and verification of NPGM-Cre mice. (**A**) Targeted alleles in the NPGM-Cre mice. (**B**) Representative micrographs of the mediobasal hypothalamus at 4 weeks after injection with AAV-EF1α-DIO-mCherry. The arrowheads indicate NPGM-immunoreactive mCherry-expressing neurons. NPGM: neurosecretory protein GM. Scale bars = 100 μm.

3.2. Cre-Dependent Overexpression of Npgm

To examine neuron-specific overexpression of *Npgm*, Cre-dependent overexpression of *Npgm* in NPGM-Cre mice was performed via stereotaxic delivery of AAV-hSyn-DIO-*Npgm*-P2A-eGFP (Figure 2A,B). *Npgm* overexpression was confirmed in the ArcLP using qRT-PCR (Student's t-test, df = 8, $t = 7.430$, $p < 0.005$, $n = 5$) and immunohistochemistry (Figure 2C,D). Notably, *Npgm* overexpression had limited effects on cumulative food intake or body mass gain (Figure 2E,F). While the mass of the iWAT was increased due to the overexpression of *Npgm* (Student's t-test, df = 8, $t = 2.620$, $p < 0.05$, $n = 5$), other adipose tissues remained unchanged (Figure 2G,H). In addition to the adipose tissues, apart from the iWAT, the liver mass and the hepatic triglyceride content were unaffected (Figure 2I,J).

Fat accumulation contributes to hyperglycemia and hyperlipidemia with chronic inflammation [37,38]. To confirm the effects of Cre-dependent *Npgm* overexpression on blood parameters, the blood glucose, triglyceride, and free fatty acids levels were evaluated. Importantly, the blood parameters were not altered due to *Npgm* overexpression (Figure 3A–C). Subsequent examination of immune cell percentages in the iWAT and eWAT revealed no changes (Figure 3D–O). These results suggest that Cre-dependent overexpression of *Npgm* induces fat accumulation without substantial effects on feeding behavior, body mass, blood parameters, or chronic inflammation in adipose tissue.

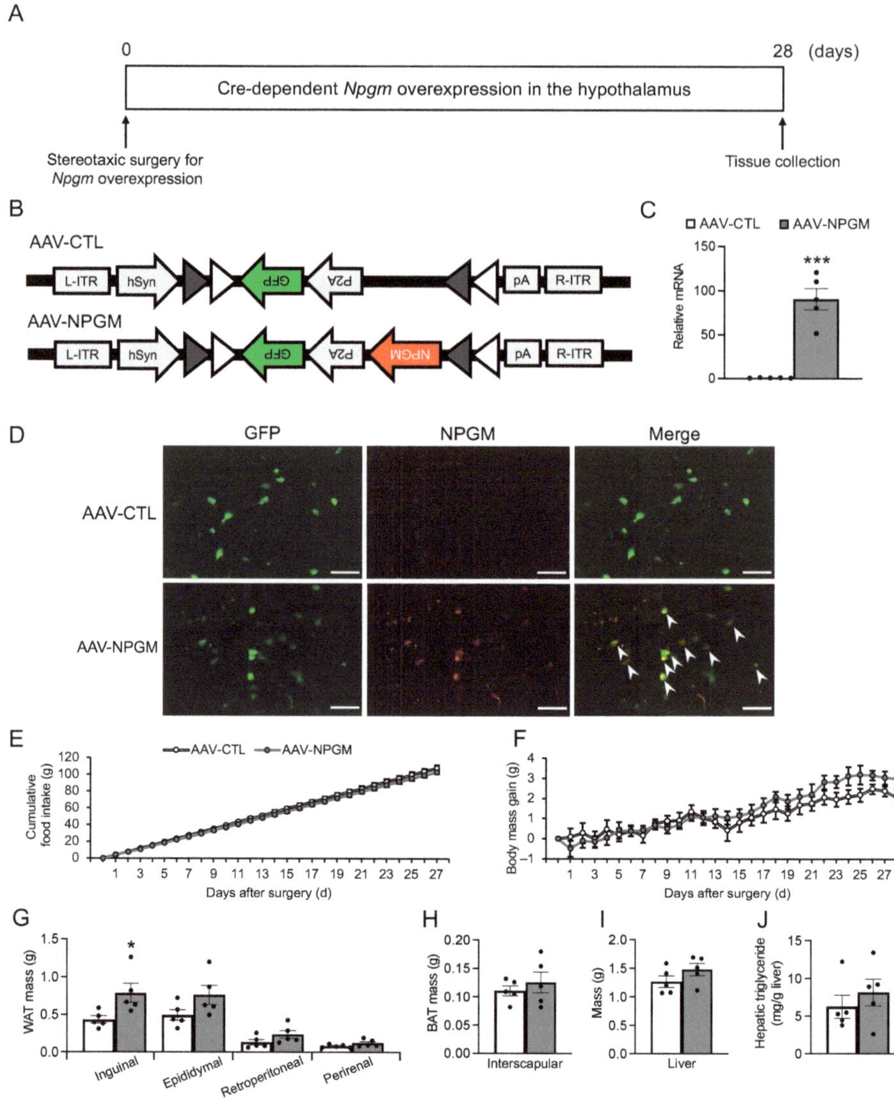

Figure 2. Cre-dependent overexpression of *Npgm* in the hypothalamus of NPGM-Cre mice. (**A**) Experimental procedure. (**B**) Construction of AAV-based vector. (**C**) mRNA expression level of *Npgm* in the mediobasal hypothalamus at the end of *Npgm* overexpression. (**D**) Representative micrograph of the mediobasal hypothalamus at 20 days after injection of AAV-CTL or AAV-NPGM. The arrowheads indicate NPGM-immunoreactive GFP-expressing neurons. (**E**) Cumulative food intake. (**F**) Body mass gain. (**G**) Mass of the inguinal, epididymal, retroperitoneal, and perirenal WAT. (**H**) Mass of the interscapular BAT. (**I**) Mass of the liver. (**J**) Content of hepatic triglycerides. Each value represents the mean ± SEM ($n = 5$). Asterisks indicate statistically significant differences (* $p < 0.05$, *** $p < 0.005$). Differences between groups were assessed using Student's t-test or two-way ANOVA with repeated measures followed by Sidak's test for multiple comparisons. NPGM: neurosecretory protein GM; AAV-CTL: AAV-based control vector; AAV-NPGM: AAV-based NPGM-precursor gene vector; WAT: white adipose tissue; BAT: brown adipose tissue. Scale bars = 100 μm.

3.3. Acute and Chronic Activation of NPGM Neurons in the Hypothalamus

Considering the established role of hypothalamic neuropeptide-expressing neurons in energy metabolism [11,15,16], NPGM neurons may also be involved in feeding behavior and body mass changes. Acute chemogenetic activation of NPGM neurons was then employed through the bilateral delivery of AAV-hSyn-DIO-hM3Dq-mCherry into the ArcLP (Figure 4A,B). After the i.p. injection of CNO, a ligand specific to hM3Dq, an increase in the number of c-fos/mCherry-double-positive cells was observed through immunostaining (Figure 4C). The CNO-injected mice exhibited an increase in food intake 1 h after injection (paired *t*-test, df = 12, $t = 2.213$, $p < 0.05$, $n = 13$) (Figure 4D). To determine the effect of chronic stimulation of NPGM neurons on body mass changes, testing was then conducted (Figure 4E). CNO was provided in the drinking water. Chronic administration of CNO slightly increased the cumulative food intake (Figure 4F). While body mass gain transiently increased 1 d after CNO administration (two-way, repeated ANOVA, time: $F_{1.639,8.193} = 5.496$, $p < 0.05$, group: $F_{1,5} = 10.46$, $p < 0.05$, interaction: $F_{3,15} = 3.162$, $p = 0.056$; 1 d: Sidak, $p < 0.01$, $n = 3$–4), the effect diminished over time (Figure 4G).

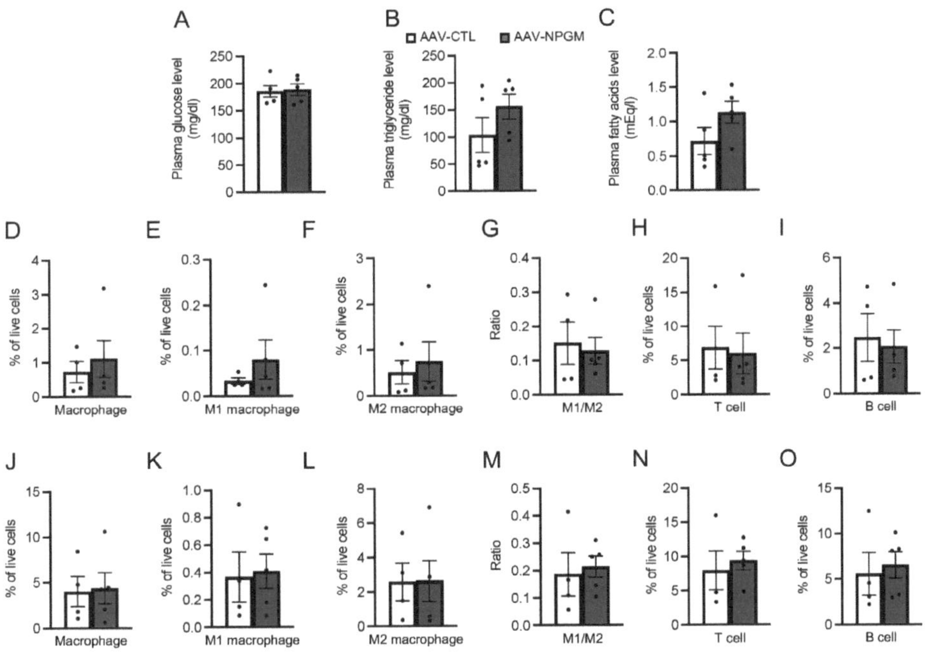

Figure 3. Effects of Cre-dependent overexpression of *Npgm* on blood parameters (**A**–**C**) and the population of immune cells in the iWAT (**D**–**I**) and eWAT (**J**–**O**). Plasma levels of (**A**) glucose, (**B**) triglyceride, and (**C**) free fatty acids. Populations of (**D**) macrophages, (**E**) M1 macrophages, (**F**) M2 macrophages, (**G**) M1/M2, (**H**) T cells, and (**I**) B cells in the iWAT. Populations of (**J**) macrophages, (**K**) M1 macrophages, (**L**) M2 macrophages, (**M**) M1/M2, (**N**) T cells, and (**O**) B cells in the eWAT. Each value represents the mean ± SEM ($n = 4$–5). Differences between groups were assessed using Student's *t*-test. NPGM: neurosecretory protein GM; AAV-CTL: AAV-based control vector; AAV-NPGM: AAV-based NPGM-precursor gene vector; iWAT: inguinal white adipose tissue; eWAT: epididymal white adipose tissue; M1 macrophages: classically activated macrophages; M2 macrophages: alternatively activated macrophages.

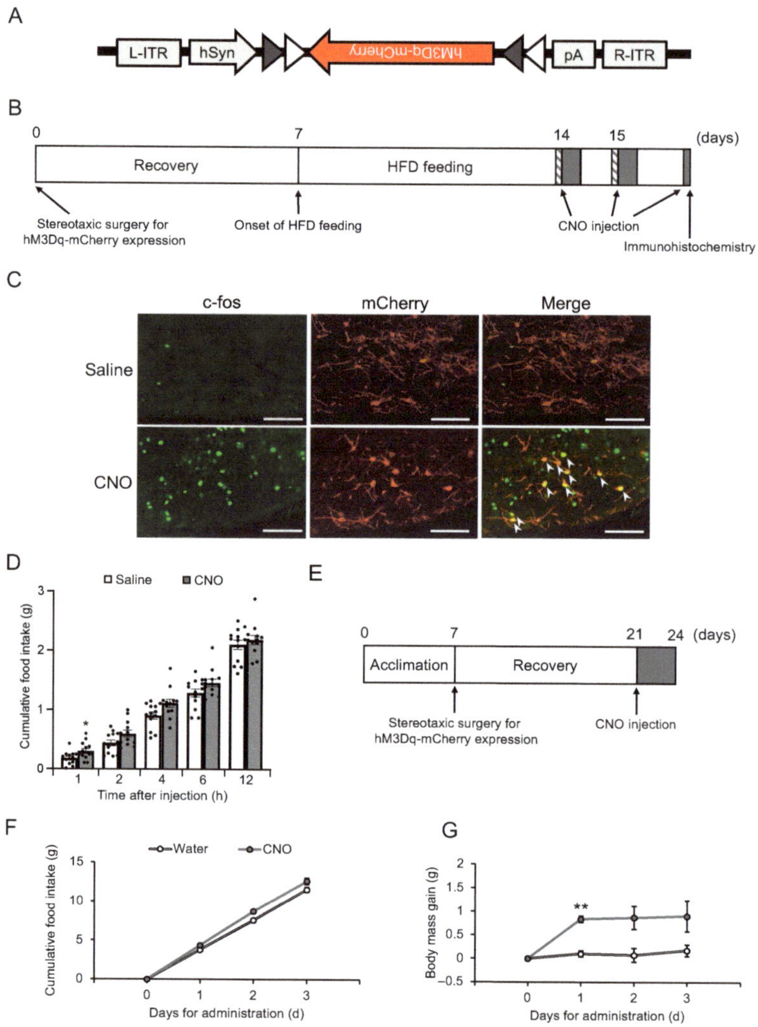

Figure 4. Acute (**B–D**) and chronic activation (**E–G**) of NPGM neurons in the hypothalamus. (**A**) Construction of AAV-based vector. (**B**) Experimental procedure to acutely activate NPGM neurons in the hypothalamus. Diagonal lines show food deprivation, and the gray zones indicate activation of NPGM neurons. (**C**) Representative micrographs of the mediobasal hypothalamus 90 min after injection of CNO in NPGM-Cre mice expressing hM3Dq. Arrowheads indicate c-fos-immunoreactive mCherry-expressing neurons. (**D**) Cumulative food intake after CNO injection. (**E**) Experimental procedure to chronically activate NPGM neurons in the hypothalamus. The gray zone indicates the activation of NPGM neurons. (**F**) Cumulative food intake after drinking CNO. (**G**) Body mass gain after drinking CNO. Each value represents the mean ± SEM (n = 3–13). Asterisks indicate statistically significant differences (* $p < 0.05$, ** $p < 0.01$). Differences between groups were assessed using the Student's t-test or two-way ANOVA with repeated measures followed by Sidak's test for multiple comparisons. NPGM: neurosecretory protein GM; CNO: clozapine-N-oxide. Scale bars = 100 μm.

3.4. Ablation of NPGM Neurons

To further evaluate the role of NPGM neurons in energy metabolism, NPGM neurons were ablated in HFD-induced obese NPGM-Cre mice using modified caspase 3

(Figure 5A,B). The AAV induces apoptosis in a Cre-dependent manner [39]. Confirmation revealed that fewer NPGM-immunoreactive cells were present than those in control mice (Figure 5C). Unexpectedly, the cumulative food intake and body mass gain were not affected by the disruption in NPGM neurons (Figure 5D,E). In addition to the change in body mass, the mass of the adipose tissues and the liver and the hepatic triglyceride content remained at control levels (Figure 5F–I).

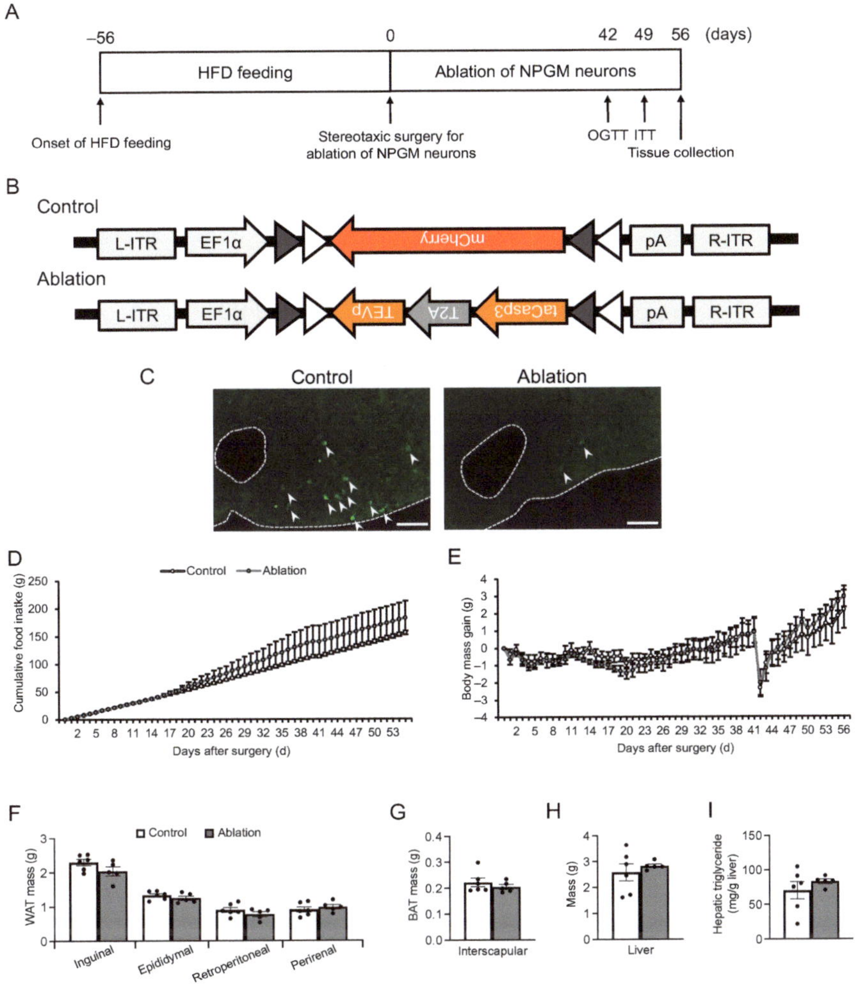

Figure 5. Ablation of NPGM neurons in the hypothalamus of NPGM-Cre mice. (**A**) Experimental procedure of the ablation of NPGM neurons in the hypothalamus. (**B**) Construction of AAV-based vector. (**C**) Representative micrograph of the mediobasal hypothalamus at 4 weeks after injection of AAVs for the control and ablation groups, respectively. The arrowheads indicate NPGM-immunoreactive neurons. (**D**) Cumulative food intake. (**E**) Body mass gain. (**F**) Mass of the inguinal, epididymal, retroperitoneal, and perirenal WAT. (**G**) Mass of the interscapular BAT. (**H**) Mass of the liver. (**I**) Content of hepatic triglyceride. Each value represents the mean ± SEM (n = 5–6). Differences between groups were assessed using Student's t-test or two-way ANOVA with repeated measures followed by Sidak's test for multiple comparisons. NPGM: neurosecretory protein GM; WAT: white adipose tissue; BAT: brown adipose tissue. Scale bars = 100 μm.

Subsequently, the effects of NPGM neuron ablation on the glucose/lipid metabolism were confirmed. Blood parameters were unchanged after NPGM neuronal ablation (Figure 6A–C). The OGTT showed a slight tendency toward increased blood glucose levels after glucose injection (not significant, 15 min: Student's t-test, df = 9, t = 1.516, p = 0.164, n = 5–6), independent of insulin secretion (Figure 6D–F). In contrast, elevated blood glucose levels were not observed during the ITT in NPGM-neuron-ablated mice (Figure 6G,H). These results suggest that the ablation of NPGM neurons may exacerbate glucose tolerance.

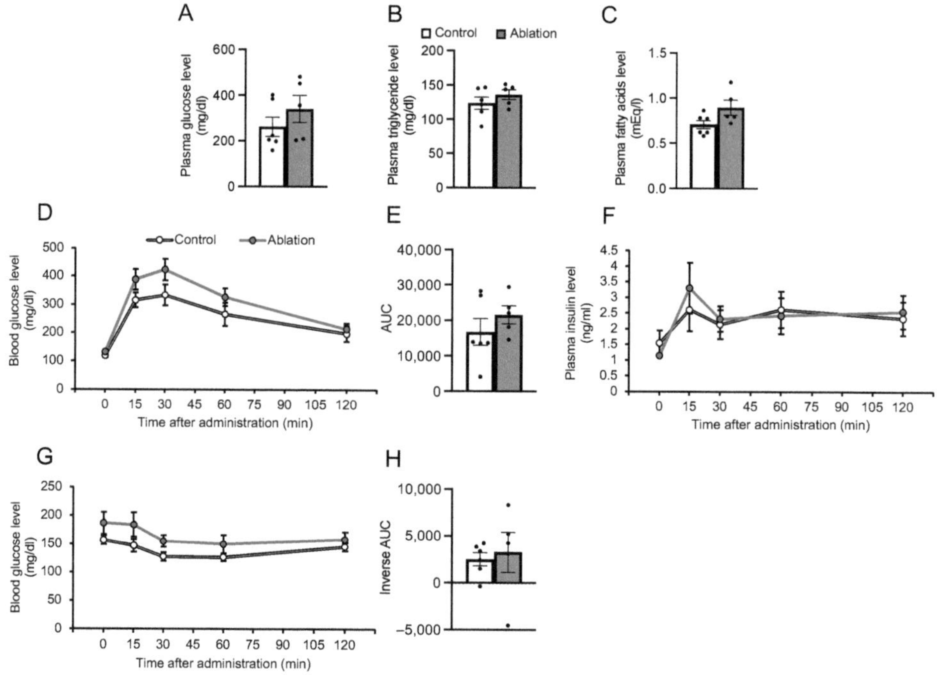

Figure 6. Effects of ablation of NPGM neurons on blood parameters (**A–C**) and glucose metabolism (**D–H**). Plasma levels of (**A**) glucose, (**B**) triglyceride, and (**C**) free fatty acids. (**D**) Blood level of glucose, (**E**) AUC for blood level of glucose, and (**F**) plasma level of insulin after oral injection of glucose. (**G**) Blood level of glucose and (**H**) AUC for blood level of glucose after i.p. injection of insulin. Each value represents the mean ± SEM (n = 5–6). Differences between groups were assessed using Student's t-test. NPGM: neurosecretory protein GM; AUC: area under the curve; i.p.: intraperitoneal.

Glucose intolerance is attributed to chronic inflammation in the adipose tissue [40]. The percentage of immune cells in the iWAT and eWAT was explored. In the iWAT of NPGM-neuron-ablated mice, the percentages of macrophages and M1 macrophages increased (macrophage: Student's t-test, df = 9, t = 3.036, p < 0.05, n = 5–6, M1 macrophage: Student's t-test, df = 9, t = 2.813, p < 0.05, n = 5–6) (Figure 7A,B). Aside from these populations, the percentage of immune cells did not change (Figure 7C–L). The data suggest that NPGM neurons may worsen glucose metabolism by exacerbating inflammatory responses in the iWAT.

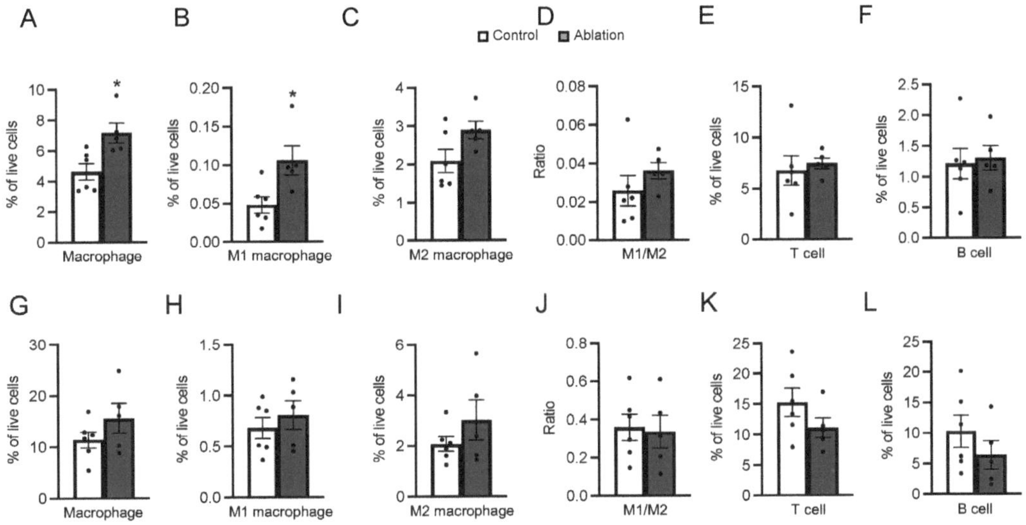

Figure 7. Effects of ablation of NPGM neurons on the population of immune cells in the iWAT (**A–F**) and eWAT (**G–L**). Populations of (**A**) macrophages, (**B**) M1 macrophages, (**C**) M2 macrophages, (**D**) M1/M2, (**E**) T cells, and (**F**) B cells in the iWAT. Populations of (**G**) macrophages, (**H**) M1 macrophages, (**I**) M2 macrophages, (**J**) M1/M2, (**K**) T cells, and (**L**) B cells in the eWAT. Each value represents the mean ± SEM (n = 5–6). Asterisks indicate statistically significant differences (* $p < 0.05$). Differences between groups were assessed using Student's t-test. NPGM: neurosecretory protein GM; iWAT: inguinal white adipose tissue; eWAT: epididymal white adipose tissue; M1 macrophages: classically activated macrophages; M2 macrophages: alternatively activated macrophages.

4. Discussion

The novel hypothalamic small protein NPGL is known to lead to overeating and fat accumulation, resulting in obesity [19–21,23]. Compared to the analysis of NPGL, the effects of the paralogous gene NPGM and its expressing neurons on energy metabolism have remained relatively unexplored in mice. In this study, a new Cre driver mouse, NPGM-Cre, was generated, and a neuron-specific approach to NPGM-expressing cells was employed to investigate the relationship between NPGM neurons and energy metabolism. Cre-dependent overexpression of *Npgm* resulted in fat accumulation in the iWAT without an increase in feeding behavior. The activation of NPGM neurons transiently stimulated feeding behavior and body mass gain. Moreover, the ablation of NPGM neurons slightly induced glucose intolerance with the progression of chronic inflammation in the iWAT. In summary, these results suggest that NPGM neurons maintain glucose homeostasis by regulating lipid storage and anti-inflammatory responses.

Although Cre-dependent overexpression of *Npgm* expanded fat depots in the iWAT, the food intake and mRNA expression levels of lipid-metabolism-related factors in the iWAT were equivalent to those in the control. The Arc, in which NPGM neurons are localized, is known to regulate lipid metabolism in the WAT under the regulation of the SNS [41]. Norepinephrine (NE) released from the axon terminus stimulates increased cAMP production via the β-adrenergic receptor [42]. cAMP triggers the phosphorylation of adipose triglyceride lipase (ATGL), hormone-sensitive lipase (HSL), perilipin, and lipolytic enzymes via activation of protein kinase A, resulting in lipolysis [42]. Furthermore, retrograde transsynaptic tracing from the iWAT revealed relatively more labeling in the suprachiasmatic nucleus (SCN), dorsomedial hypothalamus (DMH), and Arc than from the eWAT [41]. Consequently, fat accumulation independent of hyperphagia due to Cre-dependent *Npgm* overexpression may be caused by the suppression of SNS. Owing to the three Cys residues in mammalian NPGM, three patterns are expected for potential disulfide bonds [43]. However, the pattern of endogenous NPGM remains unknown. Given that the translation of overexpressed *Npgm* and the processing of precursor proteins depend on the endogenous regulatory system in NPGM neurons, the overexpressed NPGM in this study can be an appropriate form as well as an intrinsic NPGM. Future studies are required to unravel the intrinsic disulfide patterns and their effects on fat accumulation to further our understanding of endogenous NPGM-related fat accumulation.

Chronic activation of NPGM neurons via hM3Dq transiently stimulated food intake but subsequently abolished this effect. Chronic neural activation by the designer receptors exclusively activated by designer drugs (DREADD) system has provided essential insights into neuroscience [44–48]. However, a recent report has pointed out the possibility of receptor desensitization [49]. Indeed, a previous study suggested receptor desensitization during chronic treatment with ligands and recovery of receptor levels to pretreatment levels during the washout period [49,50]. In addition, a study reported that chronic activation of hM3Dq is regulated by negative feedback from inhibitory presynaptic input [45]. Hence, the transient effects on NPGM neurons may be explained by means of receptor desensitization or negative feedback regulation. Meanwhile, previous research has shown that high-dose CNO (10 mg/kg) was reversely metabolized to clozapine, a dopamine/serotonin antagonist, in mice, and recommended the use of an appropriate control group [51,52]. Considering that the dose of CNO in this study was 5 mg/kg, the reverse metabolism of CNO was not considered. Alternatively, recent studies have applied ion channels for the chronic manipulation of neurons [12,15]. Novel approaches, such as the use of ion channels, offer an elegant means of evaluating the role of NPGM neurons in energy metabolism.

The ablation of NPGM neurons using engineered caspase 3 slightly exacerbated glucose tolerance and infiltrated proinflammatory M1 macrophages in the iWAT (subcutaneous WAT) but not in the eWAT (visceral WAT). Low-grade chronic inflammation caused by macrophages in the adipose tissue is well known to be a trigger of metabolic dysfunction including insulin resistance [5]. A previous study estimated that the percentage of macrophages in the adipose tissues is over 50% in obese mice, although it is not even 10% in lean mice [53]. Macrophages in the adipose tissues are classified as M1 macrophages and M2 macrophages, at least [7]. M1 macrophages promote chronic inflammation through the production of proinflammatory cytokines such as TNF-α and interleukin-6 (IL-6), whereas M2 macrophages inhibit chronic inflammation through the secretion of anti-inflammatory cytokines including interleukin-10 (IL-10) [54,55]. Furthermore, subcutaneous and visceral WATs greatly differ in their contributions to chronic inflammation and metabolic abnormalities [56]. Indeed, previous studies have reported the infiltration of macrophages in the visceral rather than subcutaneous WATs of obese animals [57,58]. Given the possibility of metabolic regulation of the iWAT by NPGM neurons via the SNS, it is reasonable to assume that the lesions of NPGM neurons disrupted immune homeostasis in the iWAT only, resulting in limited glucose intolerance.

Cre-dependent overexpression of *Npgm* evoked fat accumulation in the iWAT without affecting the percentage of immune cells in the WAT. However, ablation of the NPGM neurons maintained the WAT mass and led to the infiltration of M1 macrophages into the iWAT. The phenotypes resulting from neuronal ablation are not necessarily identical to those caused by the pharmacological effects of neuropeptides. Melanin-concentrating hormone (MCH) has been shown to be an orexigenic neuropeptide through i.c.v. injection and analysis in KO mice [59,60]. In contrast, MCH-neuron-ablated mice displayed equivalent feeding behavior compared to the controls, suggesting that the orexigenic effects of MCH were antagonized by the anorexigenic effects of nesfatin-1 and cocaine- and amphetamine-regulated transcript (CART) that are co-expressed neuropeptides in MCH neurons [61–64]. Recent studies have demonstrated that NPGM is coexpressed with transcripts encoding neuropeptides in the Arc [26,65]. Therefore, the discrepancies in phenotypes between Cre-dependent overexpression of *Npgm* and the ablation of NPGM neurons may be accounted for by complicated processes involving other neurotransmitters.

This study has several limitations. Although Cre-dependent overexpression of *Npgm* led to fat accumulation, we only checked the plasma levels of blood glucose, triglyceride, and free fatty acids. Metabolic abnormalities are closely associated with dyslipidemia including high levels of low-density lipoprotein cholesterol and low levels of high-density lipoprotein cholesterol in obese subjects [66]. Additional measurements focused on serum cholesterol levels are required to understand the effects of NPGM on dyslipidemia. In addition, the present study employed only males. Given that the mRNA expression level of *NPGM* in females of quail is comparable to that in males [67], the sex differences in the contribution of NPGM to energy homeostasis might not be considerable. Further studies using female mice will help us to fully understand the sex differences in the effects of NPGM on energy homeostasis. Moreover, although all experiments in this study were employed using adult mice, the mRNA expression of *NPGM* gradually decreased with development in chicks [24]. Therefore, it is possible that manipulation and ablation of NPGM neurons showed minimal effects on energy homeostasis. Elucidation of the expression of *Npgm* and of the activity of NPGM neurons along with developmental stages in NPGM-Cre mice could lead to an understanding of the physiological function of NPGM and NPGM neurons. Finally, the fasting time before the OGTT and ITT was 16 h and 4 h in the present study, respectively, although a recent study revealed that a long time of food deprivation may lead to gluconeogenesis [68]. On the other hand, given that several recent studies still adopted a long time for fasting [34,36], we consider that the fasting time in these tests is controversial. Further experiments are required to reveal the effects of NPGM and NPGM neurons on gluconeogenesis.

In conclusion, NPGM neurons participate in lipid metabolism and inflammatory responses in the WAT. Whilst the relationship between the hypothalamus and adipose tissue has been widely investigated [16,69], this report will provide a stepping stone for understanding the mechanisms underlying fat accumulation and inflammatory responses in the brain. Future studies, including the chronic activation of NPGM neurons over extended periods, hold the potential to substantially advance our understanding of the intricate mechanisms underlying lipid metabolism and inflammatory response within NPGM neurons.

Author Contributions: Conceptualization, Y.N. and K.U.; methodology, Y.N., M.K., E.I.-U. and M.F.; investigation, Y.N., M.K., E.I.-U., S.M., A.O., M.F. and K.U.; writing—original draft preparation, Y.N.; writing—review and editing, Y.N. and K.U.; visualization, Y.N.; project administration, K.U.; funding acquisition, Y.N., M.K., E.I.-U. and K.U. All authors have read and agreed to the published version of the manuscript.

Funding: This work was supported by the KAKENHI Grants (JP22KJ2330 to Y.N.; JP22KJ2331 to M.K., JP22K11827 to E.I.-U.; and JP19H03258, JP20K21760, and JP22H00503 to K.U.) and the Hiroshima University Graduate School Research Fellowship (Y.N. and M.K.).

Institutional Review Board Statement: All animal experiments were conducted in accordance with the Guide for the Care and Use of Laboratory Animals prepared by Hiroshima University (Higashi-Hiroshima, Japan). The procedures were approved by the Institutional Animal Care and Use Committee of Hiroshima University (permit numbers C21-1 and C21-16).

Informed Consent Statement: Not applicable.

Data Availability Statement: The raw data supporting the findings of this manuscript will be made available by the corresponding authors, Y.N. and K.U., to any qualified researchers upon reasonable request.

Acknowledgments: We are grateful to Osamu Takeuchi (Kyoto University) and Yasuhiro Ishihara (Hiroshima University) for their experimental support.

Conflicts of Interest: The authors declare no conflict of interest.

References

1. Engin, A. The Definition and Prevalence of Obesity and Metabolic Syndrome. *Adv. Exp. Med. Biol.* **2017**, *960*, 1–17. [CrossRef]
2. Grundy, S.M. Obesity, Metabolic Syndrome, and Cardiovascular Disease. *J. Clin. Endocrinol. Metab.* **2004**, *89*, 2595–2600. [CrossRef]
3. Powell-Wiley, T.M.; Poirier, P.; Burke, L.E.; Després, J.-P.; Gordon-Larsen, P.; Lavie, C.J.; Lear, S.A.; Ndumele, C.E.; Neeland, I.J.; Sanders, P.; et al. Obesity and Cardiovascular Disease: A Scientific Statement from the American Heart Association. *Circulation* **2021**, *143*, e984–e1010. [CrossRef]
4. Leggio, M.; Lombardi, M.; Caldarone, E.; Severi, P.; D'Emidio, S.; Armeni, M.; Bravi, V.; Bendini, M.G.; Mazza, A. The Relationship between Obesity and Hypertension: An Updated Comprehensive Overview on Vicious Twins. *Hypertens. Res.* **2017**, *40*, 947–963. [CrossRef]
5. Kawai, T.; Autieri, M.V.; Scalia, R. Adipose Tissue Inflammation and Metabolic Dysfunction in Obesity. *Am. J. Physiol. Cell Physiol.* **2021**, *320*, C375–C391. [CrossRef] [PubMed]
6. Jung, U.; Choi, M.-S. Obesity and Its Metabolic Complications: The Role of Adipokines and the Relationship between Obesity, Inflammation, Insulin Resistance, Dyslipidemia and Nonalcoholic Fatty Liver Disease. *Int. J. Mol. Sci.* **2014**, *15*, 6184–6223. [CrossRef] [PubMed]
7. Fujisaka, S.; Usui, I.; Bukhari, A.; Ikutani, M.; Oya, T.; Kanatani, Y.; Tsuneyama, K.; Nagai, Y.; Takatsu, K.; Urakaze, M.; et al. Regulatory Mechanisms for Adipose Tissue M1 and M2 Macrophages in Diet-Induced Obese Mice. *Diabetes* **2009**, *58*, 2574–2582. [CrossRef] [PubMed]
8. Nawaz, A.; Aminuddin, A.; Kado, T.; Takikawa, A.; Yamamoto, S.; Tsuneyama, K.; Igarashi, Y.; Ikutani, M.; Nishida, Y.; Nagai, Y.; et al. CD206+ M2-like Macrophages Regulate Systemic Glucose Metabolism by Inhibiting Proliferation of Adipocyte Progenitors. *Nat. Commun.* **2017**, *8*, 1–16. [CrossRef] [PubMed]
9. Luquet, S.; Perez, F.A.; Hnasko, T.S.; Palmiter, R.D. NPY/AgRP Neurons Are Essential for Feeding in Adult Mice but Can Be Ablated in Neonates. *Science* **2005**, *310*, 683–685. [CrossRef] [PubMed]
10. Tang, L.; Okamoto, S.; Shiuchi, T.; Toda, C.; Takagi, K.; Sato, T.; Saito, K.; Yokota, S.; Minokoshi, Y. Sympathetic Nerve Activity Maintains an Anti-Inflammatory State in Adipose Tissue in Male Mice by Inhibiting TNF-α Gene Expression in Macrophages. *Endocrinology* **2015**, *156*, 3680–3694. [CrossRef]
11. Krashes, M.J.; Koda, S.; Ye, C.; Rogan, S.C.; Adams, A.C.; Cusher, D.S.; Maratos-Flier, E.; Roth, B.L.; Lowell, B.B. Rapid, Reversible Activation of AgRP Neurons Drives Feeding Behavior in Mice. *J. Clin. Invest.* **2011**, *121*, 1424–1428. [CrossRef] [PubMed]
12. Zhu, C.; Jiang, Z.; Xu, Y.; Cai, Z.-L.; Jiang, Q.; Xu, Y.; Xue, M.; Arenkiel, B.R.; Wu, Q.; Shu, G.; et al. Profound and Redundant Functions of Arcuate Neurons in Obesity Development. *Nat. Metab.* **2020**, *2*, 763–774. [CrossRef] [PubMed]
13. Yaswen, L.; Diehl, N.; Brennan, M.B.; Hochgeschwender, U. Obesity in the Mouse Model of Pro-Opiomelanocortin Deficiency Responds to Peripheral Melanocortin. *Nat. Med.* **1999**, *5*, 1066–1070. [CrossRef] [PubMed]
14. Cone, R.D. Anatomy and Regulation of the Central Melanocortin System. *Nat. Neurosci.* **2005**, *8*, 571–578. [CrossRef]
15. Li, H.; Xu, Y.; Jiang, Y.; Jiang, Z.; Otiz-Guzman, J.; Morrill, J.C.; Cai, J.; Mao, Z.; Xu, Y.; Arenkiel, B.R.; et al. The Melanocortin Action Is Biased toward Protection from Weight Loss in Mice. *Nat. Commun.* **2023**, *14*, 2200. [CrossRef]
16. Rashid, M.; Kondoh, K.; Palfalvi, G.; Nakajima, K.-I.; Minokoshi, Y. Inhibition of High-Fat Diet-Induced Inflammatory Responses in Adipose Tissue by SF1-Expressing Neurons of the Ventromedial Hypothalamus. *Cell Rep.* **2023**, *42*, 112627. [CrossRef]
17. Okamoto, S.; Sato, T.; Tateyama, M.; Kageyama, H.; Maejima, Y.; Nakata, M.; Hirako, S.; Matsuo, T.; Kyaw, S.; Shiuchi, T.; et al. Activation of AMPK-Regulated CRH Neurons in the PVH Is Sufficient and Necessary to Induce Dietary Preference for Carbohydrate over Fat. *Cell Rep.* **2018**, *22*, 706–721. [CrossRef] [PubMed]
18. Zhu, C.; Xu, Y.; Jiang, Z.; Tian, J.B.; Cassidy, R.M.; Cai, Z.-L.; Shu, G.; Xu, Y.; Xue, M.; Arenkiel, B.R.; et al. Disrupted Hypothalamic CRH Neuron Responsiveness Contributes to Diet-Induced Obesity. *EMBO Rep.* **2020**, *21*, e49210. [CrossRef]
19. Ukena, K.; Iwakoshi-Ukena, E.; Taniuchi, S.; Bessho, Y.; Maejima, S.; Masuda, K.; Shikano, K.; Kondo, K.; Furumitsu, M.; Tachibana, T. Identification of a cDNA Encoding a Novel Small Secretory Protein, Neurosecretory Protein GL, in the Chicken Hypothalamic Infundibulum. *Biochem. Biophys. Res. Commun.* **2014**, *446*, 298–303. [CrossRef]
20. Iwakoshi-Ukena, E.; Shikano, K.; Kondo, K.; Taniuchi, S.; Furumitsu, M.; Ochi, Y.; Sasaki, T.; Okamoto, S.; Bentley, G.E.; Kriegsfeld, L.J.; et al. Neurosecretory Protein GL Stimulates Food Intake, de Novo Lipogenesis, and Onset of Obesity. *eLife* **2017**, *6*, e28527. [CrossRef]
21. Matsuura, D.; Shikano, K.; Saito, T.; Iwakoshi-Ukena, E.; Furumitsu, M.; Ochi, Y.; Sato, M.; Bentley, G.E.; Kriegsfeld, L.J.; Ukena, K. Neurosecretory Protein GL, a Hypothalamic Small Secretory Protein, Participates in Energy Homeostasis in Male Mice. *Endocrinology* **2017**, *158*, 1120–1129. [CrossRef] [PubMed]
22. Ukena, K. Avian and Murine Neurosecretory Protein GL Participates in the Regulation of Feeding and Energy Metabolism. *Gen. Comp. Endocrinol.* **2018**, *260*, 164–170. [CrossRef] [PubMed]
23. Narimatsu, Y.; Iwakoshi-Ukena, E.; Fukumura, K.; Shikano, K.; Furumitsu, M.; Morishita, M.; Bentley, G.E.; Kriegsfeld, L.J.; Ukena, K. Hypothalamic Overexpression of Neurosecretory Protein GL Leads to Obesity in Male C57BL/6J Mice. *Neuroendocrinology* **2022**, *112*, 606–620. [CrossRef]
24. Shikano, K.; Bessho, Y.; Kato, M.; Iwakoshi-Ukena, E.; Taniuchi, S.; Furumitsu, M.; Tachibana, T.; Bentley, G.E.; Kriegsfeld, L.J.; Ukena, K. Localization and Function of Neurosecretory Protein GM, a Novel Small Secretory Protein, in the Chicken Hypothalamus. *Sci. Rep.* **2018**, *8*, 704. [CrossRef]

25. Kato, M.; Iwakoshi-Ukena, E.; Furumitsu, M.; Ukena, K. A Novel Hypothalamic Factor, Neurosecretory Protein GM, Causes Fat Deposition in Chicks. *Front. Physiol.* **2021**, *12*, 747473. [CrossRef]
26. Campbell, J.N.; Macosko, E.Z.; Fenselau, H.; Pers, T.H.; Lyubetskaya, A.; Tenen, D.; Goldman, M.; Verstegen, A.M.J.; Resch, J.M.; McCarroll, S.A.; et al. A Molecular Census of Arcuate Hypothalamus and Median Eminence Cell Types. *Nat. Neurosci.* **2017**, *20*, 484–496. [CrossRef]
27. Martinez, T.F.; Lyons-Abbott, S.; Bookout, A.L.; De Souza, E.V.; Donaldson, C.; Vaughan, J.M.; Lau, C.; Abramov, A.; Baquero, A.F.; Baquero, K.; et al. Profiling Mouse Brown and White Adipocytes to Identify Metabolically Relevant Small ORFs and Functional Microproteins. *Cell Metab.* **2023**, *35*, 166–183.e11. [CrossRef] [PubMed]
28. Truett, G.E.; Heeger, P.; Mynatt, R.L.; Truett, A.A.; Walker, J.A.; Warman, M.L. Preparation of PCR-Quality Mouse Genomic DNA with Hot Sodium Hydroxide and Tris (HotSHOT). *Biotechniques* **2000**, *29*, 52–54. [CrossRef]
29. Narimatsu, Y.; Iwakoshi-Ukena, E.; Naito, M.; Moriwaki, S.; Furumitsu, M.; Ukena, K. Neurosecretory Protein GL Accelerates Liver Steatosis in Mice Fed Medium-Fat/Medium-Fructose Diet. *Int. J. Mol. Sci.* **2022**, *23*, 2071. [CrossRef]
30. Fukumura, K.; Narimatsu, Y.; Moriwaki, S.; Iwakoshi-Ukena, E.; Furumitsu, M.; Ukena, K. Overexpression of the Gene Encoding Neurosecretory Protein GL Precursor Prevents Excessive Fat Accumulation in the Adipose Tissue of Mice Fed a Long-Term High-Fat Diet. *Molecules* **2021**, *26*, 6006. [CrossRef]
31. Narimatsu, Y.; Matsuura, D.; Iwakoshi-Ukena, E.; Furumitsu, M.; Ukena, K. Neurosecretory Protein GL Promotes Normotopic Fat Accumulation in Male ICR Mice. *Int. J. Mol. Sci.* **2022**, *23*, 6488. [CrossRef] [PubMed]
32. Livak, K.J.; Schmittgen, T.D. Analysis of Relative Gene Expression Data Using Real-Time Quantitative PCR and the $2^{-\Delta\Delta CT}$ Method. *Methods* **2001**, *25*, 402–408. [CrossRef] [PubMed]
33. Sugasawa, T.; Ono, S.; Yonamine, M.; Fujita, S.I.; Matsumoto, Y.; Aoki, K.; Nakano, T.; Tamai, S.; Yoshida, Y.; Kawakami, Y.; et al. One Week of Cdahfd Induces Steatohepatitis and Mitochondrial Dysfunction with Oxidative Stress in Liver. *Int. J. Mol. Sci.* **2021**, *22*, 5851. [CrossRef] [PubMed]
34. Al Rijjal, D.; Wheeler, M.B. A Protocol for Studying Glucose Homeostasis and Islet Function in Mice. *STAR Protoc.* **2022**, *3*, 101171. [CrossRef] [PubMed]
35. Nagy, C.; Einwallner, E. Study of In Vivo Glucose Metabolism in High-Fat Diet-Fed Mice Using Oral Glucose Tolerance Test (OGTT) and Insulin Tolerance Test (ITT). *J. Vis. Exp.* **2018**, *131*, e56672. [CrossRef]
36. Chen, H.; Liu, J.; Shi, G.-P.; Zhang, X. Protocol for in Vivo and Ex Vivo Assessment of Hyperglycemia and Islet Function in Diabetic Mice. *STAR Protoc.* **2023**, *4*, 102133. [CrossRef] [PubMed]
37. Sullivan, P.W.; Ghushchyan, V.H.; Ben-Joseph, R. The Impact of Obesity on Diabetes, Hyperlipidemia and Hypertension in the United States. *Qual. Life Res.* **2008**, *17*, 1063–1071. [CrossRef]
38. Monteiro, R.; Azevedo, I. Chronic Inflammation in Obesity and the Metabolic Syndrome. *Mediat. Inflamm.* **2010**, *2010*, 289645. [CrossRef]
39. Yang, C.F.; Chiang, M.C.; Gray, D.C.; Prabhakaran, M.; Alvarado, M.; Juntti, S.A.; Unger, E.K.; Wells, J.A.; Shah, N.M. Sexually Dimorphic Neurons in the Ventromedial Hypothalamus Govern Mating in Both Sexes and Aggression in Males. *Cell* **2013**, *153*, 896–909. [CrossRef]
40. Wu, H.; Ballantyne, C.M. Metabolic Inflammation and Insulin Resistance in Obesity. *Circ. Res.* **2020**, *126*, 1549–1564. [CrossRef]
41. Bamshad, M.; Aoki, V.T.; Adkison, M.G.; Warren, W.S.; Bartness, T.J. Central Nervous System Origins of the Sympathetic Nervous System Outflow to White Adipose Tissue. *Am. J. Physiol.* **1998**, *275*, R291–R299. [CrossRef] [PubMed]
42. Bartness, T.J.; Song, C.K. Thematic Review Series: Adipocyte Biology. Sympathetic and Sensory Innervation of White Adipose Tissue. *J. Lipid Res.* **2007**, *48*, 1655–1672. [CrossRef] [PubMed]
43. Masuda, K.; Furumitsu, M.; Taniuchi, S.; Iwakoshi-Ukena, E.; Ukena, K. Production and Characterization of Neurosecretory Protein GM Using Escherichia Coli and Chinese Hamster Ovary Cells. *FEBS Open Bio* **2015**, *5*, 844–851. [CrossRef] [PubMed]
44. Page, C.E.; Shepard, R.; Heslin, K.; Coutellier, L. Prefrontal Parvalbumin Cells Are Sensitive to Stress and Mediate Anxiety-Related Behaviors in Female Mice. *Sci. Rep.* **2019**, *9*, 19772. [CrossRef] [PubMed]
45. Ewbank, S.N.; Campos, C.A.; Chen, J.Y.; Bowen, A.J.; Padilla, S.L.; Dempsey, J.L.; Cui, J.Y.; Palmiter, R.D. Chronic Gq Signaling in AgRP Neurons Does Not Cause Obesity. *Proc. Natl. Acad. Sci. USA* **2020**, *117*, 20874–20880. [CrossRef]
46. Pozhidayeva, D.Y.; Farris, S.P.; Goeke, C.M.; Firsick, E.J.; Townsley, K.G.; Guizzetti, M.; Ozburn, A.R. Chronic Chemogenetic Stimulation of the Nucleus Accumbens Produces Lasting Reductions in Binge Drinking and Ameliorates Alcohol-Related Morphological and Transcriptional Changes. *Brain Sci.* **2020**, *10*, 109. [CrossRef]
47. Xu, J.-J.; Gao, P.; Wu, Y.; Yin, S.-Q.; Zhu, L.; Xu, S.-H.; Tang, D.; Cheung, C.-W.; Jiao, Y.-F.; Yu, W.-F.; et al. G Protein-Coupled Estrogen Receptor in the Rostral Ventromedial Medulla Contributes to the Chronification of Postoperative Pain. *CNS Neurosci. Ther.* **2021**, *27*, 1313–1326. [CrossRef]
48. Takahashi, T.M.; Sunagawa, G.A.; Soya, S.; Abe, M.; Sakurai, K.; Ishikawa, K.; Yanagisawa, M.; Hama, H.; Hasegawa, E.; Miyawaki, A.; et al. A Discrete Neuronal Circuit Induces a Hibernation-like State in Rodents. *Nature* **2020**, *583*, 109–114. [CrossRef]
49. Bikson, M.; Nitsche, M.; Claes, M.; De Groef, L.; Moons, L. The DREADDful Hurdles and Opportunities of the Chronic Chemogenetic Toolbox. *Cells* **2022**, *11*, 1110. [CrossRef]
50. Poyraz, F.C.; Holzner, E.; Bailey, M.R.; Meszaros, J.; Kenney, L.; Kheirbek, M.A.; Balsam, P.D.; Kellendonk, C. Decreasing Striatopallidal Pathway Function Enhances Motivation by Energizing the Initiation of Goal-Directed Action. *J. Neurosci.* **2016**, *36*, 5988–6001. [CrossRef]

51. Manvich, D.F.; Webster, K.A.; Foster, S.L.; Farrell, M.S.; Ritchie, J.C.; Porter, J.H.; Weinshenker, D. The DREADD Agonist Clozapine N-Oxide (CNO) Is Reverse-Metabolized to Clozapine and Produces Clozapine-like Interoceptive Stimulus Effects in Rats and Mice. *Sci. Rep.* **2018**, *8*, 3840. [CrossRef]
52. MacLaren, D.A.A.; Browne, R.W.; Shaw, J.K.; Krishnan Radhakrishnan, S.; Khare, P.; España, R.A.; Clark, S.D. Clozapine N-Oxide Administration Produces Behavioral Effects in Long-Evans Rats: Implications for Designing DREADD Experiments. *eNeuro* **2016**, *3*, ENEURO.0219-16.2016. [CrossRef] [PubMed]
53. Weisberg, S.P.; McCann, D.; Desai, M.; Rosenbaum, M.; Leibel, R.L.; Ferrante, A.W. Obesity Is Associated with Macrophage Accumulation in Adipose Tissue. *J. Clin. Invest.* **2003**, *112*, 1796–1808. [CrossRef]
54. Ohashi, K.; Parker, J.L.; Ouchi, N.; Higuchi, A.; Vita, J.A.; Gokce, N.; Pedersen, A.A.; Kalthoff, C.; Tullin, S.; Sams, A.; et al. Adiponectin Promotes Macrophage Polarization toward an Anti-Inflammatory Phenotype. *J. Biol. Chem.* **2010**, *285*, 6153–6160. [CrossRef]
55. Lumeng, C.N.; Bodzin, J.L.; Saltiel, A.R. Obesity Induces a Phenotypic Switch in Adipose Tissue Macrophage Polarization. *J. Clin. Invest.* **2007**, *117*, 175–184. [CrossRef] [PubMed]
56. Tran, T.T.; Yamamoto, Y.; Gesta, S.; Kahn, C.R. Beneficial Effects of Subcutaneous Fat Transplantation on Metabolism. *Cell Metab.* **2008**, *7*, 410–420. [CrossRef] [PubMed]
57. Strissel, K.J.; Stancheva, Z.; Miyoshi, H.; Perfield, J.W., 2nd; DeFuria, J.; Jick, Z.; Greenberg, A.S.; Obin, M.S. Adipocyte Death, Adipose Tissue Remodeling, and Obesity Complications. *Diabetes* **2007**, *56*, 2910–2918. [CrossRef]
58. Murano, I.; Barbatelli, G.; Parisani, V.; Latini, C.; Muzzonigro, G.; Castellucci, M.; Cinti, S. Dead Adipocytes, Detected as Crown-like Structures, Are Prevalent in Visceral Fat Depots of Genetically Obese Mice. *J. Lipid Res.* **2008**, *49*, 1562–1568. [CrossRef]
59. Qu, D.; Ludwig, D.S.; Gammeltoft, S.; Piper, M.; Pelleymounter, M.A.; Cullen, M.J.; Mathes, W.F.; Przypek, R.; Kanarek, R.; Maratos-Flier, E. A Role for Melanin-Concentrating Hormone in the Central Regulation of Feeding Behaviour. *Nature* **1996**, *380*, 243–247. [CrossRef]
60. Shimada, M.; Tritos, N.A.; Lowell, B.B.; Flier, J.S.; Maratos-Flier, E. Mice Lacking Melanin-Concentrating Hormone Are Hypophagic and Lean. *Nature* **1998**, *396*, 670–674. [CrossRef]
61. Lambert, P.D.; Couceyro, P.R.; McGirr, K.M.; Dall Vechia, S.E.; Smith, Y.; Kuhar, M.J. CART Peptides in the Central Control of Feeding and Interactions with Neuropeptide Y. *Synapse* **1998**, *29*, 293–298. [CrossRef]
62. Yang, S.-C.; Shieh, K.-R.; Li, H.-Y. Cocaine- and Amphetamine-Regulated Transcript in the Nucleus Accumbens Participates in the Regulation of Feeding Behavior in Rats. *Neuroscience* **2005**, *133*, 841–851. [CrossRef] [PubMed]
63. Oh-I, S.; Shimizu, H.; Satoh, T.; Okada, S.; Adachi, S.; Inoue, K.; Eguchi, H.; Yamamoto, M.; Imaki, T.; Hashimoto, K.; et al. Identification of Nesfatin-1 as a Satiety Molecule in the Hypothalamus. *Nature* **2006**, *443*, 709–712. [CrossRef] [PubMed]
64. Whiddon, B.B.; Palmiter, R.D. Ablation of Neurons Expressing Melanin-Concentrating Hormone (MCH) in Adult Mice Improves Glucose Tolerance Independent of MCH Signaling. *J. Neurosci.* **2013**, *33*, 2009–2016. [CrossRef]
65. Sugino, K.; Clark, E.; Schulmann, A.; Shima, Y.; Wang, L.; Hunt, D.L.; Hooks, B.M.; Tränkner, D.; Chandrashekar, J.; Picard, S.; et al. Mapping the Transcriptional Diversity of Genetically and Anatomically Defined Cell Populations in the Mouse Brain. *eLife* **2019**, *8*, e38619. [CrossRef] [PubMed]
66. Klop, B.; Elte, J.W.F.; Cabezas, M.C. Dyslipidemia in Obesity: Mechanisms and Potential Targets. *Nutrients* **2013**, *5*, 1218–1240. [CrossRef]
67. Kato, M.; Iwakoshi-Ukena, E.; Narimatsu, Y.; Furumitsu, M.; Ukena, K. Expression of MRNAs Encoding Hypothalamic Small Proteins, Neurosecretory Protein GL and Neurosecretory Protein GM, in the Japanese Quail, Coturnix Japonica. *bioRxiv* **2023**. [CrossRef]
68. Carper, D.; Coué, M.; Laurens, C.; Langin, D.; Moro, C. Reappraisal of the Optimal Fasting Time for Insulin Tolerance Tests in Mice. *Mol. Metab.* **2020**, *42*, 101058. [CrossRef]
69. Cardoso, F.; Klein Wolterink, R.G.J.; Godinho-Silva, C.; Domingues, R.G.; Ribeiro, H.; da Silva, J.A.; Mahú, I.; Domingos, A.I.; Veiga-Fernandes, H. Neuro-Mesenchymal Units Control ILC2 and Obesity via a Brain–Adipose Circuit. *Nature* **2021**, *597*, 410–414. [CrossRef]

biomedicines

Article

Circulating and Exosomal microRNA-33 in Childhood Obesity

Manuela Cabiati [1], Letizia Guiducci [1], Emioli Randazzo [2], Valentina Casieri [3], Giovanni Federico [2] and Silvia Del Ry [1,3,*]

[1] Laboratory of Biochemistry and Molecular Biology, Institute of Clinical Physiology, CNR, 56124 Pisa, Italy; manuela.cabiati@cnr.it (M.C.); letizia.guiducci@cnr.it (L.G.)

[2] Unit of Pediatric Endocrinology and Diabetes, Department of Clinical and Experimental Medicine, University of Pisa, 56124 Pisa, Italy; emioli.randazzo@gmail.com (E.R.); giovanni.federico@med.unipi.it (G.F.)

[3] Unit of Translational Critical Care Medicine, Scuola Superiore Sant'Anna, 56126 Pisa, Italy; valentina.casieri@santannapisa.it

* Correspondence: silvia.delry@cnr.it

Abstract: Background: MicroRNA-33 may control a wide range of different metabolic functions. Methods: This study aims to assess the miR-33a circulating profile in normal-weight (N = 20) and obese (O = 30) adolescents and to correlate its expression levels to their metabolic parameters. In a subset of subjects, we compared circulating miR-33a with exosomal miR-33a. Results: Metabolic parameters were altered in O, with initial hyperinsulinemia. Circulating miR-33a was significantly higher in O than in N ($p = 0.0002$). Significant correlations between miR-33a and auxological and metabolic indices (Insulin $p = 0.01$; Cholesterol $p = 0.01$; LDL $p = 0.01$; HbA1c $p = 0.01$) were found. Splitting our population (O + N) into two groups, according to the median value of mRNA expression miR-33a levels (0.701), irrespective of the presence or absence of obesity, we observed that those having a higher expression of miR-33a were more frequently obese (87.5% vs. 12.5%; $p < 0.0001$) and had significantly increased values of auxological and metabolic parameters. Exosomes extracted from plasma of N and O carried miR-33a, and its expression was lower in O ($p = 0.026$). No correlations with metabolic parameters were observed. Conclusion: While exosome miR-33a does not provide any advantage, circulating miR-33a can provide important indications in an initial phase of metabolic dysfunction, stratifying obese adolescents at higher cardiometabolic risk.

Keywords: microRNA-33; obesity; childhood; Real-Time PCR

Citation: Cabiati, M.; Guiducci, L.; Randazzo, E.; Casieri, V.; Federico, G.; Del Ry, S. Circulating and Exosomal microRNA-33 in Childhood Obesity. *Biomedicines* **2023**, *11*, 2295. https://doi.org/10.3390/biomedicines11082295

Academic Editors: Zaida Abad-Jiménez and Teresa Vezza

Received: 25 July 2023
Revised: 9 August 2023
Accepted: 14 August 2023
Published: 18 August 2023

1. Introduction

Metabolic syndrome (MetS), a condition strongly related to a greater risk of developing cardiovascular disease and type 2 diabetes mellitus [1], is characterized by the clustering of clinical conditions including central obesity, dyslipidemia, hypertension, and insulin resistance [1]. MetS is considered a major health problem, and recently, it has become increasingly relevant due to the exponential worldwide increase of obesity and the linked important therapeutic challenges [1,2]. Lifestyle variation, a healthy diet, and pharmacological treatment are suggested for managing this syndrome [3]; nevertheless, the mechanisms participating in metabolic dysregulation remain unclear and are intensively examined to find new therapeutic avenues.

Many studies have focused their attention on the roles of metabolic enzymes; hormones; signaling molecules; and regulatory proteins, such as transcription factors and co-factors, in metabolic diseases. Recently, an important role as modulators of metabolic homeostasis has been attributed to RNAs, i.e., microRNAs (miRNAs) [4,5]. MicroRNAs are a cluster of short, non-coding RNAs able to influence gene expression by suppressing their translation or by promoting the degradation of mRNAs [6]. MicroRNAs are among the most plentiful gene regulators in humans and have now been linked to many physiological and pathological processes, including obesity and diabetes [7–9].

The increasing information about miRNAs and their molecular actions enabled innovative engineering and applications of these biomarkers, particularly in medical therapy [10]. The use of miRNAs in medicine as new biomarkers could be facilitated by the fact that they may be carried by exosomes, which are small membrane vesicles of 40–100 nm in size [10].

The most plentiful miRNA in lipoprotein particles, generally suggested as an important regulator of lipid metabolism, is miR-33 [7,11,12]. The family of miR-33 consists of miR-33a and miR-33b, which are located in intron 16 of human SREBP-2 and SREBP-1 genes (master regulators of genes involved in cholesterol/lipid biosynthesis and trafficking), respectively [13]. MiR-33a directly targets the cholesterol transporters Abca1 and Abcg1, which are responsible for the efflux of cholesterol from the cell, underling the importance of this miRNA in cholesterol metabolism. MiR-33b is also implicated in fatty acid β-oxidation, as carnitine palmitoyltransferase (Cpt1a) [14]. It seems that metabolic regulation by miR-33 is more complex than previously known. Many studies reported an increase in Abca1 expression and plasma HDL levels [11,14–16].

A possible deleterious metabolic consequence of antagonizing miR-33 was shown by some studies in knockout mice [17,18]. Their results suggested that these metabolic abnormalities might be due to elevated SREBP-1 protein in the liver (SREBP-1 was shown to be a target of miR-33); in agreement, it was also reported that expression of SREBP-1 lipogenic target genes, such as FASN and ACC1, is elevated in the liver of mice treated regularly with miR-33-targeting antisense oligonucleotides [19]. In addition, it was found that loss of miR-33 leads to reduced insulin sensitivity in the liver, white adipose tissue, and skeletal muscle, even in mice fed on a chow diet [18]. Hyperinsulinemic–euglycemic clamp studies discovered that miR-33 knockout mice exhibited reduced glucose uptake in white adipose tissue and skeletal muscle, reduced suppression of endogenous glucose production in the liver, and reduced capacity to suppress the release of non-esterified fatty acids from white adipose tissue [18]. In miR-33 null mice fed on a high-fat diet, it was observed an increased infiltration of inflammatory cells in white adipose tissue, suggesting that complete miR-33 loss, especially in the context of a high-fat diet, may have significant harmful metabolic consequences that are comparable to obesity-associated insulin resistance, type 2 diabetes, and white adipose tissue inflammation in humans.

Thus, the loss of miR-33 promotes insulin resistance in key metabolic tissues [18]. MiR-33, in fact, targets genes involved in many important metabolic functions, including fatty acid and lipid metabolism, insulin signaling, and mitochondrial function [20–24]. The potential of miR-33 to control many different metabolic functions hints at its involvement in regulating diverse metabolic functions in many tissues. This is especially interesting if one considers that miR-33 is encoded within intronic regions of the SREBP-2 and SREBP-1 genes, which are among the most important regulators of cellular cholesterol and fatty acid metabolism, and that it has been demonstrated to be differentially regulated in many different metabolic tissues under conditions of obesity and insulin resistance [25]. Taken together, these studies clearly indicated that miR-33 is a central player in the regulation of liver metabolism, suggesting that it might have a potential role in the treatment of metabolic diseases. Since being obese in childhood increases the risk of being obese as an adult [26,27], evaluating the expression levels of miR-33 in obese children could provide important indications in those at higher risk of developing metabolic complications, such as dyslipidemia, hypertension, and impaired glucose metabolism, later in life [28]. So, the primary aim of the present investigation was to evaluate the miR-33 circulating profile in normal-weight (N) and obese (O) adolescents and to correlate its expression levels with metabolic parameters to confirm its involvement in metabolic complications. In addition, bearing in mind the possible use of miRNA in medical therapy, together with exosomes as miRNA carriers [10], we reasoned that the evaluation of miRNA-33 expression in exosomes could be helpful in the screening and surveillance of metabolic complications in obese adolescents, allowing earlier and more efficient prevention, diagnosis, and treatment. Considering this and our recent observation of the presence of miRNA-33a in the exosomes of obese adolescents [29], we established, as a secondary aim of the present study, comparing the expression levels of circulating miRNA-33a with those found in exosomal vesicles in a subset of our cohort, looking at possible correlations between miRNA-33a expression in exosomes and some metabolic indices.

2. Materials and Methods

2.1. Subjects and Plasma Collection

We enrolled 50 adolescents, 30 obese (O) and 20 age- and sex-matched normal-weight (N) subjects, as a control group. Table 1 summarizes the clinical characteristics and body composition of the population participating in the study. Obese adolescents were referred as outpatients to the Unit of Pediatric Endocrinology and Diabetes, Department of Clinical and Experimental Medicine, University of Pisa, Italy. They had primary obesity, not induced by drug or diseases, and they were not affected by diabetes or cardiac dysfunction. Obesity was defined according to the criteria of the International Task Force on Obesity in childhood, using population reference data specific for age and sex for body mass index (BMI) [30]. As also reported in our previous papers [29,31–36], the normal-weight adolescents were healthy subjects, who repeated a blood examination after an intervening disease. At the time of blood sampling, they had not taken drugs for at least one week, and their blood results, including indices of inflammation, were in the normal range. BMI was calculated using the formula weight [(Kg)/height (m)^2] [30]. We used the same National reference data [30] to calculate the BMI z-score and Height z-score. Sexual development was assessed according to Tanner's stages [37,38]. Total body fat (%) was measured using the Tanita BC-418 Segmental Body Composition Analyzer (Tanita Corporation, Tokyo, Japan) [39]. Blood pressure was measured by trained investigators, according to a standardized protocol [40].

Table 1. Demography, clinical characteristics, and body composition of study population.

	Normal-Weight Subjects	Obese Subjects	*p*
Age (years)	13.1 ± 0.16	12.2 ± 0.36	ns
Male:Female	9:11	18:12	ns
Pubertal Stage (Tanner score)	3.1 ± 0.18	3.6 ± 0.143	*0.0461*
Height (cm)	157.4 ± 0.96	155.4 ± 2.34	ns
Height z-score	0.22 ± 0.06	0.91 ± 0.2	*0.0086*
Weight (Kg)	50.3 ± 0.79	72.7 ± 3.1	*<0.0001*
BMI	20.2 ± 0.20	29.8 ± 0.79	*<0.0001*
BMI z-score	0.68 ± 0.08	2.91 ± 0.2	*<0.0001*
WC (cm)	84.3 ± 2.1	92.1 ± 1.78	*0.0093*
Fat Mass (%)	19.7 ± 1.51	36.7 ± 1.78	*<0.0001*
SBP (mmHg)	112.5 ± 1.8	108.0 ± 2.2	ns
DBP (mmHg)	62.2 ± 1.4	63.8 ± 1.7	ns
HOMA-IR	0.95 ± 0.09	2.5 ± 0.24	*<0.0001*

Table Legend: All data are expressed as mean ± SEM. BMI: Body Mass Index; WC: Waist Circumference; SBP: Systolic Blood Pressure; DBP: Diastolic Blood Pressure; HOMA-IR: HOmeostatic Model Assessment of Insulin Resistance; ns: not significant.

We collected blood samples from all the subjects by venipuncture, performed in the morning after overnight fasting. To evaluate the circulating levels of metabolic parameters, blood samples were collected into ethylenediaminetetraacetic acid (EDTA) (1 mg/mL) and lithium-heparin-containing vials [29,31–36]. They were measured by appropriate commercial kits, as previously reported [29,31–36].

The investigation conforms to principles outlined in the Declaration of Helsinki [41]. The study was approved by the local Ethics Committee, and informed consent was obtained from the parents of each subject.

2.2. *Circulating miRNA-33a Extraction and Reverse Transcription*

As previously reported [31,32], the extraction of miRNAs was conducted by using the miRNeasy Serum/Plasma Kit (Qiagen S.p.a., Milano, Italy). Briefly, 200 µL of plasma was lysed in an adequate lysis reagent and applied to silica-membrane columns. To monitor RNA recovery and reverse transcription efficiency, a Spike-In Control (C. elegans miR-39 miRNA mimic) was used as the internal control for plasma miRNA expression profiling. High-quality RNA was then eluted in a small volume (14 µL) of RNase-free water; samples were stored at −80 °C after integrity, purity, and concentration evaluation. The fraction of mature miRNAs was reverse-transcribed using the miScript II RT kit (Qiagen S.p.a., Milano, Italy), starting from a 1 µg/sample in 20 µL of final reaction volume (60 min at 37 °C, 5 min at 95 °C, and 4 °C for ∞). cDNA samples were stored, as appropriate, and diluted at +4 °C.

2.3. *Exosome Isolation and Characterization, Vesicular RNA Extraction and Reverse Transcription*

In a subset of the study population (Ssp, n = 23), we carried out exosome isolation. Table 2 summarizes the clinical characteristics of the subjects included in the subgroups.

Table 2. Demography, clinical characteristics, and body composition of Ssp.

	Normal-Weight Subjects	Obese Subjects	p
Age (years)	13.1 ± 0.2	12.7 ± 0.78	ns
Male:Female	8:8	5:2	ns
Pubertal Stage (Tanner score)	3.09 ± 0.21	3.15 ± 0.42	ns
Height (cm)	157.5 ± 0.96	156.6 ± 6.3	ns
Height z-score	0.24 ± 0.06	0.46 ± 0.4	*0.032*
Weight (Kg)	50.3 ± 0.70	74.9 ± 7.9	*<0.0001*
BMI	20.2 ± 0.20	30.2 ± 2.0	*<0.0001*
BMI z-score	0.68 ± 0.09	2.8 ± 0.23	*<0.0001*
WC (cm)	84.8 ± 2.3	89.8 ± 3.3	*0.0085*
Fat Mass (%)	19.3 ± 1.68	39.2 ± 10.8	*0.028*
SBP (mmHg)	113.9 ± 1.6	108.0 ± 4.5	ns
DBP (mmHg)	62.6 ± 1.5	58.8 ± 4.9	ns
HOMA-IR	0.95 ± 0.1	2.18 ± 0.53	*0.0036*

Table Legend: BMI: Body Mass Index; WC: Waist Circumference; SBP: Systolic Blood Pressure; DBP: Diastolic Blood Pressure; HOMA-IR: HOmeostatic Model Assessment of Insulin Resistance; ns: not significant.

As reported in a previous study of ours [29], exosomes were extracted from 600 µL of plasma using a dedicated and innovative assay (exoRNeasy mini/midi kit, QIAGEN GmbH, Hilden, Germany); before RNA extraction and in a random manner, we analyzed the morphology of exosomes with Transmission Electron Microscopy (TEM). The presence of specific exosomal proteins in our preparations was evaluated by Western blotting analysis. In particular, we used primary antibodies to detect human CD9 (monoclonal antibody, 1:1000, #SA35-08, Novus Biological, San Diego, CO, USA), human TGS101 (polyclonal antibody, 1:1000; #T5701, Sigma-Aldrich Chemical, St. Louis, MO, USA), and human Alix (monoclonal antibody, 1:1000, #2171, Cell Signaling Technology, Boston, MA, USA) as established exosomal markers, while human Calnexin (polyclonal antibody, 1:1000, ab10286, Abcam, Cambridge, UK) was used as a negative exosomal marker. Densitometric analysis of protein bands was carried out with Image J software 1.52t (National Institutes of Health, Bethesda, MD, USA). Briefly, for exosome RNA extraction, plasma samples were prefiltered using syringe filters excluding particles larger than 0.8 µm (Millipore® Membrane Filter, 0.8 µm pore size, MILLEX-AA, Merck, D); then, exosomes were isolated, adding in a 1:1 ratio of a 2× binding buffer (XBP). Next, samples were transferred into the ExoEasy membrane affinity column, to bind the exosomes to the membrane, and then centrifuged at 500× g for 2 min at RT. After discarding the flow-through, 3.5 mL of wash buffer (XWP) was added to the column, according to the starting plasma quantity, and centrifuged at 5000× g for 5 min at RT. Vesicles were lysed by adding to the spin column a phenol/guanidine-based combined lysis reagent (QIAzol), and the lysate was collected by serial centrifugations. After the lysis, elution step, and addition of chloroform, the resulting samples were separated into aqueous and organic phases by centrifugation. Ethanol was added to the aqueous phase, and samples were applied to the RNeasy MinElute spin column, washed, centrifuged, and finally eluted in RNase-free water (16 μL). The samples were then stored at −80 °C; RNA integrity, purity, and concentration were assessed by measuring absorbance at 230, 260, and 280 nm (NanoDrop Thermofisher, Waltham, MA, USA) and applying the Beer–Lambert law (expected values between 1.8 and 2.1, for protein contamination). The exosomal miRNA reverse transcription was carried out with the same procedure (miScript ® II RT kit, Qiagen, Hilden, Germany) used for circulating miRNA-33a retro-transcription.

2.4. Real-Time PCR Analysis for Circulating and Exosomal miRNA-33a

Real-time PCR reactions were carried out in duplicate using a Bio-Rad C1000™ thermal cycler system (CFX-96 Real-Time detection system, Bio-Rad). To monitor cDNA amplification, a fluorogenic DNA binding dye, EvaGreen (SsoFAST EvaGreen Supermix, Bio-Rad Laboratories Inc., Hercules, CA, USA), was used.

The miR-33a primer sequence was synthesized by Sigma Aldrich (Milan, Italy). In particular, the GenBank accession number for hsa-miR-33a was NR_029507, and the forward primer sequence (5′—3′) was CAATGTTTCCACAGTGCATCAC. The Ce_miR-39 (miRNA mimic Ce_miR-39_1 miScript Primer Assay) was employed to normalize miRNA data. All experiments were carried out according to the MIQE (Minimum Information for publication of Quantitative Real-Time PCR Experiments) guidelines [42].

2.5. Statistics

Statistical analysis was performed using Statview 5.0.1 Software for Windows (SAS Institute, Inc., Cary, NC, USA). Relative quantification was performed by the ΔΔCt method using Bio-Rad's CFX96 manager software v.3.1 (CFX-96 Real-Time PCR detection systems, Bio-Rad Laboratories Inc., Hercules, CA, USA). Skewed variables were log-transformed before statistical analysis. Differences between more than two independent groups were analyzed by Fisher's test after ANOVA, and relations between variables were assessed by linear regression analysis. The results were expressed as mean ± S.E.M., and a p-value < 0.05 was considered significant.

3. Results

3.1. Clinical Characteristics

As reported in Table 1, obese subjects showed significant differences as regards weight, BMI, BMI z-score, fat mass, HOMA-IR, and WC with respect to normal-weight subjects, and we observed the same in the subset of adolescents in whom we analyzed isolated exosomes (Table 2). Obese adolescents presented also a significantly slight increase in pubertal development as a result of their nutritional excess. Systolic and diastolic blood pressure results were similar in N and O. Moreover, as reported in Figure 1a–c, O showed higher significant levels of cholesterol, LDL, and triglycerides, while HDL values were similar between the two groups (Figure 1d).

We observed a significant reduction of glucose plasma levels in O with respect to N in the presence of higher plasma levels of insulin and Hb1ac, confirming a condition of hyperinsulinemia in obese subjects (Figure 2a–d).

We observed a similar behavior of the same parameters and analytes in the Ssp population, as reported in Figures 3 and 4.

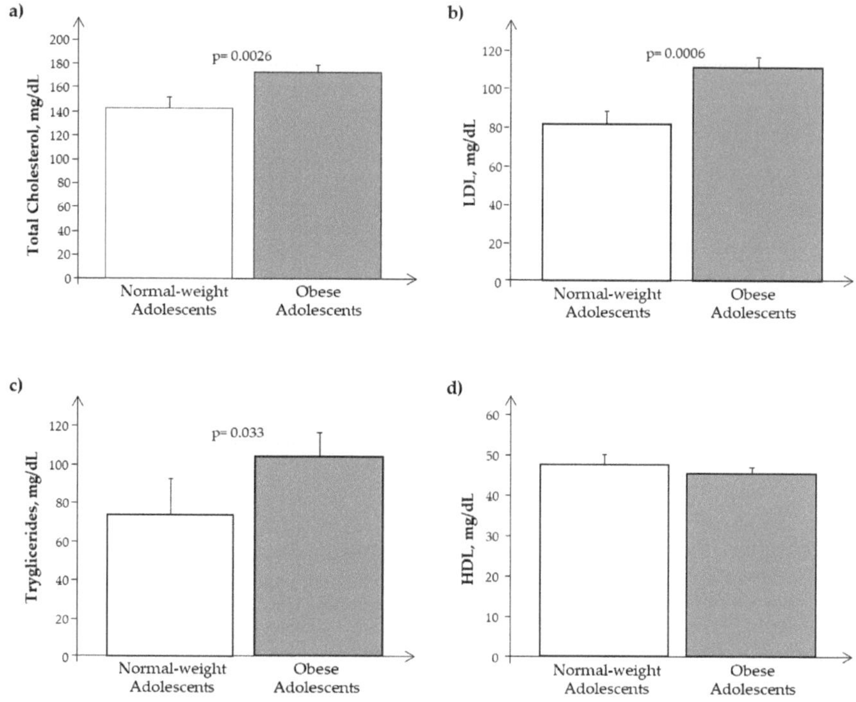

Figure 1. Plasma levels of (**a**) cholesterol, (**b**) LDL, (**c**) triglycerides, and (**d**) HDL in obese (O) with respect to normal-weight (N) adolescents.

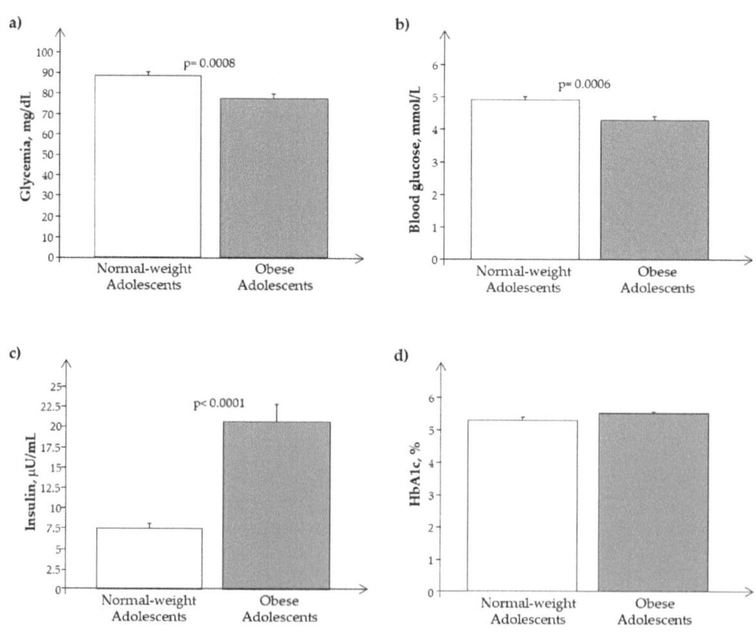

Figure 2. Plasma levels of (**a**) glycemia, (**b**) glucose, (**c**) insulin, and (**d**) Hb1ac in obese (O) with respect to normal-weight (N) adolescents.

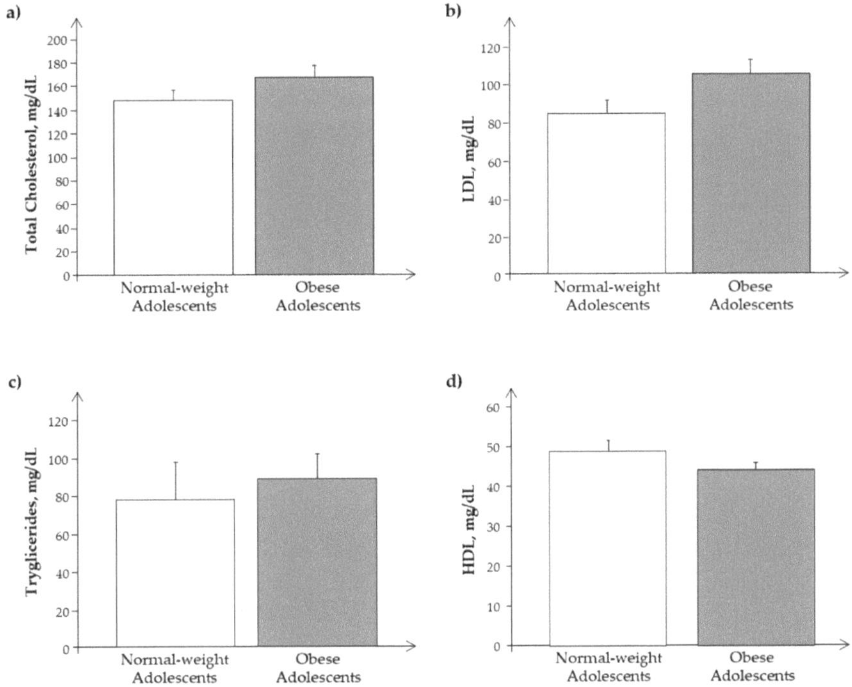

Figure 3. Plasma levels of (**a**) cholesterol, (**b**) LDL, (**c**) triglycerides, and (**d**) HDL in the subset of the study population, Ssp. (obese and normal-weight adolescents).

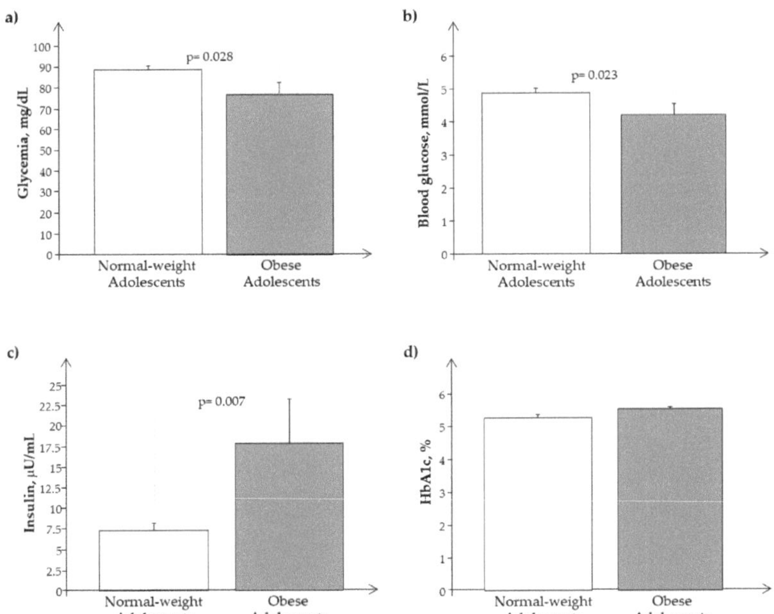

Figure 4. Plasma levels of (**a**) glycemia, (**b**) glucose, (**c**) insulin, and (**d**) Hb1ac in the subset of the study population, Ssp. (obese and normal-weight adolescents).

3.2. *miRNA-33a Expression in Plasma and Exosome Samples*

Plasma expression of miR-33a resulted in significantly higher expression in O than in N (0.9 ± 0.08 vs. 0.5 ± 0.07, respectively; $p = 0.0002$). It was interesting to note that miR-33a was significantly related to both auxological parameters and metabolic indices, such as cholesterol, LDL, insulin, and HbA1c (Figure 5a–d).

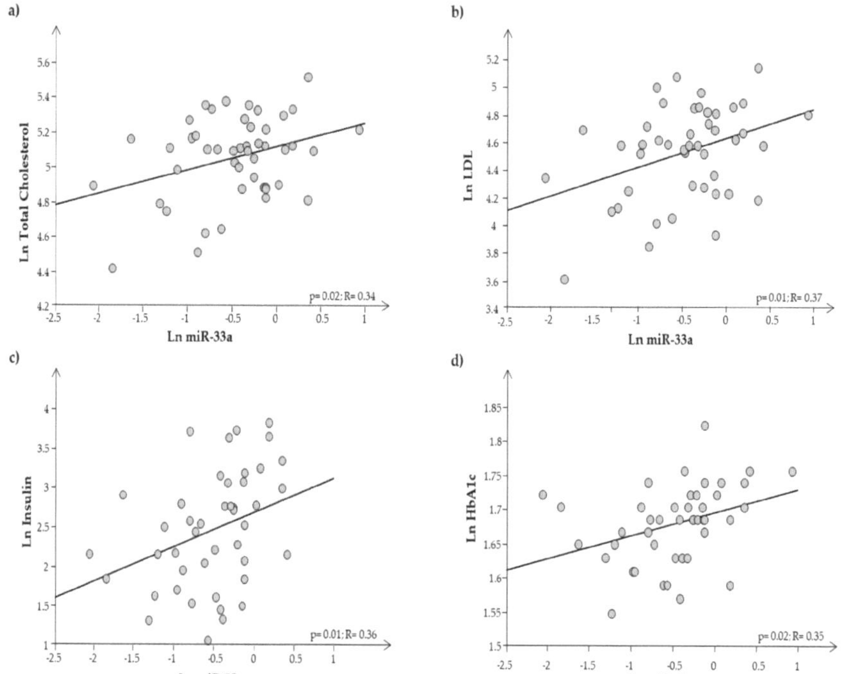

Figure 5. Regression analysis between miR-33a and (**a**) total cholesterol, (**b**) LDL cholesterol, (**c**) insulin, and (**d**) HbA1c.

After calculating the miR-33a expression levels in plasma samples in the entire population (O + N) as the median value, we used it as the cut-off (0.701) to classify our population into two groups of subjects, irrespective of the presence or absence of obesity: those having lower miR-33a expression levels (Group 1: <median value) and those with higher miR-33a expression levels (Group 2: ≥ median value). Subsequently, comparing some clinical and metabolic data between Group 1 ($n = 23$) and Group 2 ($n = 24$), we observed that those having higher miR-33a expression levels (Group 2) were more frequently obese (Group 2 vs. Group 1 = 87.5% vs. 12.5%; $p < 0.0001$) and had an increased BMI z-score (Group 2 vs. Group 1 = 2.5 ± 0.21 vs. 1.5 ± 0.2; $p = 0.0013$) and fat mass (Group 2 vs. Group 1 = 35.3 ± 1.9 vs. 21.5 ± 2.0%; $p < 0.0001$).

In comparison with Group 1, Group 2 had significantly higher levels of circulating insulin (Group 2 vs. Group 1 = 20.3 ± 2.5 vs. 10.4 ± 1.7 μU/mL; $p = 0.0005$) and HbA1c (Group 2 vs. Group 1 = 5.5 ± 0.057 vs. 5.2 ± 0.05%, $p = 0.0004$ corresponding to 37.0 ± 0.61 vs. 33.7 ± 0.64 mmol/mol, $p = 0.0005$). Also, plasma levels of total cholesterol (Group 2 vs. Group 1 = 169.7 ± 6.6 vs. 152.8 ± 7.9 mg/dL), LDL cholesterol (Group 2 vs. Group 1 = 105.9 ± 6.3 vs. 90.7 ± 6.4 mg/dL), HDL cholesterol (Group 2 vs. Group 1 = 46.6 ± 1.9 vs. 44.4 ± 2.1 mg/dL), and triglycerides (Group 2 vs. Group 1 = 95.5 ± 12.4 vs. 87.6 ± 17.3 mg/dL) resulted in being higher, even if not significantly, in Group 2 than in Group 1, while blood glucose showed an opposite behavior, being lower in Group 2 than in Group 1 (4.34 ± 0.136 vs. 4.7 ± 1.33 mmol/L, p = 0.045 corresponding to 78.47 ± 2.3 vs. 85.3 ± 2.29 mg/dL, p = 0.044).

We performed the same evaluations in exosome samples obtained from the subjects of the Ssp.

We previously reported that, following the MISEV 2018 guideline [43], TEM analysis revealed that the exoRNeasy mini/midi kit procedure isolated round-shaped intact vesicles from pre-filtered plasma samples and that Western blotting analysis, detecting proteins specifically expressed in exosomes, confirmed TEM results [29].

Exosomes extracted from plasma of both normal-weight and obese subjects carried miR-33a, but its expression levels were lower in exosomes obtained from O than from N (0.58 ± 0.29 vs. 1.33 ± 0.29, respectively; p = 0.026), confirming our previous results in obese adolescents [29]. We also observed some significant correlations between the expression of exosomal miR-33a and auxologic parameters (BMI: r = 0.49, p = 0.01; fat mass: r = 0.48, p = 0.04; HOMA: r = 0.42, p = 0.04), but not with metabolic indices. By dividing Ssp into two groups, according to the median value of miR-33a expression levels (1.085), irrespective of the presence or absence of obesity, we did not obtain any additional information.

Comparing the results obtained in plasma and in exosome samples (Figure 6a), we observed that miR-33a showed higher expression levels in exosomes than in plasma samples of normal-weight subjects (p = 0.0073), while in obese subjects, we noted the opposite, that is, an increased expression in plasma samples rather than in exosomes (p = 0.001).

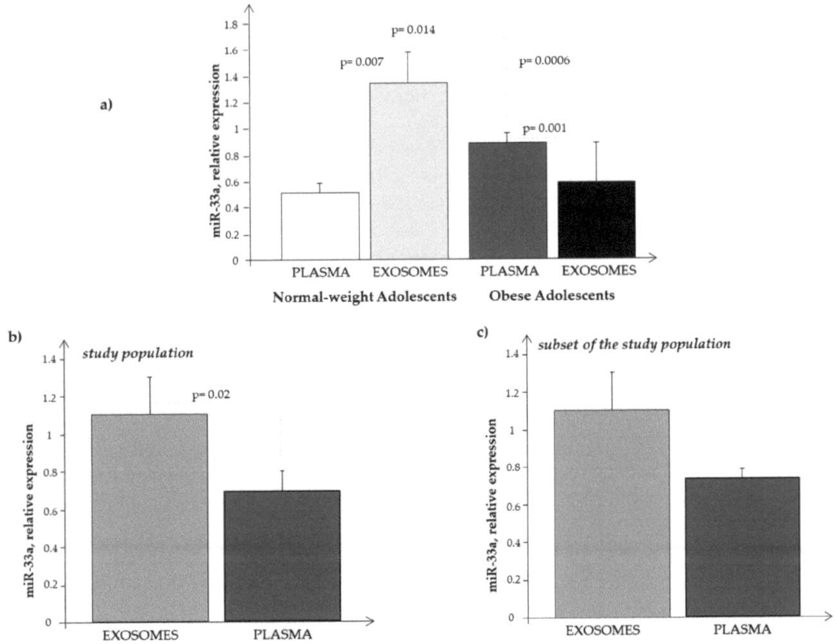

Figure 6. Expression level of miR-33a obtained in (**a**) plasma and in exosome samples in normal-weight (light-grey bar) and obese (black bar) subjects, respectively; (**b**) in plasma (light-grey bar) and in exosome (black bar) samples considering the subjects as a whole (N + O); and (**c**) in plasma (light-grey bar) and in exosome (black bar) samples considering the subjects of the Ssp as a whole (N + O).

We also compared the expression levels of circulating miR-33a with that of exosomal miR-33a observed in the whole population (O plus N subjects), and we found that the levels were significantly higher in exosomes than in plasma samples (Figure 6b). Performing the same analysis in the subjects of the Ssp, we observed the same trend (Figure 6c), even if the differences were not significant, probably due to the smaller number of subjects analyzed, confirming somehow a greater expression of miR-33a at the exosomal level.

Finally, we observed a significant negative correlation between circulating and exosomal miR-33a expression levels ($r = -0.65$, $p = 0.001$).

4. Discussion

Our obese adolescents showed a metabolic disorder, including a certain degree of hyperinsulinemia. Data from the literature reported that miR-33a has an important role in obesity, diabetes, and metabolic syndrome [7–9,44–46]. In the present study, we analyzed the circulating levels of this miRNA in a group of obese adolescents not having yet a clear metabolic syndrome, but only some metabolic alterations of the syndrome. Interestingly, they already showed a significant increase in the circulating expression levels of miR-33a as compared to N. In particular, we observed the presence of a significant direct correlation between the expression level of circulating miR-33a and total cholesterol, LDL cholesterol, insulin, and HbA1c in our cohort, suggesting the role of this biomarker in lipid regulation during adolescence. Thus, our results not only confirmed some observations previously reported in animal and human studies [7–9,44–46], but also, they highlighted the potential of the circulating expression of this miRNA as an early biomarker useful to discriminate subjects at higher metabolic risk. The use of an early marker in childhood could help in the early recognition of obese children/adolescents at higher risk of metabolic alterations that may lead to cardiometabolic diseases in adult life as a result of obesity in children/adolescents. Expression levels of circulating miR-33a correlated significantly with metabolic parameters, in addition to the auxological ones, and this was also observed after dividing the whole population (obese plus normal-weight subjects) into two groups, according to the median value of miR-33a expression levels. Despite plasma miRNAs and exosome miRNAs having been extensively studied as diagnostic or therapeutic biomarkers in several disease-related research studies, few studies determined whether miRNA expression levels were different between plasma and plasma-derived exosomes. This was the reason why we decided to compare the expression level of circulating miR-33a with that of exosomal vesicles obtained from the same subjects (considering them as a whole, O + N), and we found that miR-33a was mostly transported by exosomes (Figure 6b,c). Analyzing the same results separately in normal-weight and in obese subjects, we observed the same trend in N, while the opposite was seen in O, that is, miR-33a expression levels were higher in plasma samples than in circulating exosomes (Figure 6a).

It is rather hard to explain why circulating and exosome miRNA behave so differently; we found an inverse correlation between the expression levels of circulating and exosomal miR-33a. However, some considerations can be drawn, as we recently reported in a paper by our group [29]. It should be considered that the expression of circulating and exosome miRNAs represents the result of their regulation in two diverse compartments: plasma and cell-derived exosomes. The tendency of miRNA in plasma summarizes, in a nonspecific manner, biological mechanisms originating from several districts (cells, tissues, etc.), while the expression of miRNA in the exosome may mirror the result of a more sophisticated regulation [47]. It is known the special role that exosomes and their cargo of non-coding RNAs (primary miRNAs) play in intercellular communication [48]. Moreover, exosome biogenesis and secretion are influenced by the metabolic status of the cell, which, in turn, depends on factors such as ceramide metabolism, endoplasmic reticulum stress, autophagy, and intracellular calcium [49]. After their release, exosomes interplay with their target cells by transferring their bioactive cargo into them. As a result, this transfer triggers phenotypic changes in the recipient cells [50]. Furthermore, since exosome-mediated intra-adipose and inter-organ communication is of great significance for energy metabolism, one alternative explanation could be that this special function of exosomes in adipose tissue biology may be interrupted in obesity [50]. These reflections may help to explain, at least in part, the different results we found in examining the expression behavior of miR-33a in plasma and in exosome samples.

In the literature are well-described changes of miRNA patterns in biofluids in some diseases [51–53], including obesity, and our results on the expression of miR-33a in circulating exosomal vesicles seem to show that this evaluation does not provide additional indications on the cardiometabolic risk of our obese adolescents in adulthood.

The contribution of microRNA in glucose and lipid metabolism can provide solid indications in support of their role as key players in the regulation of complex metabolic pathways.

5. Conclusions

In adolescents, excessive intake of calories and following unbalanced diets too rich in sugars and saturated fats, in combination with stress and low levels of physical activity, can cause overweight, obesity, and metabolic alterations with a higher risk of developing cardiometabolic diseases in adult life.

The availability of a circulating biomarker, such as miR-33a, for early identification of obese adolescents at higher cardiometabolic risk provides an important tool for designing interventions to correct unhealthy lifestyles when these metabolic alterations are still reversible.

Finally, even if much remains to be understood about the role of miR-33 in conditions such as obesity, it may represent an ideal target for future therapies.

Author Contributions: Conceptualization, data curation, formal analysis: S.D.R.; methodology: S.D.R., M.C., L.G. and V.C.; patient enrolment: E.R.; writing—original draft: S.D.R.; writing—review and editing: M.C. and G.F. All authors have read and agreed to the published version of the manuscript.

Funding: This research received no external funding.

Institutional Review Board Statement: The study was conducted according to the guidelines of the Declaration of Helsinki and approved by the Institutional Review Board (Comitato Etico Pediatrico Sezione del Comitato Etico Regionale per la Sperimentazione Clinica c/o AOU Meyer, Viale Pieraccini 24, 50139 Firenze) (protocol code 3157, 23 September 2014).

Informed Consent Statement: Informed consent was obtained from the parents and from each subject as appropriate.

Data Availability Statement: The data that support the findings of this study are available in IFC-CNR.

Conflicts of Interest: The authors declare no conflict of interest. The funders had no role in the design of the study; in the collection, analyses, or interpretation of data; in the writing of the manuscript; or in the decision to publish the results.

References

1. Rochlani, Y.; Pothineni, N.V.; Kovelamudi, S.; Mehta, J.L. Metabolic syndrome: Pathophysiology, management, and modulation by natural compounds. *Ther. Adv. Cardiovasc. Dis.* **2017**, *11*, 215–225. [CrossRef] [PubMed]
2. Prasad, H.; Ryan, D.A.; Celzo, M.F.; Stapleton, D. Metabolic syndrome: Definition and therapeutic implications. *Postgrad Med* **2012**, *124*, 21–30. [CrossRef] [PubMed]
3. Gharipour, M.; Kelishadi, R.; Khosravi, A.; Shirani, S.; Masjedi, M.; Sarrafzadegan, N. The impact of a community trial on the pharmacological treatment in the individuals with the metabolic syndrome: Findings from the Isfahan Healthy Heart Program, 2001–2007. *Arch. Med. Sci.* **2012**, *8*, 1009–1017. [CrossRef]
4. Bartel, D.P. Metazoan MicroRNAs. *Cell* **2018**, *173*, 20–51. [CrossRef] [PubMed]
5. Rottiers, V.; Naar, A.M. MicroRNAs in metabolism and metabolic disorders. *Nat. Rev. Mol. Cell Biol.* **2012**, *13*, 239–250. [CrossRef]
6. Udali, S.; Guarini, P.; Moruzzi, S.; Choi, S.W.; Friso, S. Cardiovascular epigenetics: From DNA methylation to microRNAs. *Mol. Asp. Med.* **2012**, *34*, 883–901. [CrossRef]
7. Alrob, O.A.; Khatib, S.; Naser, S.A. MicroRNAs 33, 122, and 208: A potential novel targets in the treatment of obesity, diabetes, and heart-related diseases. *J. Physiol. Biochem.* **2017**, *73*, 307–314. [CrossRef]
8. Aryal, B.; Singh, A.K.; Rotllan, N.; Price, N.; Fernández-Hernando, C. MicroRNAs and lipid metabolism. *Curr. Opin. Lipidol.* **2017**, *28*, 273–280. [CrossRef]
9. Vienberg, S.; Geiger, J.; Madsen, S.; Dalgaard, L.T. MicroRNAs in metabolism. *Acta Physiol.* **2017**, *219*, 346–361. [CrossRef]

10. Hayes, J.; Peruzzi, P.P.; Lawler, S. MicroRNAs in cancer: Biomarkers, functions and therapy. *Trends Mol. Med.* **2014**, *20*, 460–469. [CrossRef]
11. Najafi-Shoushtari, S.H.; Kristo, F.; Li, Y.; Shioda, T.; Cohen, D.E.; Gerszten, R.E.; Naar, A.M. MicroRNA-33 and the SREBP host genes cooperate to control cholesterol homeostasis. *Science* **2010**, *328*, 1566–1569. [CrossRef] [PubMed]
12. Yang, Z.; Cappello, T.; Wang, L. Emerging role of microRNAsin lipid metabolism. *Acta Pharm. Sin. B* **2015**, *5*, 145–150. [CrossRef] [PubMed]
13. Goedeke, L.; Vales-Lara, F.M.; Fenstermaker, M.; Cirera-Salinas, D.; Chamorro-Jorganes, A.; Ramirez, C.M.; Mattison, J.A.; de Cabo, R.; Suarez, Y.; Fernandez-Hernando, C. A regulatory role for microRNA 33* in controlling lipid metabolism gene expression. *Mol. Cell Biol.* **2013**, *33*, 2339–2352. [CrossRef] [PubMed]
14. Horie, T.; Ono, K.; Horiguchi, M.; Nishi, H.; Nakamura, T.; Nagao, K.; Kinoshita, M.; Kuwabara, Y.; Marusawa, H.; Iwanaga, Y.; et al. MicroRNA-33 encoded by an intron of sterol regulatory element-binding protein 2 (Srebp2) regulates HDL in vivo. *Proc. Natl. Acad. Sci. USA* **2010**, *107*, 17321–17326. [CrossRef]
15. Marquart, T.J.; Allen, R.M.; Ory, D.S.; Baldan, A. miR-33 links SREBP-2 induction to repression of sterol transporters. *Proc. Natl. Acad. Sci. USA* **2010**, *107*, 12228–12232. [CrossRef]
16. Rayner, K.J.; Suarez, Y.; Davalos, A.; Parathath, S.; Fitzgerald, M.L.; Tamehiro, N.; Fisher, E.A.; Moore, K.J.; Fernandez-Hernando, C. MiR-33 contributes to the regulation of cholesterol homeostasis. *Science* **2010**, *328*, 1570–1573. [CrossRef]
17. Horie, T.; Nishino, T.; Baba, O.; Kuwabara, Y.; Nakao, T.; Nishiga, M.; Usami, S.; Izuhara, M.; Sowa, N.; Yahagi, N.; et al. MicroRNA-33 regulates sterol regulatory element-binding protein 1 expression in mice. *Nat. Commun.* **2013**, *4*, 2883. [CrossRef]
18. Price, N.L.; Singh, A.K.; Rotllan, N.; Goedeke, L.; Wing, A.; Canfrán-Duque, A.; Diaz-Ruiz, A.; Araldi, E.; Baldán, Á.; Camporez, J.P.; et al. Genetic Ablation of miR-33 Increases Food Intake, Enhances Adipose Tissue Expansion, and Promotes Obesity and Insulin Resistance. *Cell. Rep.* **2018**, *22*, 2133–2145. [CrossRef]
19. Goedeke, L.; Salerno, A.; Ramírez, C.M.; Guo, L.; Allen, R.M.; Yin, X.; Langley, S.R.; Esau, C.; Wanschel, A.; Fisher, E.A.; et al. Long-term therapeutic silencing of miR-33 increases circulating triglyceride levels and hepatic lipid accumulation in mice. *EMBO Mol. Med.* **2014**, *6*, 1133–1141. [CrossRef]
20. Dávalos, A.; Goedeke, L.; Smibert, P.; Ramírez, C.M.; Warrier, N.P.; Andreo, U.; Cirera-Salinas, D.; Rayner, K.; Suresh, U.; Pastor-Pareja, J.C.; et al. miR-33a/b contributes to the regulation of fatty acid metabolism and insulin signaling. *Proc. Natl. Acad. Sci. USA* **2011**, *108*, 9232–9237. [CrossRef]
21. Gerin, I.; Clerbaux, L.A.; Haumont, O.; Lanthier, N.; Das, A.K.; Burant, C.F.; Leclercq, I.A.; MacDougald, O.A.; Bommer, G.T. Expression of miR-33 from an SREBP2 intron inhibits cholesterol export and fatty acid oxidation. *J. Biol. Chem.* **2010**, *285*, 33652–33661. [CrossRef] [PubMed]
22. Karunakaran, D.; Thrush, A.B.; Nguyen, M.A.; Richards, L.; Geoffrion, M.; Singaravelu, R.; Ramphos, E.; Shangari, P.; Ouimet, M.; Pezacki, J.P.; et al. Macrophage mitochondrial energy status regulates cholesterol efflux and is enhanced by anti-mir33 in atherosclerosis. *Circ. Res.* **2015**, *117*, 266–278. [CrossRef] [PubMed]
23. Ouimet, M.; Ediriweera, H.; Gundra, U.M.; Sheedy, F.J.; Ramkhelawon, B.; Hutchison, S.B.; Rinehold, K.; van Solingen, C.; Fullerton, M.D.; Cecchini, K.; et al. MicroRNA-33-dependent regulation of macrophage metabolism directs immune cell polarization in atherosclerosis. *J. Clin. Investig.* **2015**, *125*, 4334–4348. [CrossRef] [PubMed]
24. Ouimet, M.; Ediriweera, H.; Afonso, M.S.; Ramkhelawon, B.; Singaravelu, R.; Liao, X.; Bandler, R.C.; Rahman, K.; Fisher, E.A.; Rayner, K.J.; et al. microRNA-33 regulates macrophage autophagy in atherosclerosis. *Arter. Thromb. Vasc. Biol.* **2017**, *37*, 1058–1067. [CrossRef] [PubMed]
25. Horton, J.D.; Goldstein, J.L.; Brown, M.S. SREBPs: Activators of the complete program of cholesterol and fatty acid synthesis in the liver. *J. Clin. Investig.* **2002**, *109*, 1125–1131. [CrossRef] [PubMed]
26. Raitakari, O.T.; Juonala, M.; Viikari, J.S. Obesity in childhood and vascular changes in adulthood: Insights into the Cardiovascular Risk in Young Finns Study. *Int. J. Obes.* **2005**, *29*, S101e4. [CrossRef]
27. Bastien, M.; Poirier, P.; Lemieux, I.; Després, J.P. Overview of epidemiology and contribution of obesity to cardiovascular disease. *Prog. Cardiovasc. Dis.* **2014**, *56*, 369e81. [CrossRef]
28. Nathan, B.M.; Moraan, A. Metabolic complications of obesity in childhood and adolescence: More than just diabetes. *Curr. Opin. Endocrinol. Diabetes. Obes.* **2008**, *15*, 21–29. [CrossRef]
29. Cabiati, M.; Randazzo, E.; Guiducci, L.; Falleni, A.; Cecchettini, A.; Casieri, V.; Federico, G.; Del Ry, S. Evaluation of exosomal coding and non-coding RNA signature in obese adolescents. *Int. J. Mol. Sci.* **2022**, *24*, 139. [CrossRef]
30. Cole, T.J.; Bellizzi, M.C.; Flegal, K.M.; Diettz, W.H. Establishing a standard definition for child overweight and obesity worldwide: International survey. *BMJ* **2000**, *320*, 1240–1243. [CrossRef]
31. Cabiati, M.; Fontanini, M.; Giacomarra, M.; Politano, G.; Randazzo, E.; Peroni, D.; Federico, G.; Del Ry, S. Screening and identification of putative Long Non-Coding RNA in childhood obesity: Evaluation of their transcriptional levels. *Biomedicines* **2022**, *10*, 529. [CrossRef] [PubMed]
32. Cabiati, M.; Randazzo, E.; Salvadori, C.; Peroni, D.; Federico, G.; Del Ry, S. Circulating microRNAs associated with C-type natriuretic peptide in childhood obesity. *Peptides* **2020**, *133*, 170387. [CrossRef] [PubMed]
33. Del Ry, S.; Cabiati, M.; Bianchi, V.; Randazzo, E.; Peroni, D.; Clerico, A.; Federico, G. C-type natriuretic peptide plasma levels and whole blood mRNA expression show different trends in adolescents with different degree of endothelial dysfunction. *Peptides* **2020**, *124*, 170218. [CrossRef] [PubMed]

34. Ry, S.D.; Cabiati, M.; Bianchi, V.; Caponi, L.; Maltinti, M.; Caselli, C.; Kozakova, M.; Palombo, C.; Morizzo, C.; Marchetti, S.; et al. C-type natriuretic peptide is closely associated to obesity in Caucasian adolescents. *Clin. Chim. Acta.* **2016**, *460*, 172–177.
35. Ry, S.D.; Cabiati, M.; Bianchi, V.; Caponi, L.; Cecco, P.D.; Marchi, B.; Randazzo, E.; Caselli, C.; Prescimone, T.; Clerico, A.; et al. Mid-regional-pro-adrenomedullin plasma levels are increased in obese adolescents. *Eur. J. Nutr.* **2016**, *55*, 1255–1260.
36. Ry, S.D.; Cabiati, M.; Bianchi, V.; Storti, S.; Caselli, C.; Prescimone, T.; Clerico, A.; Saggese, G.; Giannessi, D.; Federico, G. C-type natriuretic peptide plasma levels are reduced in obese adolescents. *Peptides* **2013**, *50*, 50–54.
37. Marshall, W.A.; Tanner, J.M. Variations in pattern of pubertal changes in girls. *Arch. Dis. Child.* **1969**, *44*, 291–303. [CrossRef]
38. Marshall, W.A.; Tanner, J.M. Variations in the pattern of pubertal changes in boys. *Arch. Dis. Child.* **1970**, *45*, 13–23. [CrossRef]
39. McCarthy, H.D.; Cole, T.J.; Fry, T.; Jebb, S.A.; Prentice, A.M. Body fat reference curves for children. *Int. J. Obes.* **2006**, *30*, 598–602. [CrossRef]
40. National High Blood Pressure Education Program Working Group on High Blood Pressure in Children and Adolescents. The fourth report on the diagnosis, evaluation and treatment of high blood pressure in children and adolescents. *Pediatrics* **2004**, *114*, 555–576. [CrossRef]
41. Rickham, P.P. Human experimentation. Code of ethics of the world medical association. Declaration of Helsinki. *Br. Med. J.* **1964**, *2*, 177. [PubMed]
42. Bustin, S.A.; Benes, V.; Garson, J.A.; Hellemans, J.; Huggett, J.; Kubista, M.; Mueller, R.; Nolan, T.; Pfaffl, M.W.; Shipley, G.L.; et al. The MIQE guidelines: Minimum information for publication of quantitative real-time PCR experiments. *Clin. Chem.* **2009**, *55*, 611–622. [CrossRef] [PubMed]
43. Théry, C.M.; Witwer, K.W.; Aikawa, E.; Alcaraz, M.J.; Anderson, J.D.; Andriantsitohaina, R.; Antoniou, A.; Arab, T.; Archer, F.; Atkin-Smith, G.K.; et al. Minimal information for studies of extracellular vesicles 2018 (MISEV2018): A position statement of the International Society for Extracellular Vesicles and update of the MISEV2014 guidelines. *J. Extracell. Vesicles* **2018**, *7*, 1535750. [CrossRef] [PubMed]
44. Zhang, X.; Zhao, H.; Sheng, Q.; Liu, X.; You, W.; Lin, H.; Liu, G. Regulation of microRNA-33, SREBP and ABCA1 genes in a mouse model of high cholesterol. *Arch. Anim. Breed.* **2021**, *64*, 103–108. [CrossRef]
45. Gharipour, M.; Sadeghi, M. Pivotal role of microRNA-33 in metabolic syndrome: A systematic review. *ARYA Atheroscler.* **2013**, *9*, 372–376.
46. Ramírez, C.M.; Goedeke, L.; Fernández-Hernando, C. Micromanaging metabolic syndrome. *Cell Cycle* **2011**, *10*, 3249–3252. [CrossRef]
47. Endzeliņš, E.; Berger, A.; Melne, V.; Bajo-Santos, C.; Soboļevska, K.; Ābols, A.; Rodriguez, M.; Šantare, D.; Rudņickiha, A.; Lietuvietis, V.; et al. Detection of circulating miRNAs: Comparative analysis of extracellular vesicle-incorporated miRNAs and cell-free miRNAs in whole plasma of prostate cancer patients. *BMC Cancer* **2017**, *17*, 730. [CrossRef]
48. Raposo, G.; Stahl, P.D. Extracellular vesicles: A new communication paradigm? *Nat. Rev. Mol. Cell Biol.* **2019**, *20*, 509–510. [CrossRef]
49. Kita, S.; Maeda, N.; Shimomura, I. Interorgan communication by exosomes, adipose tissue, and adiponectin in metabolic syndrome. *J. Clin. Investig.* **2019**, *129*, 4041–4049. [CrossRef]
50. Maligianni, I.; Yapijakis, C.; Bacopoulou, F.; Chrousos, G. The potential role of exosomes in child and adolescent obesity. *Children* **2021**, *3*, 196. [CrossRef]
51. Mitchell, P.S.; Parkin, R.K.; Kroh, E.M.; Fritz, B.R.; Wyman, S.K.; Pogosova-Agadjanyan, E.L.; Peterson, A.; Noteboom, J.; O'Briant, K.C.; Allen, A.; et al. Circulating micro-RNAs as stable blood-based markers for cancer detection. *Proc. Natl. Acad. Sci. USA* **2008**, *105*, 10513–10518. [CrossRef] [PubMed]
52. Ortega, F.J.; Mercader, J.M.; Catalan, V.; Moreno-Navarrete, J.M.; Pueyo, N.; Sabater, M.; Gomez-Ambrosi, J.; Anglada, R.; Fernandez-Formoso, J.A.; Ricart, W.; et al. Targeting the circulating microRNA signature of obesity. *Clin. Chem.* **2013**, *59*, 781–792. [CrossRef] [PubMed]
53. Yang, Z.; Chen, H.; Si, H.; Li, X.; Ding, X.; Sheng, Q.; Chen, P.; Zhang, H. Serum miR-23a, a potential biomarker for diagnosis of pre-diabetes and type 2 diabetes. *Acta Diabetol.* **2014**, *51*, 823–831. [CrossRef] [PubMed]

biomedicines

MDPI

Article

Effect of Bariatric Surgery on Plasma Cell-Free Mitochondrial DNA, Insulin Sensitivity and Metabolic Changes in Obese Patients

Larysa V. Yuzefovych [1], Viktor M. Pastukh [1], Madhuri S. Mulekar [2], Kate Ledbetter [3], William O. Richards [3] and Lyudmila I. Rachek [1,*]

[1] Department of Pharmacology, College of Medicine, University of South Alabama, Mobile, AL 36688, USA; yuzefovych@southalabama.edu (L.V.Y.); vpastukh@southalabama.edu (V.M.P.)
[2] Department of Mathematics and Statistics, College of Art and Science, University of South Alabama, Mobile, AL 36688, USA; mmulekar@southalabama.edu
[3] Department of Surgery, College of Medicine, University of South Alabama, Mobile, AL 36688, USA; kledbetter@health.southalabama.edu (K.L.); brichards@health.southalabama.edu (W.O.R.)
* Correspondence: lrachek@southalabama.edu; Tel.: +1-(251)-460-6960

Abstract: While improvement of mitochondrial function after bariatric surgery has been demonstrated, there is limited evidence about the effects of bariatric surgery on circulatory cell-free (cf) mitochondrial DNA (mtDNA) and intracellular mtDNA abundance. Plasma and peripheral blood mononuclear (PBM) cells were isolated from healthy controls (HC) and bariatric surgery patients before surgery and 2 weeks, 3 months, and 6 months after surgery. At baseline, the plasma level of short cf-mtDNA (*ND6*, ~100 bp) fragments was significantly higher in obese patients compared to HC. But there was no significant variation in mean *ND6* values post-surgery. A significant positive correlation was observed between preop plasma *ND6* levels and HgbA1c, *ND6* and HOMA-IR 2 weeks post-surgery, and mtDNA content 6 months post-surgery. Interestingly, plasma from both HC and obese groups at all time points post-surgery contains long (~8 kb) cf-mtDNA fragments, suggesting the presence of near-intact and/or whole mitochondrial genomes. No significant variation was observed in mtDNA content post-surgery compared to baseline data in both PBM and skeletal muscle samples. Overall, bariatric surgery improved insulin sensitivity and other metabolic parameters without significant changes in plasma short cf-mtDNA levels or cellular mtDNA content. Our study provides novel insights about possible molecular mechanisms underlying the metabolic effects of bariatric surgery and suggests the development of new generalized approaches to characterize cf-mtDNA.

Keywords: circulating cell-free mitochondrial DNA; obesity; insulin resistance; type 2 diabetes

Citation: Yuzefovych, L.V.; Pastukh, V.M.; Mulekar, M.S.; Ledbetter, K.; Richards, W.O.; Rachek, L.I. Effect of Bariatric Surgery on Plasma Cell-Free Mitochondrial DNA, Insulin Sensitivity and Metabolic Changes in Obese Patients. *Biomedicines* **2023**, *11*, 2514. https://doi.org/10.3390/biomedicines11092514

Academic Editors: Teresa Vezza and Zaida Abad-Jiménez

Received: 7 July 2023
Revised: 21 August 2023
Accepted: 23 August 2023
Published: 12 September 2023

1. Introduction

Obesity is associated with insulin resistance (IR), and this represents a major risk factor for metabolic syndrome (MetS), type 2 diabetes (T2D), and cardiovascular disorders such as coronary artery disease and heart failure [1,2]. It has been demonstrated that mitochondrial dysfunction contributes to obesity-related IR [3–5]. Previously, we and others showed that bariatric surgery rapidly improves mitochondrial respiration in morbidly obese patients [6,7]. Furthermore, even moderate weight loss and physical activity result in improvements in mitochondrial function, which results in improvements in insulin sensitivity [8]. With the increasing use of bariatric surgery for the treatment of severe obesity, it is critical to understand the molecular mechanisms responsible for the notable improvements seen after the procedure.

Overall, the publications regarding the effect of bariatric surgery and/or surgically-induced weight loss on cf-mtDNA fragments or mtDNA integrity are scarce [9–13]. Considering the growing importance of mtDNA analysis in obesity and IR research, the goal of this study was to evaluate plasma levels of cf-mtDNA and intracellular mtDNA content in relation to the metabolic changes before and after bariatric surgery in obese patients. Additionally, since we previously demonstrated that increased levels of plasma cf-mtDNA correlate with the degree of IR in obese T2D patients [14], in the current study we determined whether cf-mtDNA and intracellular mtDNA content vary when obese patients are grouped by presence or absence of T2D or MetS and compared to healthy controls (HC).

2. Materials and Methods

2.1. Subjects

We recruited 13 morbidly obese patients with T2D ($n = 7$), MetS ($n = 4$), and without T2D or MetS ($n = 2$) who fulfilled the National Institutes of Health criteria for bariatric surgery (body mass index (BMI) > 35 kg/m^2). Ten patients underwent laparoscopic sleeve gastrectomy (LSG), and three patients underwent laparoscopic Roux-en-Y gastric bypass (LRYGB). Eight volunteer HC (subjects without obesity (body mass index < 30 kg/m^2) were recruited from the general community. The presence of T2D in patients was confirmed based on a detailed clinical examination using the American Diabetes Association criteria for T2D: fasting glucose $\geq$126 mg/dL, HgbA1c $\geq$ 6.5%. The Adult Treatment Panel (ATP III) diagnostic criteria were used to establish the presence of the MetS in patients as described [15]. Post-bariatric surgery, patients were taken off medications.

All subjects were sedentary. All human studies were conducted in accordance with NIH guidelines, complied with the Declaration of Helsinki, and were approved by the Institutional Review Board of the University of South Alabama (protocol #1253267, "Effect of bariatric surgery mtDNA DAMP's, inflammation & diabetic outcome"). All human subjects gave informed consent.

2.2. Metabolic Parameters and Muscle Biopsy

Each subject had a medical history, a physical examination, and blood sampling. From each bariatric surgery patient, blood was collected before surgery (BS) and 2 weeks (2 W), 3 months (3 M), and 6 months (6 M) after bariatric surgery. Metabolic parameters were measured using standard kits. Homeostasis model assessment for insulin resistance index (HOMA-IR) was calculated using the formula: HOMA-IR = [glucose (mg/dL) $\times$ insulin (µIU)/mL]/405, with fasting values. For most obese patients prior to surgery and at some points after, samples of skeletal muscle (~100 mg) were obtained from the vastus lateralis by percutaneous needle biopsy under sterile conditions with local anesthesia (lidocaine, 1%) and were snap frozen in liquid nitrogen for DNA and protein analysis.

2.3. Isolation of Plasma and PBM Cells

On the day of the blood sampling and/or muscle biopsy, subjects reported to the laboratory after an overnight fast (12 h). Peripheral blood (16 mL) was collected into two sterile density gradient tubes (Vacutainer with Ficoll-Hypaque solution, Becton Dickinson, Franklin Lakes, NJ, USA). Blood was fractionated by centrifugation at 1500$\times$ g for 30 min at 21 °C with a swinging bucket rotor, as previously described with modifications [14]. The plasma (upper) fraction was separated from PBM cells, carefully transferred into a 15 mL conical tube, and centrifuged a second time at 1800$\times$ g for 10 min to eliminate cell and large debris contamination. Then, it was aliquoted and stored at −80 °C until analyzed. The mononuclear-containing buffy coat fraction was suspended in 10 mL of phosphate-buffered saline (PBS; pH 7.4, Thermo Fisher, Waltham, MA, USA) and collected by centrifugation at 700$\times$ g for 10 min at ambient temperature as described previously [6]. This wash step was repeated once, and the mononuclear-containing buffy coat fraction was suspended in 0.1 mL of PBS (pH 7.4) and frozen at −70 °C.

2.4. Isolation of Plasma cf-DNA

Plasma cf-DNA was isolated using a DNeasy Blood and Tissue kit (Qiagen, Germantown, MD, USA) from aliquots of plasma samples that had been frozen at −80 °C, as we described previously with some modifications [14]. Extraction was performed with 1 mL of plasma sequentially loaded on a single column (200 μL 5 times), and the DNA was eluted with 50 μL of deionized water. The cf-DNA was stored at −20 °C for further analysis.

2.5. Analysis of Plasma cf-mtDNA

First, we performed qRT-PCR as we described previously, with a slight modification [16], using primers to amplify a 104 bp sequence within the mitochondrial *ND6* gene. To assess the concentration of mtDNA fragments (ng/mL) in plasma samples, we constructed a standard calibration curve using a PCR product from the whole *ND6* region using total DNA isolated from HUVEC cells. Equal amounts of eluant (5 μL) were used for qRT-PCR (SsoAdvanced™ Universal SYBR® Green Supermix, #1725271, Hercules, CA, USA), according to the manufacturer's protocol. All primers were designed by Beacon Designer 8.2 software (PREMIER Biosoft International, Palo Alto, CA, USA). Sequences of primers are provided in Supplemental Table S1. For amplification of long PCR fragments, Platinum™ SuperFi™ PCR Master (Invitrogen, Thermo Fisher Scientific, Waltham, MA, USA) was used according to the manufacturer's recommendations, with 1 μL of DNA and two sets of overlapping mitochondrial primers, Mito 1 and Mito 2 (500 nM/each, Supplemental Table S1). The thermal cycling conditions were as follows: (1) Mito1—95 °C (2:00) + [95 °C (0:30) + 60 °C (0:30) + 68 °C (8:00)] × 30 cycles + 68 °C (5:00) + 4 °C (∞). (2) Mito2—95 °C (2:00) + [95 °C (0:30) + 55 °C (0:30) + 68 °C (8:00)] × 30 cycles + 68 °C (5:00) + 4 °C (∞). Equal amounts of PCR products were loaded in the 0.8% agarose gel with reference to the DNA standards: 1 Kb DNA ladder (Promega, Fitchburg, WI, USA) and 1 Kb DNA ladder (Invitrogen, Thermo Fisher Scientific, Waltham, MA, USA). A total of 1 ng of DNA isolated from HUVEC cells was used as a positive control, and a sample without DNA but containing water was used as a negative control. Positive and negative controls were included in each run.

2.6. Analysis of mtDNA Content

Total DNA was isolated from PBM cells from HC and each obese patient at baseline and 2 weeks, 3 months, and 6 months after bariatric surgery using the DNeasy Blood and Tissue Kit (Qiagen, Germantown, MD, USA). Using the same kit, we isolated total DNA from skeletal muscle samples obtained from obese patients. 20 ng of total DNA was used for qRT-PCR (Luna® Universal qPCR Master Mix, #M3003L, NEB, Ipswich, MA, USA) according to the manufacturer's protocol and PCR conditions as previously described [16]. The ratio of mtDNA to nDNA indicates the relative mtDNA content per cell in the tissue. The mtDNA/nDNA ratio was assessed by the qRT-PCR method using primers specific for the mitochondrial *ND1* gene [16] and normalizing for the amount of nDNA used in each reaction by using primers specific for *18S rRNA* [14].

2.7. Statistical Analyses

All data collected were analyzed using JMP Pro 16.2.0 (a product of SAS Inc., Cary, NC, USA). A significance level of 0.05 was used to determine the significance of the results. This is an unbalanced block design because of missing data for different patients at different times. Mean outcomes for four-time sets (BS, 2 W, 3 M, and 6 M) were compared using the ANOVA technique with patient as a blocking factor, i.e., measurements over different times were matched by patient. Statistically significant variation in measurements among patients and data collected for the same patients over time justifies the use of patients as a blocking factor. Post hoc comparisons for four-time sets were conducted using Tukey's HSD. Mean outcomes for T2D-Mets groups and IR were compared using ANOVA, and post hoc comparisons with BS were conducted using Dunnett's method. Mean outcomes for two bariatric surgery procedure groups (LSG or LRYGB) were compared using a *t*-test after accounting for differences in standard deviations, if any.

3. Results

3.1. Metabolic Parameters

Initially, we recruited 14 morbidly obese patients but excluded patient number 13 for not following instructions for postoperative blood draws. Patient number 13 ate prior to the fasting blood draws, making the outcomes of the blood tests invalid. As a result, we had 13 morbidly obese patients (9 females and 4 males) with T2D (n = 7), MetS (n = 4), and without T2D or MetS (n = 2), and eight HC (6 females and 2 males) subjects without obesity. The study has two groups for which separate statistical analysis was performed: (1) HC-Obese patients, BS; and (2) Obese patients at all time points. Clinical and metabolic characteristics of study groups are shown in Table 1 for Obese patients at all time points and in Supplemental Table S2 for HC and Obese BS patients.

Table 1. Summary of clinical, metabolic, plasma cf-mtDNA data, and mtDNA content of obese bariatric surgery patients before surgery and at all time points post-surgery.

Obese Only	Time	n	Mean	Std Dev	R^2	p (Patient)	p (Time)			
BMI	BS	13	49.90	7.91	0.88	<0.0001	<0.0001	A		
	2 W	12	44.49	7.59					B	
	3 M	10	39.54	6.99						C
	6 M	10	36.61	6.62						C
Glucose	BS	13	125.31	43.64	0.78	<0.0001	0.1317			
	2 W	12	117.17	72.08						
	3 M	10	89.30	9.68						
	6 M	10	85.30	20.24						
HgbA1c	BS	13	7.04	1.73	0.89	0.0023	0.0151	A		
	2 W	0								
	3 M	8	5.39	0.31					B	
	6 M	7	5.30	0.29					B	
Insulin	BS	12	50.29	42.57	0.56	0.3424	0.0034	A		
	2 W	12	16.35	10.14					B	
	3 M	10	10.58	4.24					B	
	6 M	10	14.40	14.90					B	
HOMA-IR	BS	12	18.66	22.50	0.59	0.0780	0.0126	A		
	2 W	12	4.52	3.07					B	
	3 M	10	2.40	1.16					B	
	6 M	10	3.58	5.01				A	B	
Triglycerides	BS	13	159.85	77.95	0.92	0.0002	0.0026	A		
	2 W	0								
	3 M	8	99.50	13.77					B	
	6 M	7	91.57	31.97					B	
Total Cholesterol	BS	13	167.23	31.93	0.75	0.0213	0.9287			
	2 W	0								
	3 M	8	168.13	27.53						
	6 M	7	169.00	28.61						
HDL	BS	13	40.08	6.91	0.91	0.0002	0.0113		B	
	2 W	0								
	3 M	8	43.88	11.70					B	
	6 M	7	51.57	14.59				A		
LDL	BS	13	95.08	29.76	0.69	0.0773	0.8800			
	2 W	0								
	3 M	8	104.44	20.80						
	6 M	7	99.14	12.62						
Plasma *ND6*	BS	6	0.06	0.03	0.66	0.0443	0.8539			
	2 W	9	0.07	0.07						
	3 M	7	0.08	0.06						
	6 M	8	0.07	0.08						
Skeletal muscle *ND1/18SrRNA*	BS	13	1.74	1.06	0.69	0.3104	0.0700			
	2 W	0								
	3 M	9	1.14	0.42						
	6 M	2	1.34	0.61						
PBM *ND1/18S rRNA*	BS	13	0.97	0.28	0.39	0.3311	0.3759			

Table 1. *Cont.*

Obese Only	Time	*n*	Mean	Std Dev	R^2	*p* (Patient)	*p* (Time)
	2 W	12	0.85	0.32			
	3 M	10	0.87	0.46			
	6 M	11	1.16	0.44			

BS—before surgery. The *n* is variable because of missing data for different patients at different times. Levels not connected by the same letter are significantly different.

On average, HC (47.50 ± 9.01 years) were slightly older than Obese, BS patients (41.54 ± 12.78 years), but the difference was not significant ($p = 0.2266$, Supplemental Table S2). Obese, BS patients had significantly higher BMI ($p < 0.0001$), fasting plasma glucose ($p = 0.0102$), HgbA1c ($p = 0.0023$), insulin ($p = 0.0052$), HDL ($p = 0.0036$), and triglyceride ($p = 0.0023$) levels compared to HC (Supplemental Table S2). Additionally, they were more insulin-resistant, as shown by a significantly increased HOMA-IR ($p = 0.0237$) in the Obese BS group. On average, total cholesterol and LDL for the two groups were not significantly different (Supplemental Table S2).

As can be seen by the missed data in Table 1, some parameters were not measured in the Obese group at all time points due to the missed patient visits, particularly at the time of the COVID-19 pandemic. Some parameters (HgbA1c, triglycerides, HDL, and LDL) were not measured at the 2 W time point, which is indicated as $n = 0$. Mean outcomes for clinical and metabolic data over four time points (BS, 2 W, 3 M, and 6 M) were compared using the ANOVA technique with patients as a blocking factor, i.e., measurements over different times were matched by patient. Significant patient-to-patient variation was observed in the following metabolic parameters, indicating blocking (or matching) by patient was justified (Table 1): BMI ($p < 0.0001$), glucose ($p < 0.0001$), HgbA1c ($p < 0.0023$), triglycerides ($p = 0.0002$), total cholesterol ($p = 0.0213$), HDL ($p = 0.0002$), and plasma *ND6* ($p = 0.0443$).

Bariatric surgery, on average, reduced BMI over time (Table 1). Mean BMI differed significantly at least two different times ($p < 0.0001$, power = 1.0). In fact, the mean BMI decreased as time advanced. Post hoc analysis using Tukey's test shows that mean BMI dropped significantly from BS (49.90) to 2 W (44.49) and then to 3 M (39.54). Although, the 6 M (36.61) mean BMI was lower than that for 3 M (39.54), the difference was not statistically significant, but the 6 M BMI was significantly different from the BS and 2 W BMIs. Although the mean glucose level decreased steadily from 125.31 at BS to 85.30 at 6 M, the difference was not statistically significant ($p = 0.1317$). Mean HgbA1c differed significantly for at least two different times ($p = 0.0151$, power = 0.99); in fact, it decreased over time. Post hoc analysis using Tukey's test showed a significant drop from BS (7.04) to 3 M (5.39) and 6 M (5.30), but a decrease from 3 M (5.39) to 6 M (5.30) was not statistically significant. Insulin levels did not differ significantly from patient to patient ($p = 0.3424$); however, the mean insulin level differed significantly for at least two different times ($p = 0.0034$, power = 0.751). In fact, it decreased over time from BS to 3 M (BS: 50.29, 2 W: 16.35, and 3 M: 10.58) only to increase to 14.40 at 6 M. Post hoc analysis using Tukey's test showed a significant decrease in mean insulin level from BS to 2 W, 3 M, and 6 M; however, the decrease from 2 W to 3 M was not significant, nor was the increase from 3 M to 6 M. HOMA-IR levels did not differ significantly from patient to patient ($p = 0.0780$), but mean the HOMA-IR level differed significantly at least two different times ($p = 0.0126$, power = 0.77). In fact, it decreased from BS (18.66) to 2 W (4.52). The drop from BS to 3 M (2.40) was also significant. But it then increased to 6 M (3.58), though not significantly. Post hoc analysis using Tukey's test showed a significant decrease in mean HOMA-IR level from BS to 2 W; however, later changes in HOMA-IR level over time were not statistically significant. HOMA-IR levels among patients at BS varied the most, decreasing with time only to increase at 6 M.

Triglyceride levels varied significantly from patient to patient ($p = 0.0002$). Mean triglyceride levels differed significantly at least two different times ($p = 0.0026$, power = 1.0). In fact, mean triglyceride levels decreased as time advanced. Post hoc analysis using Tukey's test shows that mean triglyceride levels at both 3 M (99.50) and 6 M (91.57) were significantly lower than BS (159.85); however, the decrease from 3 M to 6 M was not statistically significant. Total cholesterol levels differed significantly by patient ($p = 0.0213$), but no significant differences in mean cholesterol levels were observed over time ($p = 0.9287$, BS: 167.23, 3 M: 168.13, and 6 M: 169.00). Significant variation in HDL levels was observed among patients ($p = 0.0002$). Mean HDL levels differed significantly for at least two time periods ($p = 0.0113$, power = 1.0). A steady increase in HDL levels with time was observed. Post hoc analysis using Tukey's test indicates a slight but not statistically significant increase in mean HDL level from BS (40.08) to 3 M (43.88). The increase in mean HDL level from BS to 6 M (51.57) as well as from 3 M to 6 M is significant. No significant variation in LDL levels was observed among patients ($p = 0.0773$) or at three different times ($p = 0.8800$).

Comparison of clinical and metabolic parameters among types of bariatric surgery (LSG or LRYGB groups) showed that the 3 M mean insulin level was significantly lower among LRYGB patients than LSG patients (5.80 vs. 11.77, $p = 0.0040$, Supplemental Table S3). Also, the 3 M mean HOMA-IR level was significantly lower among LRYGB patients than LSG patients (1.18 vs. 2.71, $p = 0.0066$, Supplemental Table S3).

3.2. *Effect of Obesity and Bariatric Surgery on Plasma cf-mtDNA*

First, we analyzed short cf-mtDNA fragments (104 bp) within the mitochondrial *ND6* gene using qRT-PCR. Analysis of *ND6* cf-mtDNA in plasma was performed separately for two groups: (1) HC-Obese, BS group, and (2) Obese group, at all times (BS, 2 W, 3 M, and 6 M). Upon comparing the level of short cf-mtDNA in plasma from HC and Obese BS patients, we found that the level of plasma cf-mtDNA *ND6* fragments was five times greater in Obese BS (0.05 ± 0.03) compared to HC (0.01 ± 0.01, $p = 0.0026$, Figure 1A). Additionally, when patients in the Obese BS group were classified as MetS, T2D, or none, we found that mean plasma *ND6* levels were significantly higher in samples of T2D than HC (Dunnett's method for comparison with HC, $p = 0.0098$, Figure 1B, Supplemental Table S4). Similarly, a significantly higher mean plasma *ND6* level was observed in the IR in Obese BS patients compared to HC (Dunnett's method for comparison with HC, $p = 0.0119$, Figure 1C, Supplemental Table S5).

Comparison of plasma *ND6* levels among HC/T2D/nonT2D groups resulted in significant differences in the mean levels of at least two groups (Figure 1D, ANOVA, $p = 0.0085$). Post hoc analysis using Dunnett's method showed that mean plasma *ND6* levels of T2D patients were significantly higher than HC (Figure 1D, $p = 0.0048$); however, those of non-T2D patients were not significantly different than HC (Figure 1D, $p = 0.1177$).

A significant positive correlation was observed between HgbA1c and plasma *ND6* ($r = 0.73$, $n = 18$, $p = 0.0006$, Figure 2A). However, when HC and obese patient data were separated, a negative correlation was observed for HC, which was not statistically significant probably due to the small sample size ($r = -0.55$, $n = 7$, $p = 0.1967$, Supplemental Figure S1), and a significant positive correlation was observed for Obese BS patients ($r = 0.63$, $n = 11$, $p = 0.0362$, Supplemental Figure S1).

Post-surgery analysis of plasma *ND6* levels in obese patients showed that ND6 values varied significantly among patients ($p = 0.0443$, Table 1). Although mean plasma cf-mtDNA *ND6* levels at three post-operative points were slightly higher compared to baseline, the differences were not significant ($p = 0.8539$, Supplemental Figure S2). Interestingly, comparison of the Obese MetS, T2D, and without MetS (designated as "none") groups showed that mean plasma *ND6* levels were higher in T2D patients' samples at all time points except at 3 M after bariatric surgery compared to those in the other two groups (Supplemental Table S6).

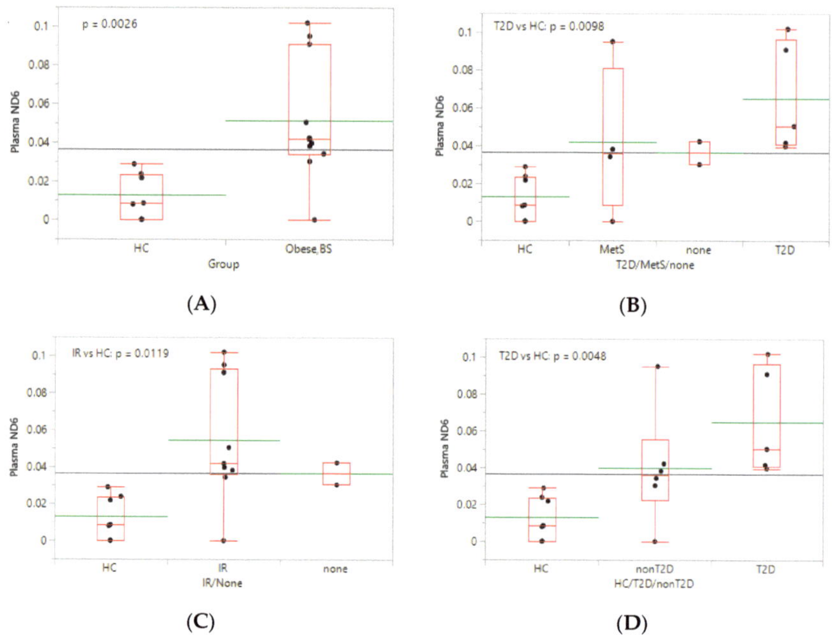

Figure 1. Effect of (**A**) obesity, (**B**) MetS, (**C**) IR, and (**D**) T2D on plasma *ND6* level. The data are presented as boxplots, with the mean marked by green lines.

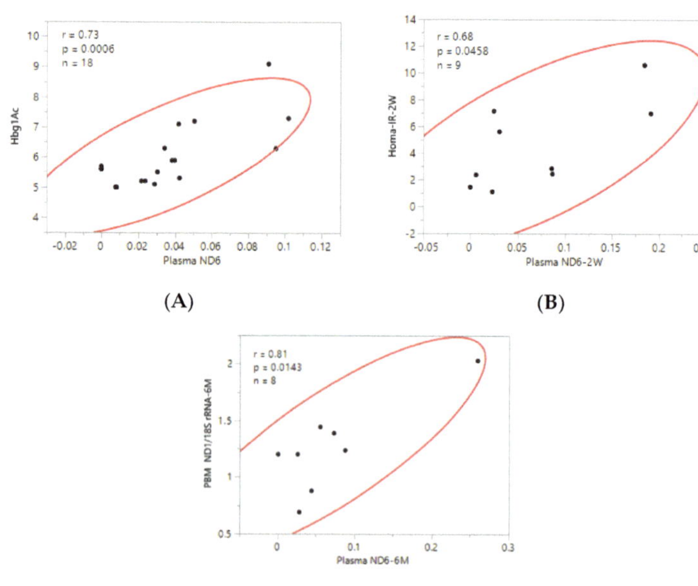

Figure 2. A significant positive correlation as measured by Pearson's correlation coefficient was observed between plasma levels of *ND6* and (**A**) pre-surgery HgbA1c in the HC-Obese, BS group of patients (n = 18); (**B**) HOMA-IR measured 2 W post-surgery (n = 9); and (**C**) PBMN mtDNA content measured 6 M post-surgery (n = 8). Red ellipse is the 95% bivariate normal ellipse.

A significant positive correlation was observed between plasma levels of plasma *ND6* and HOMA-IR, 2 W post-surgery (r = 0.68, p = 0.0458, Figure 2B), and PBMN *ND1/18S rRNA*, 6 M post-surgery (r = 0.81, p = 0.0143, Figure 2C). No significant correlation was observed between BMI and plasma *ND6* for any of the 4 time periods.

Second, we analyzed the presence of long cf-mtDNA fragments using two pairs of overlapping mitochondrial sequences covering the whole mtDNA genome for the identification of long mtDNA fragments. The location of sequences and size of the amplified fragments are shown on the scheme presented in Supplemental Figure S3. Long-range PCR analysis of samples from the Obese group at all time points after BS revealed that plasma from most samples contains long amplicons (~8 kb), suggesting the presence of near intact and/or whole mitochondrial genomes (Figure 3A,B). Although the majority of HC samples also showed long amplicons in their plasma, in some HC (#8 for Mito1 and #5 and 8 for Mito2), bands for long amplicons were not even observed, suggesting either the absence or a very low quantity of the long fragments in their pool of plasma long cf-mtDNA (Figure 3A,B). Although the inferences about differences between controls and obese samples should be qualified by the limitation that the data for long PCR products is not quantitative, the abundance of long PCR products was noticeably greater in samples from obese patients (Figure 3A,B) compared to samples from the HC group, at least for the Mito1 primers dataset. Therefore, not only were the short *ND6* fragments accumulated in plasma from obese patients, but also the complete mitochondrial genome, as determined by long-range PCR amplifications of 2 amplicons (~8 kb each, Figure 3A,B).

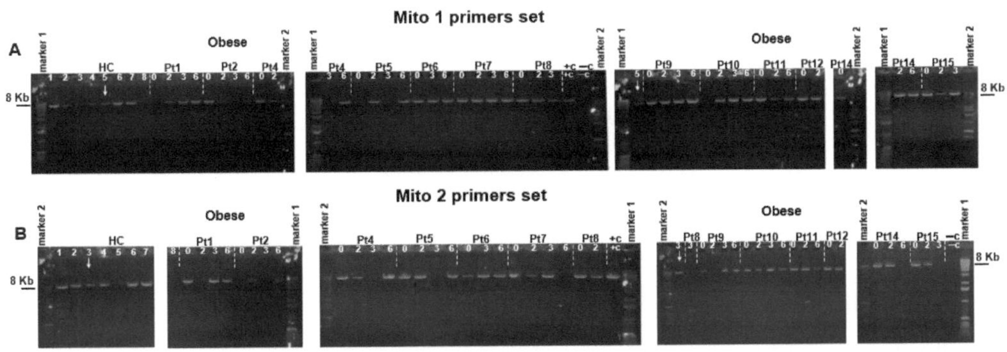

Figure 3. Representative gel electrophoresis of long-range PCR analysis. Using total DNA isolated from plasma as a template showed that plasma from HC and obese patients (all time points) contained (**A**) Mito1 (~8.4 kb) or (**B**) Mito2 (~8 kb) long overlapping amplicons that covered the full length mitochondrial genome. 1 ng of total DNA isolated from HUVEC cells was used as a positive control (+c), and a sample without DNA but containing water was used as a negative control (−c). No signal was detected while using the negative control. To compare the level of amplicons between obese and HC, a PCR sample from HC #5 was loaded on every gel analysis for long-range PCR with the Mito1 set of primers (indicated with an arrow above the gel). A sample from HC #3 was also loaded in every gel for comparison while using the Mito2 primer set, which is indicated with an arrow above the gel. Marker 1-1 Kb DNA ladder (Invitrogen) and Marker 2-1 Kb DNA ladder (Promega).

3.3. *Effect of Obesity and Bariatric Surgery on mtDNA Content in PBM Cells and Skeletal Muscle*

Next, we evaluated relative mtDNA content (mtDNA/nDNA ratio) using total DNA isolated from PBM or skeletal muscle. First, we used total DNA isolated from PBM cells from HC and each obese patient at baseline and at the time points indicated above after bariatric surgery. Although not statistically significant, the *ND1/18S rRNA* ratio in PBM cells on average was observed to be higher among Obese BS patients (1.0 ± 0.31) compared to HC (0.76 ± 0.32) (p = 0.1013, Supplemental Figure S4A). When patients in the Obese BS group were further classified as MetS (n = 4), T2D (n = 7), and none (n = 2), they all had a higher mean *ND1/18S rRNA* ratio in PBM cells than HC, but the differences were not statistically significant, probably due to small sample sizes (Supplemental Figure S4B, Supplemental Table S4). Interestingly, when Obese BS patients were classified as IR vs. none, their mean PBM cell *ND1/18S rRNA* ratio, although not significant, was higher compared to HC, with the "none" group having the highest mean (p = 0.0976, ANOVA, Supplemental Figure S4C, Supplemental Table S5). It is important to note that the "none" group had only two patients, and the IR group had two outliers on the higher end. Comparison of mean PBM cell *ND1/18S rRNA* levels by HC/T2D/non T2D groups resulted in no significant differences (p = 0.2021, ANOVA, Supplemental Figure S4D). Also, note an outlier in the T2D group (Supplemental Figure S4D). The mean PBM cell *ND1/18S rRNA* level was observed to be higher in obese non-T2D patients compared to HC, but the difference was not significant (Supplemental Figure S4D).

Comparison of PBM *ND1/18S rRNA* levels in obese patients at BS and three different post-surgery times showed no significant variation among patients (p = 0.3311) or over the different times it was measured (p = 0.3759, Table 1 and Figure 4A). Additionally, we analyzed total DNA isolated from skeletal muscle samples obtained from some obese patients at various time points post-surgery. As shown in Figure 4B and Table 1, no significant variation in the mean skeletal muscle *ND1/18S rRNA* ratio was observed among patients (p = 0.3104) or among three different times (p = 0.07). Additionally, the 2 W mean PBM *ND1/18S rRNA* level was significantly lower among LRYGB patients than LSG patients (0.52 vs. 0.96, p = 0.0025, Supplemental Table S3).

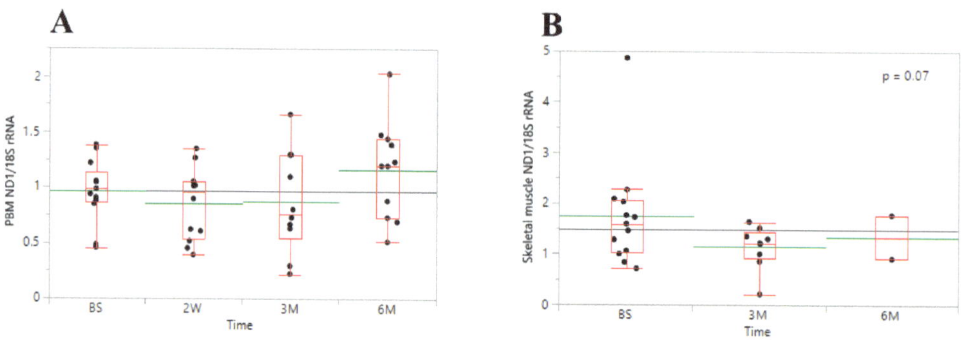

Figure 4. Effect of bariatric surgery on mtDNA abundance. (**A**)—mtDNA content in PBM prior to and follow-up post-surgery (n = 10–13) and (**B**)—mtDNA content in skeletal muscle prior to surgery and follow-up post-surgery (n = 2–13). The data are presented as boxplots, with the mean marked by green lines.

4. Discussion

Our study demonstrated that while bariatric surgery reduced BMI, it did not significantly change plasma short cf-mtDNA levels or mtDNA content in PBM cells or skeletal muscle. Surprisingly, we found that plasma from both HC and obese patients contained long (~8 kb) cf-mtDNA fragments, which cover near intact and/or whole mitochondrial genomes.

Metabolic improvement after bariatric surgery going beyond mere weight loss, and given the increasing use of bariatric surgery, it is critical to understand the physiologic basis of its metabolic benefits, including resolution of T2D. Bariatric surgery has variable effects on mitochondria, reviewed in [17], including improvement in mitochondrial respiration in PBM cells [6], liver [13], and skeletal muscle [6,7]. Additionally, bariatric surgery induced alterations in skeletal muscle expression of genes involved in calcium/lipid metabolism and mitochondrial function, which were associated with subsequent distinct methylation patterns at 52 weeks after surgery [18]. Furthermore, a recent study demonstrated that laparoscopic Roux-en-Y gastric bypass augmented autophagy/mitophagy markers and decreased mitochondrial membrane potential in leukocytes, improving mitochondrial turnover as a consequence [19].

Despite the intensive studies on mitochondrial physiology, only a few studies have examined circulatory or urine cf-mtDNA [9–11] and mtDNA/nDNA content [12,13] in obese patients after bariatric surgery. Previously, it has been shown that bariatric surgery after 6 months reduced urinary mt*ND1* and mt*COX3* copy numbers, as well as serum mt*COX3* copy numbers, only in patients with obesity with T2D [10]. Additionally, the same group demonstrated that bariatric surgery reduces the elevated urinary mtND1 copy number in obese patients with an initial high baseline mt*ND1* copy number 6 months after surgery [10]. Also, a study in obese Asian Indian patients with T2D demonstrated a significant reduction in circulatory mtDNA copy number at 6 and 12 months post-bariatric surgery [11]. In contrast with these previous reports, we did not observe a reduced level of short cf-mtDNA post-surgery. Simultaneously, we observed a significant increase in short plasma *ND6* cf-mtDNA in obese patients compared to HC (Figure 1A). Additionally, we found that the mean plasma *ND6* level was significantly higher in samples of T2D than HC (Figure 1B), which is consistent with our [14] and others previous reports for an increase in cf-mtDNA in patients with T2D [20]. Similarly, we observed a significantly higher level of plasma short cf-mtDNA in the IR in Obese BS compared to HC (Figure 1C). These findings are in line with a recent study of healthy adolescents that showed that circulating cf-DNA is associated with the risk of metabolic syndrome, not obesity per se [21]. Furthermore, this study found an association between cf-mtDNA and one of the markers of oxidative stress, advanced oxidation protein products [21]. Recently, the use of circulatory cf-DNA and cf-mtDNA has received a lot of attention due to their attractive diagnostic, prognostic, and risk assessment potential. In light of this, blood levels of cf-DNA have been found to be correlated with IR and other obesity-related cardiometabolic risk factors in clinically healthy individuals, leading to a proposal to use cf-DNA as an auxiliary risk marker for cardiometabolic disease [22]. Nevertheless, studies to determine cf-mtDNA as a potential biomarker for predicting the outcome of bariatric surgery are in the preliminary stages.

Analysis of plasma *ND6* cf-mtDNA in obese patients after surgery did not show significant changes over the time after surgery (Supplemental Figure S2), which was different from the previous reports indicating a decrease in cf-mtDNA copy number post-surgery [9–11]. The discrepancy between our results and results from previous studies can be influenced by several potential factors, including source (serum, plasma, or urine) and approaches for cf-DNA isolation, analysis, and interpretation of results. As opposed to our study, in which we performed absolute quantification of cf-mtDNA in plasma, in all previous related studies, the authors measured the relative cf-mtDNA amount, which was calculated as a ratio of cf-mtDNA/cf-nDNA in either serum or urine. Because of the relative nature of the analysis, we believe that the above-mentioned previous studies do not support the selective release/accumulation of cf-mtDNA associated with obesity or T2D and its selective decrease with weight loss after bariatric surgery. Additionally, we believe that the initial and especially post-surgery BMI can be a contributing factor to the discrepancy in the analysis as well. Also, it needs to be mentioned that even after profound weight loss our patients were still severely obese, with a BMI ~36.6 at 6 M post-surgery. We need to point out that both pre- and post-operative BMI for our patients were much greater than in related publications [9–11], which most likely contributed to the outcomes of our study. It must also be highlighted that gender differences could affect our findings. Indeed, the above-mentioned related studies have had a higher male percentage: ~41% (9) and ~37% [10,11] vs. ~31% male in our study. Interestingly, similar to our study, the decrease in body weight and fat percentage did not result in changes in cf-DNA levels six months after bariatric surgery [23]. Additionally, we need to mention that some data for 3 M and 6 M post-surgery were missing for different patients at different times, partially due to the COVID-19 pandemic-related missed visits (Table 1). This may also affect the outcomes for the dynamic for plasma *ND6* level. Regarding the association between cf-mtDNA level and markers of T2D in HC and Obese patients, we observed a significant positive correlation between HgbA1c and plasma *ND6* (Figure 2A). However, when HC and obese patient data were separated, no correlation between *ND6* and HgbA1c was observed for HC (Supplemental Figure S1), but we note that 100% of the HC had a normal HgbA1c and *ND6* level ≤ 0.3 ng/mL, and all of the obese patients at baseline had a *ND6* ≥ 0.4 ng/mL (Figure 2A). This suggests that plasma levels of *ND6* above 0.4 ng/mL are pathologic and associated with increasing levels of HgbA1c.

Analysis of long-range PCR showed that plasma from HC and Obese patients (at all time points after bariatric surgery) contained long overlapping amplicons that covered the full-length mitochondrial genome (Figure 3A,B). Therefore, not only were the short *ND6* fragments accumulated in plasma from obese patients but also the complete mitochondrial genome. The latter may be an adaptive mechanism to preserve mitochondrial function and bioenergetics (i.e., via restoration of OXPHOS), since cells may share/transfer whole healthy mitochondria and/or mtDNA to each other when they recover from the stress of obesity and diabetes. Previous studies showed that isolated mitochondria can be transferred to any cell type via simple coincubation or brief centrifugation in vitro [24,25], reviewed in [26]. Furthermore, the injection of autologous or non-autologous mitochondria has been effective in treating injury and various diseases [27,28], reviewed in [29]. In line with this, a recent paper has shown that exogenous healthy mitochondria are preferentially trafficked to cells and tissues in which mitochondria are damaged [30], which has implications for the delivery of therapeutic agents to injured or diseased sites. Whether this process happened after bariatric surgery needs further investigation. As we mentioned above, even after 6 M post-surgery, patients were still morbidly obese, although their metabolic parameters, such as HOMA-IR and HgbA1c, improved significantly. While it needs to be interpreted with limitations noted in the Results part, we can speculate that the observed increase in long cf-mtDNA may represent an adaptive response in order to compensate for the cells' bioenergetic needs that ultimately provide improvement in metabolic parameters and IR. The main unanswered scientific question here is: does bariatric surgery stimulate adaptive responses through mitochondria and/or mtDNA transfer? In this context, most likely, the possibility of using quantitative analysis of cf-mtDNA as an indicator of bariatric surgery results prognostic needs further investigation, with greater sample size and improvement and standardization of methods of detection.

To the best of our knowledge, this is the first report in which the accumulation of long plasma mtDNA fragments was analyzed in obesity and bariatric surgery research. Numerous previous studies have demonstrated mitochondrial transfer [24,25], reviewed in [26], between mammalian cells in culture and in vivo in its application as a therapeutic tool for treating injury and various diseases [27,28], reviewed in [29]. The study by Dache et al. reported that blood preparation with resting platelets contains respiratory-competent cell-free mitochondria in their normal physiological state [31]. Additionally, the authors showed that normal and tumor-cultured cells are able to secrete their mitochondria [31]. Furthermore, a few studies demonstrated a horizontal transfer of mtDNA [32,33]. Sansone et al. demonstrated that the horizontal transfer of mtDNA from extracellular vesicles acts as an oncogenic signal, promoting an exit from dormancy in hormonal therapy-resistant breast cancer [32]. A more recent study demonstrated that chimeras composed of cells with wild-type and mutant mtDNA exhibited increased trafficking of wild-type mtDNA to mutant cells, suggesting that horizontal mtDNA transfer may be a compensatory mechanism to restore compromised mitochondrial function [33]. A recent review discussed that circulating cf-mtDNA in its physiological forms in human blood is unlikely to be pro-inflammatory [34], and its relevance and physiological role remain to be established. Likewise, the origin and specificity of our findings of plasma long cf-mtDNA fragments for bariatric surgery-related improvement in metabolic parameters and IR are unknown and will require further testing.

Interestingly, mtDNA content tended to be slightly increased in PBM cells (Figure 4A), but decreased in skeletal muscle at 6 M after surgery compared to BS (Figure 4B), suggesting a tissue-specific effect of bariatric surgery on mtDNA content. Our finding is in line with a recent study showing that obesity has an opposing association with mitochondrial respiration in adipose and liver tissue with no overall association with NAFLD severity; however, bariatric surgery increased hepatic mtDNA/nDNA content 12 months after bariatric surgery, which appeared to be driven primarily by a substantial increase in mitochondrial biogenesis [13]. Additionally, it was reported that the copy number of mtDNA in various fat stores was higher in obese patients with T2D than in obese patients without diabetes or in control subjects [35]. This finding [35] was different from our study since mtDNA content in PBM cells was slightly increased in non-diabetics compared to both HC and T2D patients, although the difference was not significant (Supplemental Figure S4D). When compared to HC, mtDNA content in PBM cells was also slightly greater, regardless of the patient's T2D or MetS status (Supplemental Figure S4B). A significant positive correlation was observed between plasma levels of plasma *ND6* and mtDNA content in PBM cells 6 M post-surgery (Figure 2C).

A limitation of this study was the relatively small sample size, the missing data for different patients at different times, and the short time of follow-up post-surgery. Future studies with a larger sample size and designed to address limits and methodological issues related to the isolation and analysis of cf-mtDNA are needed to generalize these results. Regarding technical limitations, we need to highlight that the current methods of cf-mtDNA quantitation in plasma are time-consuming, lack reproducibility, and do not indicate selective release and/or accumulation of cf-mtDNA in plasma. Variations in DNA isolation methods, normalization of cf-mtDNA to cf-nDNA, or generation of a standard curve altogether limit the reproducibility or clinical utility of mtDNA analysis in previous reports. In this context, a recent report using droplet digital PCR provides absolute DNA quantification without the need for a standard curve. Importantly, this method is more sensitive than conventional qRT-PCR at low concentrations of cf-mtDNA, making it feasible to analyze DNA in plasma directly without a DNA isolation step [36].

In conclusion, we found that bariatric surgery decreased weight and improved insulin sensitivity and other metabolic parameters in obese patients without significant changes in plasma short cf-mtDNA levels or cellular mtDNA content. However, a significant correlation between preop *ND6* levels and HgbA1c and *ND6* levels and insulin resistance (HOMA-IR) at 2 weeks post-op suggests that one of the potential drivers of insulin resistance may be short cf-mtDNA fragments such as *ND6*. These findings, along with the presence of long cf-mtDNA fragments, may underlie the clinical importance of cf-mtDNA as a mechanism explaining the effects of bariatric surgery on the improvement of insulin sensitivity. Further studies need to test the structure, potential characteristics, determinants, and origin of cf-mtDNA associated with metabolic improvements following surgery.

Supplementary Materials: The following supporting information can be downloaded at: https://www.mdpi.com/article/10.3390/biomedicines11092514/s1, Figure S1: Correlation between HgbA1c and plasma *ND6* levels prior to surgery with patients separated as HC and Obese, BS (n = 7, HC; n = 11, Obese, BS); Figure S2: Effect of bariatric surgery on plasma cf-mtDNA. *ND6* level in obese patients prior to surgery and at the indicated time points follow up post-surgery. Data are presented as boxplots with mean marked by green lines, n = 6–9; Figure S3: Schematic of location of primers and long overlapping mtDNA sequences (Mito1 and Mito2) amplified whole mitochondrial genome during long-range PCR; Figure S4: Effect of obesity, MetS and T2D on the mtDNA abundance in PBM cells. Data are presented as boxplots with mean marked by green lines. (**A**) mtDNA content in PBM cells from Obese patients and HC (n = 8–13). (**B–D**) mtDNA abundance in PBM cells demonstrated by group (**B**) data in Obese, BS group were plotted according to the MetS, T2D or none (n = 2–8); (**C**) data in Obese, BS group were plotted according to the IR or none (no IR); n = 2–8 subjects per group; (**D**) data in Obese, BS group were plotted according to the T2D or non-T2D (n = 6–8); Table S1: Primers for mtDNA and nDNA; Table S2: Summary of clinical, metabolic, cf-mtDNA data, and mtDNA content of patients, HC-Obese, BS group; Table S3: Comparison of clinical and metabolic parameters, plasma cf-mtDNA data, and mtDNA content among type of bariatric surgery; Table S4: Comparison of *ND6* and PBM *ND1/18SrRNA* in HC and Obese, BS by factor: MetS, T2D or none; Table S5: Comparison of *ND6* and PBM *ND1/18SrRNA* in HC and Obese, BS group where

Obese group is classified as IR or none; Table S6: Comparison of plasma *ND6* in Obese group by MetS, T2D, or none.

Author Contributions: Conceptualization, W.O.R. and L.I.R.; Data curation, K.L. and L.I.R.; Formal analysis, V.M.P., M.S.M. and L.I.R.; Funding acquisition, M.S.M. and L.I.R.; Investigation, L.V.Y., V.M.P., W.O.R. and L.I.R.; Writing—original draft, L.V.Y. and L.I.R.; Writing—review and editing, L.V.Y., V.M.P., M.S.M., K.L., W.O.R. and L.I.R. All authors have read and agreed to the published version of the manuscript.

Funding: This study was supported in part by University of South Alabama College of Medicine Intramural Grants Program, Award #807 (L.I.R.). Statistical analysis reported in this publication by Dr. Mulekar was supported by the National Center for Advancing Translational Sciences of the National Institutes of Health under award number UL1TR001417.

Institutional Review Board Statement: The study was conducted in accordance with the Declaration of Helsinki and approved by the Institutional Review Board of the University of South Alabama (protocol #1253267, "Effect of bariatric surgery mtDNA DAMP's, inflammation & diabetic outcome", date of approval 12 June 2018, project identification code 18-197).

Informed Consent Statement: Informed consent was obtained from all subjects involved in the study.

Data Availability Statement: All relevant data are within the manuscript and its Supplementary Files.

Acknowledgments: Part of this study was presented in the abstract form at the 83rd Scientific Sessions of the American Diabetes Association, San Diego, California, 23–26 June 2023.

Conflicts of Interest: The authors declare no conflict of interest. The funders had no role in the design of the study; in the collection, analyses, or interpretation of data; in the writing of the manuscript, or in the decision to publish the results.

Abbreviations

BMI, body mass index; cf, cell-free; BS, before surgery; HC, healthy controls; HDL, high-density lipoprotein; HOMA-IR, homeostasis model assessment for insulin resistance; LDL, low-density lipoprotein; LSG, laparoscopic sleeve gastrectomy; LRYGB, laparoscopic Roux-en-Y gastric bypass; MetS, metabolic syndrome; mtDNA, mitochondrial DNA; n-number of patients; nDNA, nuclear DNA; PBM, peripheral blood mononuclear; PBS, phosphate- buffered saline; Pt, patients; T2D, type 2 diabetes.

References

1. Reaven, G.M. Insulin resistance: The link between obesity and cardiovascular disease. *Med. Clin. N. Am.* **2011**, *95*, 875–892. [CrossRef] [PubMed]
2. Hruby, A.; Manson, J.A.E.; Qi, L.; Malik, V.S.; Rimm, E.B.; Sun, Q.; Walter, C.W.; Frank, B.H. Determinants and consequences of obesity. *Am. J. Public Health* **2016**, *106*, 1656–1662. [CrossRef]
3. Bonnard, C.; Durand, A.; Peyrol, S.; Chanseaume, E.; Chauvin, M.A.; Morio, B.; Vidal, H.; Rieusset, J. Mitochondrial dysfunction results from oxidative stress in the skeletal muscle of diet-induced insulin-resistant mice. *J. Clin. Investig.* **2008**, *118*, 789–800. [CrossRef]
4. Anderson, E.J.; Lustig, M.E.; Boyle, K.E.; Woodlief, T.L.; Kane, D.A.; Lin, C.T.; Price, J.W., III; Kang, L.; Rabinovitch, P.S.; Szeto, H.H.; et al. Mitochondrial H_2O_2 emission and cellular redox state link excess fat intake to insulin resistance in both rodents and humans. *J. Clin. Investig.* **2009**, *119*, 573–581. [CrossRef]
5. Hoehn, K.L.; Salmon, A.B.; Hohnen-Behrens, C.; Turner, N.; Hoy, A.J.; Maghzal, G.J.; Stocker, R.; Van Remmen, H.; Kraegen, E.W.; Cooney, G.J.; et al. Insulin resistance is a cellular antioxidant defense mechanism. *Proc. Natl. Acad. Sci. USA* **2009**, *106*, 17787–17792. [CrossRef] [PubMed]
6. Nijhawan, S.; Richards, W.; O'Hea, M.F.; Audia, J.P.; Alvarez, D.F. Bariatric surgery rapidly improves mitochondrial respiration in morbidly obese patients. *Surg. Endosc.* **2013**, *27*, 4569–4573. [CrossRef]
7. Fernstrom, M.; Bakkman, L.; Loogna, P.; Rooyackers, O.; Svensson, M.; Jakobsson, T.; Brandt, L.; Lagerros, Y.T. Improved Muscle Mitochondrial Capacity Following Gastric Bypass Surgery in Obese Subjects. *Obes. Surg.* **2016**, *26*, 1391–1397. [CrossRef]
8. Toledo, F.G.; Menshikova, E.V.; Ritov, V.B.; Azuma, K.; Radikova, Z.; DeLany, J.; Kelley, D.E. Effects of physical activity and weight loss on skeletal muscle mitochondria and relationship with glucose control in type 2 diabetes. *Diabetes* **2007**, *56*, 2142–2147. [CrossRef]

9. Lee, H.; Oh, S.; Yang, W.; Park, R.; Kim, H.; Jeon, J.S.; Noh, H.; Han, D.C.; Cho, K.W.; Kim, Y.J.; et al. Bariatric surgery reduces elevated urinary mitochondrial DNA copy number in patients with obesity. *J. Clin. Endocrinol. Metab.* **2019**, *104*, 2257–2266. [CrossRef] [PubMed]
10. Seo, M.; Kim, H.; Noh, H.; Jeon, J.S.; Byun, D.W.; Kim, S.H.; Kim, H.J.; Suh, K.; Park, H.K.; Kwon, S.H. Effect of bariatric surgery on circulating and urinary mitochondrial DNA copy numbers in obesity with or without diabetes. *BMJ Open Diabetes Res. Care* **2020**, *8*, e001372. [CrossRef]
11. Chandru, S.; Prabhu, P.; Balasubramanyam, M.; Subhashini, R.; Tiwaskar, M.; Pramodkumar, T.A.; Pradeepa, R.; Anjana, R.M.; Mohan, V. Beneficial Primary Outcomes of Metabolic Surgery with Changes in Telomere Length and Mitochondrial DNA in Obese Asian Indians with Dysglycemia. *J. Assoc. Physicians Ind.* **2021**, *69*, 11–12.
12. Skuratovskaia, D.; Litvinova, L.; Vulf, M.; Zatolokin, P.; Popadin, K.; Mazunin, I. From normal to obesity and back: The associations between mitochondrial DNA copy number, gender, and body mass index. *Cells* **2019**, *8*, 430. [CrossRef]
13. Pedersen, J.S.; Rygg, M.O.; Chrøis, K.; Sustarsic, E.G.; Gerhart-Hines, Z.; Wever Albrechtsen, N.J.; Serizawa, R.R.; Kristiansen, V.B.; Basse, A.L.; Boilesen, A.E.B.; et al. Influence of NAFLD and bariatric surgery on hepatic and adipose tissue mitochondrial biogenesis and respiration. *Nat. Commun.* **2022**, *13*, 2931. [CrossRef] [PubMed]
14. Yuzefovych, L.V.; Pastukh, V.M.; Ruchko, M.V.; Simmons, J.D.; Richards, W.O.; Rachek, L.I. Plasma mitochondrial DNA is elevated in obese type 2 diabetes patients and correlates positively with insulin resistance. *PLoS ONE* **2019**, *14*, e0222278. [CrossRef] [PubMed]
15. Huang, P.L. A comprehensive definition for metabolic syndrome. *Dis. Model Mech.* **2009**, *2*, 231–237. [CrossRef]
16. Simmons, J.D.; Lee, Y.L.; Mulekar, S.; Kuck, J.L.; Brevard, S.B.; Gonzalez, R.P.; Gillespie, M.N.; Richards, W.O. Elevated levels of plasma mitochondrial DNA DAMPs are linked to clinical outcome in severely injured human subjects. *Ann. Surg.* **2013**, *258*, 591–596. [CrossRef]
17. Dankel, S.N.; Staalesen, V.; Bjørndal, B.; Berge, R.K.; Mellgren, G.; Burri, L. Tissue-specific effects of bariatric surgery including mitochondrial function. *J. Obes.* **2011**, *2011*, 435245. [CrossRef]
18. Gancheva, S.; Ouni, M.; Jelenik, T.; Koliaki, C.; Szendroedi, J.; Toledo, F.G.S.; Markgraf, D.F.; Pesta, D.H.; Mastrototaro, L.; De Filippo, E.; et al. Dynamic changes of muscle insulin sensitivity after metabolic surgery. *Nat. Commun.* **2019**, *10*, 4179. [CrossRef] [PubMed]
19. Abad-Jiménez, Z.; López-Domènech, S.; García-Gargallo, C.; Vezza, T.; Gómez-Abril, S.Á.; Morillas, C.; Díaz-Pozo, P.; Falcón, R.; Bañuls, C.; Víctor, V.M.; et al. Roux-en-Y Gastric Bypass Modulates AMPK, Autophagy and Inflammatory Response in Leukocytes of Obese Patients. *Biomedicines* **2022**, *10*, 430. [CrossRef]
20. Liu, J.; Zou, Y.; Tang, Y.; Xi, M.; Xie, L.; Zhang, Q.; Gong, J. Circulating cell-free mitochondrial deoxyribonucleic acid is increased in coronary heart disease patients with diabetes mellitus. *J. Diabetes Investig.* **2016**, *7*, 109–114. [CrossRef]
21. Celec, P.; Janovičová, Ĺ.; Gurecká, R.; Koborová, I.; Gardlík, R.; Šebeková, K. Circulating extracellular DNA is in association with continuous metabolic syndrome score in healthy adolescents. *Physiol. Genom.* **2021**, *53*, 309–318. [CrossRef]
22. Jylhävä, J.; Lehtimäki, T.; Jula, A.; Moilanen, L.; Kesäniemi, Y.A.; Nieminen, M.S.; Kähönen, M.; Hurme, M. Circulating cell free DNA is associated with cardiometabolic risk factors: The health 2000 survey. *Atherosclerosis* **2014**, *233*, 268–271. [CrossRef] [PubMed]
23. Camuzi Zovico, P.V.; Gasparini Neto, V.H.; Venâncio, F.A.; Soares Miguel, G.P.; Graça Pedrosa, R.; Kenji Haraguchi, F.; Barauna, V.G. Cell-free DNA as an obesity biomarker. *Physiol. Res.* **2020**, *69*, 515–520. [CrossRef] [PubMed]
24. Kesner, E.E.; Saada-Reich, A.; Lorberboum-Galski, H. Characteristics of Mitochondrial Transformation into Human Cells. *Sci. Rep.* **2016**, *6*, 26057. [CrossRef]
25. Kim, M.J.; Hwang, J.W.; Yun, C.K.; Lee, Y.; Choi, Y.S. Delivery of exogenous mitochondria via centrifugation enhances cellular metabolic function. *Sci. Rep.* **2018**, *8*, 3330. [CrossRef]
26. Berridge, M.V.; McConnell, M.J.; Grasso, C.; Bajzikova, M.; Kovarova, J.; Neuzil, J. Horizontal transfer of mitochondria between mammalian cells: Beyond co-culture approaches. *Curr. Opin. Genet. Dev.* **2016**, *38*, 75–82. [CrossRef]
27. Alexander, J.F.; Seua, A.V.; Arroyo, L.D.; Ray, P.R.; Wangzhou, A.; Heibeta-Luckemann, L.; Schedlowski, M.; Price, T.J.; Kavelaars, A.; Heijnen, C.J. Nasal administration of mitochondria reverses chemotherapy-induced cognitive deficits. *Theranostics* **2021**, *11*, 3109–3130. [CrossRef] [PubMed]
28. Chang, J.C.; Wu, S.L.; Liu, K.H.; Chen, Y.H.; Chuang, C.S.; Cheng, F.C.; Su, H.L.; Wei, Y.H.; Kuo, S.J.; Liu, C.S. Allogeneic/xenogeneic transplantation of peptide-labeled mitochondria in Parkinson's disease: Restoration of mitochondria functions and attenuation of 6-hydroxydopamine-induced neurotoxicity. *Transl. Res.* **2016**, *170*, 40–56.e3. [CrossRef]
29. Valenti, D.; Vacca, R.A.; Moro, L.; Atlante, A. Mitochondria Can Cross Cell Boundaries: An Overview of the Biological Relevance, Pathophysiological Implications and Therapeutic Perspectives of Intercellular Mitochondrial Transfer. *Int. J. Mol. Sci.* **2021**, *22*, 8312. [CrossRef]
30. Lee, S.-E.; Kang, Y.C.; Kim, Y.; Kim, S.; Yu, S.-H.; Park, J.H.; Kim, I.-H.; Kim, H.-Y.; Han, K.; Lee, H.K.; et al. Preferred Migration of Mitochondria toward Cells and Tissues with Mitochondrial Damage. *Int. J. Mol. Sci.* **2022**, *23*, 15734. [CrossRef]
31. Al Amir Dache, Z.; Otandault, A.; Tanos, R.; Pastor, B.; Meddeb, R.; Sanchez, C.; Arena, G.; Lasorsa, L.; Bennett, A.; Grange, T.; et al. Blood contains circulating cell-free respiratory competent mitochondria. *FASEB J.* **2020**, *34*, 3616–3630. [CrossRef]

32. Sansone, P.; Savini, C.; Kurelac, I.; Chang, Q.; Amato, L.B.; Strillacci, A.; Stepanova, A.; Iommarini, L.; Mastroleo, C.; Daly, L.; et al. Packaging and transfer of mitochondrial DNA via exosomes regulate escape from dormancy in hormonal therapy-resistant breast cancer. *Proc. Natl. Acad. Sci. USA* **2017**, *114*, E9066–E9075. [CrossRef] [PubMed]
33. Gutierrez, N.M.; Mikhalchenko, A.; Ma, H.; Koski, A.; Li, Y.; Van Dyken, C.; Tippner-Hedges, R.; Yoon, D.; Liang, D.; Hayama, T.; et al. Horizontal mtDNA transfer between cells is common during mouse development. *iScience* **2022**, *25*, 103901. [CrossRef] [PubMed]
34. Picard, M.; Shirihai, O.S. Mitochondrial signal transduction. *Cell Metab.* **2022**, *34*, 1620–1653. [CrossRef]
35. Litvinova, L.; Zatolokin, P.; Vulf, M.; Mazunin, I.; Skuratovskaia, D. The relationship between the mtDNA copy number in insulin-dependent tissues and markers of endothelial dysfunction and inflammation in obese patients. *BMC Med. Genom.* **2019**, *12*, 41. [CrossRef]
36. Hepokoski, M.L.; Odish, M.; Lam, M.T.; Coufal, N.G.; Rolfsen, M.L.; Shadel, G.S.; Moyzis, A.G.; Sainz, A.G.; Takiar, P.G.; Patel, S.; et al. Absolute quantification of plasma mitochondrial DNA by droplet digital PCR marks COVID-19 severity over time during intensive care unit admissions. *Am. J. Physiol. Lung Cell. Mol. Physiol.* **2022**, *323*, L84–L92. [CrossRef] [PubMed]

biomedicines

MDPI

Article

HOMA-IR as a Predictor of PAI-1 Levels in Women with Severe Obesity

Fabiana Martins Kattah [1,*,†], Milijana Janjusevic [2,†], Nayra Figueiredo [3], Emilly Santos Oliveira [1], Glaucia Carielo Lima [1], Ana Raimunda Dâmaso [4], Lila Missae Oyama [4], Alessandra Lucia Fluca [2], Paulo Reis Eselin de Melo [5], Maria Aderuza Horst [1], Aneta Aleksova [2,‡] and Flávia Campos Corgosinho [1,3,*,‡]

1 Faculty of Nutrition, Federal University of Goias, Goiânia 74605-080, Brazil; emillysantos@discente.ufg.br (E.S.O.); glauciacarielo@ufg.br (G.C.L.); aderuza@ufg.br (M.A.H.)
2 Cardiothoracovascular Department, Azienda Sanitaria Universitaria Giuliano Isontina (ASUGI), Department of Medical Surgical and Health Science, University of Trieste, 34149 Trieste, Italy; mjanjusevic@units.it (M.J.); alessandralucia.fluca@units.it (A.L.F.); aaleksova@units.it or aaleksova@gmail.com (A.A.)
3 Faculty of Medicine, Federal University of Goias, Goiania 74605-080, Brazil; nayrafgrdo@hotmail.com
4 Paulista Medicine School, Federal University of São Paulo, São Paulo 04023-062, Brazil; ana.damaso@unifesp.br (A.R.D.); lmoyama@unifesp.br (L.M.O.)
5 Alberto Rassi Hospital, Goiânia 74110-010, Brazil; prcirurgia@icloud.com
* Correspondence: fabianakattah@discente.ufg.br (F.M.K.); flaviacorgosinho@ufg.br (F.C.C.); Tel.: +55-(62)-3209-6270 (F.M.K. & F.C.C.)
† These authors contributed equally to this work.
‡ These authors contributed equally to this work.

Abstract: Background: Obesity is a chronic inflammatory disorder that increases the risk of cardiovascular diseases (CVDs). Given the high CVD mortality rate among individuals with obesity, early screening should be considered. Plasminogen activator inhibitor (PAI-1), a cytokine that links obesity and CVDs, represents a promising biomarker. However, PAI-1 is not part of the clinical routine due to its high cost. Therefore, it is necessary to find good predictors that would allow an indirect assessment of PAI-1. Methods: This study enrolled 47 women with severe obesity (SO). The obtained anthropometric measurements included weight, height, neck (NC), waist (WC), and hip circumference (HC). Blood samples were collected to analyse glucose and lipid profiles, C-reactive protein, liver markers, adiponectin, and PAI-1 (determined by ELISA immunoassay). Homeostasis model assessment-adiponectin (HOMA-AD), homeostasis model assessment of insulin resistance (HOMA-IR), quantitative insulin sensitivity check index (QUICKI), triglyceride–glucose index (TyG), and atherogenic index of plasma (AIP) were calculated. The women were grouped according to PAI-1 levels. The data were analysed using IBM SPSS Statistics, version 21. The significance level for the analysis was set at 5%. Results: Women with SO who have higher levels of PAI-1 have lower values of high-density lipoprotein cholesterol (HDL) ($p = 0.037$) and QUICKI (0.020) and higher values of HOMA-AD (0.046) and HOMA-IR (0.037). HOMA-IR was demonstrated to be a good predictor of PAI-1 in this sample ($B = 0.2791$; $p = 0.017$). Conclusions: HOMA-IR could be used as a predictor of PAI-1 levels, pointing out the relevance of assessing glycaemic parameters for the prevention of CVDs in women with SO.

Keywords: PAI-1; cardiovascular disease; inflammation; severe obesity; HOMA-IR

Citation: Martins Kattah, F.; Janjusevic, M.; Figueiredo, N.; Santos Oliveira, E.; Carielo Lima, G.; Dâmaso, A.R.; Oyama, L.M.; Fluca, A.L.; de Melo, P.R.E.; Aderuza Horst, M.; et al. HOMA-IR as a Predictor of PAI-1 Levels in Women with Severe Obesity. *Biomedicines* **2024**, *12*, 1222. https://doi.org/10.3390/biomedicines12061222

Academic Editors: Teresa Vezza and Zaida Abad-Jiménez

Received: 23 April 2024
Revised: 24 May 2024
Accepted: 27 May 2024
Published: 31 May 2024

1. Introduction

Obesity, a health-threatening condition characterised by an excessive accumulation of body fat [1], is increasing in prevalence worldwide across all ages, genders, nationalities, and socioeconomic status [2]. Obesity is associated with an increase in cardiometabolic risk through effects on cardiovascular structure, the promotion of a pro-inflammatory state with alterations in cytokine secretion patterns, and the emergence of other metabolic disorders [3]. More precisely, obesity is associated with a worse lipid panel, increased blood pressure, impaired plasma glucose levels, type 2 diabetes mellitus (T2DM), liver dysfunction, and low levels of cardiorespiratory fitness parameters, factors that contribute to cardiovascular diseases (CVDs) [4].

One of the cytokines whose expression is elevated in obesity and increases the risk of CVDs is the plasminogen activator inhibitor (PAI-1), a regulator of fibrinolysis that acts on thrombogenic pathways [5]. The main activity of PAI-1 is to inhibit both the tissue and urokinase plasminogen activators, which are responsible for the cleavage of plasminogen to plasmin [6]. The processes of fibrinogenesis and fibrinolysis are important in both intravascular and extravascular physiology and the pathology of CVDs [7]. However, the role of PAI-1 is not limited to the control of fibrinolysis, as it is also involved in the control of tissue remodelling, angiogenesis, inflammation, and extracellular matrix degradation [6]. By controlling this plethora of mechanisms, PAI-1 has previously been reported to participate in the pathophysiology of several metabolic syndromes, including obesity and insulin resistance.

The interplay between PAI-1 and glycaemic control appears to be related to hyperinsulinemia promoting PAI-1 expression as well as reducing the rate of mRNA degradation of PAI-1, supporting protein production [8,9]. Furthermore, it has been hypothesised that insulin resistance decreases the activity of the PI3-K/Akt pathway while upregulating the mitogen-activated protein kinase/extracellular signal-regulated kinase (MAPK/ERK) pathway, favouring the release of inflammatory markers, among them PAI-1 [10]. Also, given that individuals with atherosclerotic plaque have increased PAI-1 expression [11] and that PAI-1 concentrations are associated with the severity of subclinical atherosclerosis in patients with obesity, PAI-1 quantification may be an additional tool to identify patients with obesity at higher risk of developing CVDs [11].

However, despite PAI-1 being a good predictor, the high cost of measuring this adipokine prevents its use in routine clinical practice. Therefore, it is important to explore other parameters that can be predictors of PAI-1. The atherogenic index of plasma (AIP) is independently correlated with a higher incidence of coronary heart disease [12] and appears to be associated with PAI-1 levels in individuals with severe obesity; however, further studies are needed to confirm this possible relationship [11]. The triglyceride (TG)–high-density lipoprotein cholesterol (HDL-c) ratio, which is a component of the AIP equation, could be a parameter for the identification of individuals with severe obesity at risk of developing metabolic syndrome (MS) [13]. Furthermore, although the triglyceride–glucose index (TyG) is also used as a cardiometabolic risk marker and may be used as a marker of atherosclerosis [14], no studies have evaluated its correlation with PAI-1 levels. In addition, the homeostasis model assessment of insulin resistance (HOMA-IR) and quantitative insulin sensitivity check index (QUICKI) are usually performed to determine insulin resistance as a TyG, but little is known about the relationship between those markers and PAI-1 values as well. Thus, this study aimed to verify the correlation of PAI-1 with cardiometabolic risk markers in women with severe obesity, seeking to evaluate predictors of PAI-1 levels. Thus, the hypothesis of this study is that some of those indexes can predict PAI-1 in order to better screen patients with higher cardiovascular risk.

2. Material and Methods

2.1. Participants

The study population consisted of 47 women with a body mass index (BMI) above 40 kg/m^2 (severe obesity) who were hospitalised for bariatric surgery at the Hospital Estadual Geral de Goiânia Dr Alberto Rassi (HGG), Goiânia, GO, Brazil. BMI was calculated by dividing the person's weight by the square of their height (in metres) [15]. The research team obtained a list of patients eligible for bariatric surgery from the Obesity Surgery Control Program (PCCO). During the first outpatient consultation at the HGG, the researcher explained the aim of the project to patients who met the inclusion criteria, and informed consent forms were signed in duplicate. Data collected for this study included age, date of birth, medication use, the presence of comorbidities such as T2DM, hypertension, thyroid dysfunction, and anthropometric measurements. In addition, blood samples were collected from all participants.

Non-inclusion criteria were participants younger than 20 years or older than 59 years with acute inflammatory, infectious, or neoplastic diseases, genetic syndromes, rheumatic and autoimmune diseases, fibromyalgia, chronic alcohol consumption (>30 g/d), or abuse of illicit/psychotropic drugs.

This study was carried out according to the principles of the Declaration of Helsinki and was approved by the Ethics Committees of the Universidade Federal de Goiás and the hospital (3,251,178 and 961/19, respectively).

2.2. Anthropometric Measurements

The anthropometric assessment was given by the mean value of two measurements of weight, height, hip, waist, and neck circumference. Weight was measured using the Lider scale with a maximum capacity of 200 kg, with the volunteer standing in the centre of the scale wearing light clothes and no shoes. Height was measured with the patient in an upright position, barefoot, looking forward, and with arms outstretched at the sides, using a Fillizola scale available at the hospital. BMI was calculated by dividing the person's weight by the square of their height (in metres), and obesity was classified as grade I (30 to 34.99 kg/m^2), grade II (35 to 39.99 kg/m^2), and grade III ($\geq$40 kg/m^2). The waist circumference was measured at the level of the umbilical line while the volunteer was standing. The neck circumference was measured below the level of the cricoid cartilage. All measurements were taken by a trained researcher.

2.3. Blood Analysis

Blood sampling was performed by peripheral vein puncture of the forearm by trained nurses after a 12 h overnight fast. The collection was carried out in the laboratory of Atalaia Medicina Diagnóstica, Goiânia—GO. Biochemical analysis was performed with a colorimetric enzymatic method, specific for each dose (insulin, blood glucose, glycated haemoglobin A1c (HbA1c), lipid profile, and ultra-high sensitive C-reactive protein (hs-CRP), according to the laboratory). For additional analysis, EDTA tubes containing samples from each participant were transported in a thermal box to the Clinical Nutrition Research Laboratory and Sports (LABINCE), located at the Faculty of Nutrition (FANUT) of the Federal University of Goiás (UFG). After centrifugation, the serum was stored at −80 °C until use. PAI-1 and adiponectin levels were determined by enzyme-linked immunosorbent assay (ELISA) using a commercial kit (R&D Systems, Minnesota, EUA) according to the manufacturer's instructions performed at the Laboratory of Nutrition Physiology of the Federal University of São Paulo (UNIFESP).

2.4. Index Calculation

The AIP was calculated from the logarithm of the triglyceride–high-density lipoprotein cholesterol ratio (TG/HDL-c ratio) [12]. The HOMA-IR was obtained with the formula: fasting insulin (μUI/L) × blood glucose (mg/dL)/22.5 [16]. The homeostasis model assessment-beta (HOMA-beta) and QUICKI were calculated from blood glucose and insulin values as reported in the literature [16,17]. The homeostasis model assessment-adiponectin (HOMA-AD) was calculated as the product of fasting insulin (μUI/L) and blood glucose (mg/dL), divided by adiponectin (mg/mL) [18]. TyG was calculated by Ln (fasting triglycerides (mg/dL) × fasting blood glucose (mg/dL)/2) [19]. As there is no reference value for PAI-1 in severe obesity in the literature, the median value of PAI-1 in our cohort (21 ng/mL) was used for study purposes. Values less than 4 μg/mL for adiponectin were considered hypoadiponectinemia [20].

2.5. Statistical Analysis

Statistical analysis was performed using the Statistical Package for the Social Sciences (SPSS), version 24.0. The data were first assessed for normality using the Shapiro–Wilk test. Non-normally distributed variables were standardised using the Z-score and are presented as the mean ± standard deviation. Volunteers were grouped according to the median PAI-1 value, as is usually performed when there is a lack of a cut-off literature value for the marker of interest [21,22]. Pearson's or Spearman's correlation analysis was performed to assess correlations between the studied variables, as appropriate. To compare the difference in means, the *t*-test for independent samples was used. The chi-square test was performed to compare the frequencies of pathologies in the groups. A bivariate logistic regression model was performed after the comparison of averages. The calculation of post hoc sample power was performed using GPower (version 3.0) to determine the effect size. We considered a binomial distribution, an odds ratio of 1.32; Pr(Y = 1) H0: 0.05; α err prob: 0.05; R^2 other X: 0.232; X parm π of 0.5; and a total sample of 47 individuals, obtaining a sample power of 5.26%. A p value ≤ 0.05 was considered statistically significant.

3. Results

3.1. Characteristics of the Population

Forty-seven women were enrolled in this study. At baseline, the mean age was 40 ± 8.3 years old, and all volunteers had a BMI compatible with severe obesity. Neck and abdominal circumferences were higher than recommended, indicating a high cardiovascular risk. Moreover, the mean blood pressure classifies this sample as hypertension stage 1 [23]. Mean blood glucose values, insulin, HOMA-IR, HOMA-beta, and HbA1c also indicate the presence of cardiometabolic risk and insulin resistance. The mean AIP was 0.47 ± 0.21. The baseline characteristics of the patients are presented in Table 1.

Table 1. Descriptive analyses of women with severe obesity.

Variable	Total Sample	The group with Lower Levels of PAI-1 (≤21 ng/mL)	The Group with Higher Levels of PAI-1 (>21 ng/mL)	*p*-Value
Age (years)	40.08 ± 8.30	41.64 ± 8.27	38.46 ± 8.31	0.186
Height (m)	1.59 ± 0.06	1.58 ± 0.06	1.60 ± 0.06	0.276
Weight (kg)	122.68 ± 18.86	123.84 ± 14.98	121.57 ± 22.23	0.486
BMI (kg/m^2)	48.23 ± 6.77	49.34 ± 5.87	47.16 ± 7.51	0.172
NC (cm)	41.58 ± 3.18	41.68 ± 3.28	41.88 ± 0.16	0.832
WC (cm)	130.77 ± 12.38	132.41 ± 10.46	129.17 ± 14.02	0.376
HC (cm)	144.09 ± 13.22	145.00 ± 11.64	143.20 ± 14.77	0.480
Fasting Glucose (mg/dL)	117.72 ± 50.57	110.39 ± 37.42	126.15 ± 62.38	0.312
Insulin (μUI/mL)	26.36 ± 12.76	23.33 ± 10.76	29.70 ± 14.17	0.107
HOMA-IR	7.37 ± 3.90	6.19 ± 3.51	8.65 ± 4.06	0.037 *
HOMA-beta	262.18 ± 182.90	240.36 ± 132.3	286.18 ± 184.5	0.697
QUICKI	0.295 ± 0.022	0.303 ± 0.023	0.287 ± 0.019	0.020 *
HOMA-AD	472.21 ± 368.02	362.07 ± 274	587.87 ± 422.6	0.046 *
TG (mg/dL)	145.34 ± 67.14	140.75 ± 49.8	150.85 ± 84.5	0.482

Table 1. *Cont.*

Variable	Total Sample	The group with Lower Levels of PAI-1 (≤21 ng/mL)	The Group with Higher Levels of PAI-1 (>21 ng/mL)	*p*-Value
TC (mg/dL)	176.14 ± 31.28	179.42 ± 31.5	172.20 ± 49, 31.2	0.453
HDL-c (mg/dL)	45.86 ± 9.78	48.58 ± 9.87	42.60 ±8.82	0.042 *
LDL-c (mg/dL)	105.01± 28.65	105.90 ± 31.9	103.95 ± 24.9	0.811
VLDL-c (mg/dL)	25.26 ± 8.80	24.79 ± 6.51	25.80 ± 11.02	0.568
ALT (U/L)	18.12 ± 8.23	15.86 ± 5.80	20.60 ± 9.83	0.062
AST (U/L)	17.43 ± 7.03	15.77 ± 4.09	19.25 ± 9.02	0.108
hsCRP (mg/dL)	1.28 ± 1.05	1.34 ± 0.95	1.20 ± 1.16	0.653
HbA1c (%)	6.37 ± 1.37	6.26 ± 1.30	6.49 ± 1.47	0.627
EAG (mg/dL)	136.09 ± 39.40	133.18 ± 37.35	139.30 ± 42.27	0.621
Adiponectin (μg/mL)	7.84 ± 5.25	8.60 ± 6.69	7.05 ± 3.10	0.245
PAI-1 (ng/mL)	22.95 ± 12.09	15.58 ± 4.44	27.24 ± 5.36	<0.0001 *
AIP	0.47 ± 0.21	0.44 ± 0.19	0.50 ± 0.23	0.389
TyG	4.16 ± 0.24	4.14 ± 0.20	4.19 ± 0.27	0.547
Systolic Blood Pressure	140.46 ± 15.5	139.42 ± 16.14	141.5 ± 15.3	0.731
Diastolic Blood Pressure	87.67 ± 9.55	87. 42 ± 9.12	87.9 ± 10.3	0.931
Pathology (%)	72.3	76	68.2	0.550
Hypertension (%)	55.4	56	54.5	0.920
T2DM (%)	23.4	16	31.8	0.201
Thyroidopathy (%)	12.8	84	90.9	0.479
Insulin Resistance (%)	78.7	76	81.8	0.368

AIP: atherogenic index of plasma (≤0.21); ALT: alanine aminotransferase (<33 U/L); AST: aspartate aminotransferase (< 32 U/L); BMI: body mass index; EAG: estimated average glucose; adiponectin (>4 μg/mL); HbA1c: glycated haemoglobin (>5.7%); HC: hip circumference; glucose (between 70 and 99 mg/dL); HDL-c: high-density lipoprotein (>40 mg/dL); HOMA-AD: homeostasis model assessment-adiponectin (>0.504); HOMA-beta: homeostasis model assessment of β-cell function (154 ± 73); HOMA-IR: homeostasis model assessment of insulin resistance (<2.71); hs-CRP: ultra-high sensitive C-reactive protein (<1 mg/dL); insulin (between 2.6 and 24.9 μIU/mL); LDL-c: low-density lipoprotein (<130 mg/dL); NC: neck circumference (≤34 cm); PAI-1: plasminogen activator inhibitor-1; QUICKI: quantitative insulin sensitivity check index (≥0.321); T2DM: type 2 diabetes mellitus; TC: total cholesterol (<190 mg/dL); TG: triglycerides (<150 mg/dL); TyG: triglyceride–glucose index (<4.55); VLDL-c: very-low-density lipoprotein; WC: waist circumference (<80 cm). * $p \leq 0.05$.

The cytokine PAI-1 did not correlate with any variable in the overall sample. AIP correlated with TG (r = 0.904, $p < 0.0001$), HDL-c (r = −0.494, $p = 0.001$), very-low-density lipoprotein (VLDL) (r = 0.900, $p < 0.0001$), alanine aminotransferase (ALT) (r = 0.522, $p < 0.0001$), adiponectin (r = −0.352, $p = 0.019$), and TyG (r = 0.792, $p < 0.0001$) (Table 2).

Table 2. Correlations between AIP and metabolic variables in the total sample.

		TG	HDL-c	VLDL	ALT	Adiponectin	TyG
AIP	r	0.904	−0.494	0.900	0.522	−0.352	0.792
	p	<0.0001 *	0.001 *	<0.0001 *	<0.0001 *	0.019 *	<0.0001 *

AIP: atherogenic index of plasma (≤0.21); ALT: alanine aminotransferase; HDL-c: high-density lipoprotein; TG: triglycerides; TyG: triglyceride–glucose index (<4.55); VLDL-c: very-low-density lipoprotein. * $p \leq 0.05$.

Furthermore, TyG correlated with blood glucose (r = 0.547, $p < 0.0001$), HOMA-beta (r = −0.461, $p = 0.002$), TG (r = 0.833, $p < 0.0001$), ALT (r = 0.522, $p < 0.0001$), HbA1c (r = 0.352, $p = 0.019$), estimated average glucose (EAG) (r = 0.446, $p = 0.003$), adiponectin (r = −0.431, p = 0.004), QUICKI (r = −0.384, p = 0.012), and HOMA-AD (r = 0.363, p = 0.020) (Table 3).

Table 3. Correlations between TyG and metabolic variables in the total sample.

		Glycaemia	HOMA-beta	TG	ALT	HbA1c	EAG	Adiponectin	HOMA-AD	QUICKI
TyG	r	0.547	−0.461	0.833	0.522	0.352	0.446	−0.431	0.363	−0.384
	p	<0.0001 *	0.002 *	<0.0001 *	<0.0001 *	0.019 *	0.003 *	0.004 *	0.020 *	0.012 *

ALT: alanine aminotransferase; EAG: estimated average glucose; HbA1c: glycated haemoglobin; HOMA-AD: homeostasis model assessment-adiponectin; HOMA-beta: homeostasis model assessment of β-cell function; QUICKI: quantitative insulin sensitivity check index; TG: triglycerides; TyG: triglyceride–glucose index (<4.55). * $p \leq 0.05$.

When stratifying individuals based on median PAI-1 concentration (21 ng/mL), patients with higher PAI-1 values had significantly higher HOMA-IR (p= 0.037, Cohen's d = −0.67) and HOMA-AD (p = 0.046, Cohen's d = −0.65) levels compared to the individuals with lower PAI-1 blood concentration, while the QUICKI (p= 0.022, Cohen's d = 0.75) and HDL-c (Cohen's d = 0.64) were significantly lower in the same group, indicating worse cardiometabolic and insulin features, as shown on Table 1.

There was no difference between the groups for the incidence of hypertension, T2DM, hypercholesterolemia, thyroidopathy, insulin resistance, and hypoadiponectinemia. When comparing groups based on the use of hypoglycaemic and antihypertensive drugs, we did not observe any difference between the groups in terms of PAI-1 levels (Table S1).

In the group with a higher PAI-1 value, there was a positive correlation between PAI-1 and hs-CRP (r = 0.674, p = 0.002) (Table 4). In this group, AIP remained correlated only with ALT (r = 0.517, p = 0.020).

Table 4. Significant correlations in the group with higher levels of PAI-1.

		hs-CRP	ALT	HOMA-beta	HOMA-AD
PAI-1	r	0.674	−0.216	−0.087	−0.063
	p	0.002 *	0.360	0.072	0.797
AIP	r	−0.337	0.517	−0.316	0.330
	p	0.147	0.020 *	0.188	0.155
TyG	r	0.054	0.531	0.659	0.470
	p	0.820	0.016 *	0.002 *	0.036 *

AIP: atherogenic index of plasma (≤0.21); hs-CRP: ultra-high sensitive C-reactive protein; ALT: alanine aminotransferase; HOMA-beta: homeostasis model assessment of β-cell function; HOMA-AD: homeostasis model assessment-adiponectin; TyG: triglyceride–glucose index (<4.55). * $p \leq 0.05$.

It was possible to observe a negative correlation between TyG and QUICKI only in the group with lower PAI-1 values (r = −0.459, p = 0.032) (Table 5).

Table 5. Significant correlations in the group with lower levels of PAI-1.

		QUICKI	Hs-CRP	HOMA-beta	HOMA-AD
TyG	r	−0.459	−0.145	−0.152	0.251
	p	0.032 *	0.519	0.500	0.271

HOMA-AD: homeostasis model assessment-adiponectin; HOMA-beta: homeostasis model assessment of β-cell function; hs-CRP: ultra-high sensitive C-reactive protein; QUICKI: quantitative insulin sensitivity check index; TyG: triglyceride–glucose index. * $p \leq 0.05$.

3.2. PAI-1 Predictors

In the regression analysis, it was found that there is an additional risk of 32.1% of belonging to the group with higher cardiovascular risk (higher PAI-1) with an increase in one unit in HOMA-IR. The results showed HOMA-IR as a predictor of PAI-1 ($p = 0.017$) (Table 6).

Table 6. Bivariate logistic regression model for PAI-1 in women with severe obesity.

		B	p-Value	OR (95% for Exp (B))
Step 1	HDL-c	−0.052	0.256	0.95 (0.87–1.04)
	HOMA-AD	0.000	0.862	1 (0.996–1.004)
	QUICKI	−9.030	0.835	0.000 (0.000–1.03)
	HOMA-IR	0.247	0.431	1.280 (0.692–2.367)
	Constant	3.321	0.815	27.677
Step 2	HDL-c	−0.053	0.232	0.948 (0.869–1.035)
	QUICKI	−9.117	0.834	0.000 (0.00–9.15)
	HOMA-IR	0.217	0.404	1.243 (0.746–2.070)
	Constant	3.471	0.807	32.158
Step 3	HDL-c	−0.055	0.203	0.946 (0.869–1.030)
	HOMA-IR	0.266	0.027 *	1.305 (1.030–1.654)
	Constant	0.525	0.806	1.690
Step 4	HOMA-IR	0.279	0.017 *	1.321 (1.052–1.659)
	Constant	−2.069	0.018	0.126

HDL-c: high-density lipoprotein; HOMA-AD: homeostasis model assessment-adiponectin; HOMA-IR: homeostasis model assessment of insulin resistance; QUICKI: quantitative insulin sensitivity check index; Binomial logistic regression, * $p \leq 0.05$.

4. Discussion

It is beyond doubt that inflammation plays a pivotal role in the onset of obesity and the development of cardiovascular diseases. The role of PAI-1 in inflammation and CVDs is well established; however, it is important to note that the analysis of PAI-1 levels can be challenging and expensive, which further contributes to the limited use of this important marker in clinical practice. For the first time, we were able to show HOMA-IR, a very common biochemical marker, as a predictor of PAI-1 in women with severe obesity. A previous study by Basurto et al. [24], which included individuals with normal weight, overweight, and subjects with obesity, observed the influence of HOMA-IR on PAI-1 concentration; however, the study did not include individuals with a BMI > 40 kg/m.

In addition to the hypothesis that higher PAI-1 values are promoted by hyperinsulinemia and insulin resistance [8–10], the relationship between PAI-1 and glycaemia is also demonstrated by the effects of common hypoglycaemic drugs such as pioglitazone, troglitazone, and metformin, which reduce serum levels of PAI-1 [25–27] as well as its activity [28]. We did not find a statistically significant effect of glucose-lowering drugs on PAI-1 levels in our cohort. Nevertheless, controlling PAI-1 levels through pharmacological intervention could have multifactorial beneficial effects for patients with obesity-related T2DM.

In fact, the present study demonstrated that women with severe obesity and higher levels of PAI-1 presented lower values of QUICKI and higher levels of HOMA-AD and HOMA-IR, reinforcing the relationship between PAI-1 and glycaemic traits and emphasising its importance in clinical practice. Corroborating with our data, previous studies have demonstrated the relationship between PAI-1 and glycaemic metabolic parameters [24,29,30], showing that metabolically unhealthy individuals have higher levels of PAI-1 independently of BMI [24]. Mendivil et al. [29] observed a positive correlation between PAI-1 levels and insulin resistance parameters in subjects at high risk of developing T2DM. In addition, a study of 295 individuals aged between 18 and 45 years, including eutrophic, overweight, and patients with obesity (without a diagnosis of T2DM or hypertension), found that PAI-1 was a negative predictor of QUICKI [30]. The exact mechanisms that precisely define PAI-1 involvement in glycaemic control are not clear yet [31]. Some hypotheses postulate that PAI-1 has deleterious effects on some proteins, such as the insulin receptor, transforming growth factor beta, and peroxisome proliferator-activated receptor gamma γ, promoting insulin resistance [32]. Therefore, more studies are needed to elucidate the underlying mechanisms.

Although we did not observe significant differences in adiponectin levels between the groups stratified according to low and high PAI-1 levels, we did observe that patients with higher PAI-1 presented higher HOMA-AD, an important index combining insulin resistance and adiponectin. A study including individuals with T2DM and MS observed that PAI-1 mediates the downregulation of adiponectin [33]. Changes in adiponectin levels can alter the insulin signalling cascade, as adiponectin binds to leucine zipper 1 (APPL1) [34].

It is important to note that PAI-1 can also contribute to CVDs by altering cholesterol homeostasis [35]. In fact, our study showed that individuals with higher PAI-1 values had lower HDL-c levels, a lipoprotein with cardioprotective effects. Corroboratingly, Basurto et al. evaluated individuals based on metabolic assessment and observed a negative correlation of PAI-1 with HDL-c in women with normal weight, overweight, and obesity [24]. In addition, a systematic review and meta-analysis also suggested a causal effect of PAI-1 on HDL-c [36]. Together, our data support the importance of PAI-1 in cardiovascular health by influencing both the glycaemic and lipid markers related to endothelial dysfunction.

Moreover, it is well established that high levels of PAI-1 and hs-CRP are associated with insulin resistance and microvascular dysfunction and may contribute to CVDs [30]. Our data pointed out that in the group of patients with a higher PAI-1 value, there is a strong and positive correlation between PAI-1 and hs-CRP. A correlation between PAI-1 and CRP has previously been observed in individuals with DM, demonstrating the association between these markers in individuals with DM and carriers of the 4G polymorphism in the PAI-1 gene [37].

Regarding other cardiometabolic indexes, high mean AIP values (0.47 ± 0.21) were observed in our cohort, which also supports the increased cardiovascular risk being defined as AIP > 0.21 by Holmes et al. [38]. However, we did not find any associations between PAI-1 and AIP and TyG, which could be explained by the small sample size, as we expected given the data from a study of individuals with severe obesity that demonstrated an association between PAI-1 and AIP [11].

Concerning the association between the TyG and PAI-1, to our knowledge, there are no studies investigating the topic, although the association between PAI-1 concentrations and high concentrations of glucose and triglycerides (parameters of the index) has been demonstrated [24]. Nevertheless, we were able to show a positive correlation between TyG and HOMA-AD, and this correlation remained significant only in the group with higher values of PAI-1. We also observed a negative correlation between TyG and QUICKI, parameters that evaluate insulin resistance and sensibility, respectively, in the group with lower levels of PAI-1, reinforcing PAI-1 as a factor influencing glycaemic parameters. In addition, we demonstrated the correlations between AIP and TyG, which represent a cardiovascular risk factor.

Finally, our data showed a positive correlation between AIP and ALT. This finding is consistent with a large-scale study conducted in China with 7838 participants that showed AIP to be an independent risk predictor for fatty liver [39]. PAI-1 expression is known to be significantly higher in patients with non-alcoholic fatty liver disease (NAFLD), suggesting that NAFLD independently contributes to PAI-1 secretion [40]. Furthermore, an association has been found between plasma levels of PAI-1 and the ALT/aspartate aminotransferase (AST) ratio in individuals with severe obesity [41]. Corroboratingly, our study showed a correlation of AIP with ALT only in the group with high PAI-1 values, suggesting that clinicians should recognise the increased risk of CVDs in NAFLD patients.

This is the first study comparing women with severe obesity based on PAI-1 levels. It is important to note that there is a lack of PAI-1 cut-off value in the literature, which led us to group the patients according to the median values, as previously performed when no cut-off value had been established. Taking together our findings, we suggest that in a homogeneous sample (no differences in anthropometric assessments), individuals with higher levels of PAI-1 had a worse cardiometabolic profile, and HOMA-IR might be a useful tool to screen patients with higher PAI-1. The limitations of this study are the cross-sectional design, small sample size, and absence of a control group. Further studies with a larger sample size, especially in severe obesity, are needed to confirm these results and set up a cut-off value for this important cytokine.

5. Conclusions

Women with severe obesity and higher PAI-1 levels have an increased cardiometabolic risk, as indicated by higher HOMA-IR and HOMA-AD values, lower QUICKI, and lower HDL-c concentrations. Finally, HOMA-IR could be used as a predictor of PAI-1 levels, highlighting the importance of assessing glycaemic parameters in the prevention of CVDs in women with severe obesity.

Supplementary Materials: The following supporting information can be downloaded at: https://www.mdpi.com/article/10.3390/biomedicines12061222/s1, Table S1: Comparison between individuals based on treatment with hypoglycaemic and antihypertensive drugs.

Author Contributions: Conceptualisation, F.M.K., M.J., N.F., P.R.E.d.M., A.A. and F.C.C.; methodology, F.M.K., N.F., E.S.O., L.M.O., P.R.E.d.M., A.A. and F.C.C.; validation, L.M.O., M.A.H. and F.C.C.; formal analysis, F.M.K., L.M.O., A.A. and F.C.C.; investigation, F.M.K., N.F. and P.R.E.d.M.; resources, M.A.H. and F.C.C.; data curation, F.M.K., M.J., L.M.O. and F.C.C.; writing—original draft preparation, F.M.K., M.J., A.A. and F.C.C.; writing—review and editing, F.M.K., M.J., G.C.L., A.R.D., A.L.F., M.A.H., A.A. and F.C.C.; visualisation, F.M.K.; supervision, M.A.H., A.A. and F.C.C.; project administration, M.A.H., A.A. and F.C.C.; funding acquisition, F.M.K. and M.A.H. All authors have read and agreed to the published version of the manuscript.

Funding: This research was funded by the "Conselho Nacional de Desenvolvimento Científico e Tecnológico" (CNPq) grant number [no434159/2018-2], "Fundação de Apoio à Pesquisa"—FUNAPE—UFG [n.01/2022], and "Fundação de Amparo à Pesquisa do Estado de Goiás"—FAPEG (student scholarship).

Institutional Review Board Statement: This study was conducted in accordance with the Declaration of Helsinki and approved by the Research Ethics Committee of the Federal University of Goiás (no 3.251.178, 8 April 2019) and the State Hospital of Goiânia, Dr. Alberto Rassi (no 961/19, 17 June 2019).

Informed Consent Statement: Informed consent was obtained from all subjects involved in this study.

Data Availability Statement: The original contributions presented in this study are included in this article and Supplementary Materials; further inquiries can be directed to the corresponding authors.

Acknowledgments: The authors would like to thank the undergraduate students Arthur Ferreira Laureano and Giullia Ferreira de Paula, who contributed to the execution of this project. Ana Dâmaso (n. 305240/2021-8) has been financially supported by the Conselho Nacional de Desenvolvimento Científico e Tecnológico CNPq. Also, we want to thank all the volunteers who agreed to participate in this study.

Conflicts of Interest: The authors declare no conflicts of interest.

References

1. OMS. Obesity and Overweight. Available online: https://www.who.int/news-room/fact-sheets/detail/obesity-and-overweight (accessed on 1 June 2022).
2. Chooi, Y.C.; Ding, C.; Magkos, F. The epidemiology of obesity. *Metabolism* **2019**, *92*, 6–10. [CrossRef] [PubMed]
3. Landecho, M.F.; Tuero, C.; Valenti, V.; Bilbao, I.; de la Higuera, M.; Fruhbeck, G. Relevance of Leptin and Other Adipokines in Obesity-Associated Cardiovascular Risk. *Nutrients* **2019**, *11*, 2664. [CrossRef] [PubMed]
4. Ortega, F.B.; Lavie, C.J.; Blair, S.N. Obesity and Cardiovascular Disease. *Circ. Res.* **2016**, *118*, 1752–1770. [CrossRef] [PubMed]
5. Sillen, M.; Declerck, P.J. A Narrative Review on Plasminogen Activator Inhibitor-1 and Its (Patho)Physiological Role: To Target or Not to Target? *Int. J. Mol. Sci.* **2021**, *22*, 2721. [CrossRef] [PubMed]
6. Sillen, M.; Declerck, P.J. Targeting PAI-1 in Cardiovascular Disease: Structural Insights Into PAI-1 Functionality and Inhibition. *Front. Cardiovasc. Med.* **2020**, *7*, 622473. [CrossRef] [PubMed]
7. Kaji, H. Adipose Tissue-Derived Plasminogen Activator Inhibitor-1 Function and Regulation. *Compr. Physiol.* **2016**, *6*, 1873–1896. [CrossRef] [PubMed]
8. Alessi, M.C.; Juhan-Vague, I.; Kooistra, T.; Declerck, P.J.; Collen, D. Insulin stimulates the synthesis of plasminogen activator inhibitor 1 by the human hepatocellular cell line Hep G2. *Thromb. Haemost.* **1988**, *60*, 491–494. [CrossRef] [PubMed]

9. Fattal, P.G.; Schneider, D.J.; Sobel, B.E.; Billadello, J.J. Post-transcriptional regulation of expression of plasminogen activator inhibitor type 1 mRNA by insulin and insulin-like growth factor 1. *J. Biol. Chem.* **1992**, *267*, 12412–12415. [CrossRef] [PubMed]
10. Altalhi, R.; Pechlivani, N.; Ajjan, R.A. PAI-1 in Diabetes: Pathophysiology and Role as a Therapeutic Target. *Int. J. Mol. Sci.* **2021**, *22*, 3170. [CrossRef]
11. Carmona-Maurici, J.; Cuello, E.; Sanchez, E.; Minarro, A.; Rius, F.; Bueno, M.; de la Fuente, M.C.; Kissler, J.J.O.; Vidal, T.; Maria, V.; et al. Impact of bariatric surgery on subclinical atherosclerosis in patients with morbid obesity. *Surg. Obes. Relat. Dis.* **2020**, *16*, 1419–1428. [CrossRef]
12. Li, L.; Shi, Z.; Ma, L.; Lu, Y. Analysis of the correlation between plasma coagulation factor VII, PAI-1, and uric acid with insulin resistance and macrovascular complications in elderly patients with type 2 diabetes. *Ann. Palliat. Med.* **2021**, *10*, 664–671. [CrossRef]
13. Figueiredo, N.; de Oliveira Queiroz, M.; Lopes, K.L.S.; Oliveira, L.; Dâmaso, A.R.; de Melo, P.R.E.; Abreu, V.d.S.; Mota, J.F.; Horst, M.A.; Corgosinho, F.C. Triglyceride-to-High-Density-Lipoprotein-Cholesterol Ratio as a Predictor of Metabolic Syndrome According to Stage of Life at Obesity Onset in Women with Severe Obesity—A Pilot Study. *Obesities* **2022**, *2*, 361–371. [CrossRef]
14. da Silva, A.; Caldas, A.P.S.; Hermsdorff, H.H.M.; Bersch-Ferreira, A.C.; Torreglosa, C.R.; Weber, B.; Bressan, J. Triglyceride-glucose index is associated with symptomatic coronary artery disease in patients in secondary care. *Cardiovasc. Diabetol.* **2019**, *18*, 89. [CrossRef] [PubMed]
15. WHO. *Obesity: Preventing and Managing the Global Epidemic: Report of a WHO Consultation*; WHO Technical Report Series; WHO: Geneva, Switzerland, 2000; Volume 894, i–xii, 1–253.
16. Song, Y.; Manson, J.E.; Tinker, L.; Howard, B.V.; Kuller, L.H.; Nathan, L.; Rifai, N.; Liu, S. Insulin sensitivity and insulin secretion determined by homeostasis model assessment and risk of diabetes in a multiethnic cohort of women: The Women's Health Initiative Observational Study. *Diabetes Care* **2007**, *30*, 1747–1752. [CrossRef] [PubMed]
17. Katz, A.; Nambi, S.S.; Mather, K.; Baron, A.D.; Follmann, D.A.; Sullivan, G.; Quon, M.J. Quantitative insulin sensitivity check index: A simple, accurate method for assessing insulin sensitivity in humans. *J. Clin. Endocrinol. Metab.* **2000**, *85*, 2402–2410. [CrossRef]
18. Vilela, B.S.; Vasques, A.C.; Cassani, R.S.; Forti, A.C.; Pareja, J.C.; Tambascia, M.A.; Geloneze, B.; BRAMS Investigators. The HOMA-Adiponectin (HOMA-AD) Closely Mirrors the HOMA-IR Index in the Screening of Insulin Resistance in the Brazilian Metabolic Syndrome Study (BRAMS). *PLoS ONE* **2016**, *11*, e0158751. [CrossRef]
19. Li, H.; Zuo, Y.; Qian, F.; Chen, S.; Tian, X.; Wang, P.; Li, X.; Guo, X.; Wu, S.; Wang, A. Triglyceride-glucose index variability and incident cardiovascular disease: A prospective cohort study. *Cardiovasc. Diabetol.* **2022**, *21*, 105. [CrossRef] [PubMed]
20. Kishida, K.; Funahashi, T.; Shimomura, I. Adiponectin as a routine clinical biomarker. *Best Pract. Res. Clin. Endocrinol. Metab.* **2014**, *28*, 119–130. [CrossRef]
21. Kravchychyn, A.C.P.; Campos, R.; Ferreira, Y.A.M.; Vicente, S.; Corgosinho, F.C.; Oyama, L.M.; Thivel, D.; Tock, L.; Dâmaso, A.R. Higher increase degree of FGF21 post long-term interdisciplinary weight loss therapy preserves the free fat mass and rest metabolic rate in adolescents with obesity. *Arch. Endocrinol. Metab.* **2020**, *64*, 479–482. [CrossRef]
22. Tustumi, F. Choosing the most appropriate cut-point for continuous variables. *Rev. Col. Bras. Cir.* **2022**, *49*, e20223346. [CrossRef]
23. Whelton, P.K.; Carey, R.M.; Aronow, W.S.; Casey, D.E., Jr.; Collins, K.J.; Dennison Himmelfarb, C.; DePalma, S.M.; Gidding, S.; Jamerson, K.A.; Jones, D.W.; et al. 2017 ACC/AHA/AAPA/ABC/ACPM/AGS/APhA/ASH/ASPC/NMA/PCNA Guideline for the Prevention, Detection, Evaluation, and Management of High Blood Pressure in Adults: A Report of the American College of Cardiology/American Heart Association Task Force on Clinical Practice Guidelines. *Hypertension* **2018**, *71*, e13–e115. [CrossRef] [PubMed]
24. Basurto, L.; Sanchez, L.; Diaz, A.; Valle, M.; Robledo, A.; Martinez-Murillo, C. Differences between metabolically healthy and unhealthy obesity in PAI-1 level: Fibrinolysis, body size phenotypes and metabolism. *Thromb. Res.* **2019**, *180*, 110–114. [CrossRef] [PubMed]
25. Kruszynska, Y.T.; Yu, J.G.; Olefsky, J.M.; Sobel, B.E. Effects of troglitazone on blood concentrations of plasminogen activator inhibitor 1 in patients with type 2 diabetes and in lean and obese normal subjects. *Diabetes* **2000**, *49*, 633–639. [CrossRef] [PubMed]
26. Trost, S.; Pratley, R.; Sobel, B. Impaired fibrinolysis and risk for cardiovascular disease in the metabolic syndrome and type 2 diabetes. *Curr. Diabetes Rep.* **2006**, *6*, 47–54. [CrossRef] [PubMed]
27. Hanefeld, M.; Marx, N.; Pfutzner, A.; Baurecht, W.; Lubben, G.; Karagiannis, E.; Stier, U.; Forst, T. Anti-inflammatory effects of pioglitazone and/or simvastatin in high cardiovascular risk patients with elevated high sensitivity C-reactive protein: The PIOSTAT Study. *J. Am. Coll. Cardiol.* **2007**, *49*, 290–297. [CrossRef] [PubMed]
28. Pandolfi, A.; Cetrullo, D.; Polishuck, R.; Alberta, M.M.; Calafiore, A.; Pellegrini, G.; Vitacolonna, E.; Capani, F.; Consoli, A. Plasminogen activator inhibitor type 1 is increased in the arterial wall of type II diabetic subjects. *Arterioscler. Thromb. Vasc. Biol.* **2001**, *21*, 1378–1382. [CrossRef] [PubMed]
29. Mendivil, C.O.; Robles-Osorio, L.; Horton, E.S.; Hamdy, O.; Caballero, A.E. Young Hispanics at risk of type 2 diabetes display endothelial activation, subclinical inflammation and alterations of coagulation and fibrinolysis. *Diabetol. Metab. Syndr.* **2013**, *5*, 37. [CrossRef] [PubMed]
30. Cheng, C.; Daskalakis, C. Association of Adipokines with Insulin Resistance, Microvascular Dysfunction, and Endothelial Dysfunction in Healthy Young Adults. *Mediat. Inflamm.* **2015**, *2015*, 594039. [CrossRef] [PubMed]

31. Alessi, M.C.; Juhan-Vague, I. PAI-1 and the metabolic syndrome: Links, causes, and consequences. *Arterioscler. Thromb. Vasc. Biol.* **2006**, *26*, 2200–2207. [CrossRef]
32. Batiha, G.E.; Al-Kuraishy, H.M.; Al-Maiahy, T.J.; Al-Buhadily, A.K.; Saad, H.M.; Al-Gareeb, A.I.; Simal-Gandara, J. Plasminogen activator inhibitor 1 and gestational diabetes: The causal relationship. *Diabetol. Metab. Syndr.* **2022**, *14*, 127. [CrossRef]
33. Nawaz, S.S.; Siddiqui, K. Plasminogen activator inhibitor-1 mediate downregulation of adiponectin in type 2 diabetes patients with metabolic syndrome. *Cytokine X* **2022**, *4*, 100064. [CrossRef] [PubMed]
34. Fang, H.; Judd, R.L. Adiponectin Regulation and Function. *Compr. Physiol.* **2018**, *8*, 1031–1063. [CrossRef] [PubMed]
35. Levine, J.A.; Oleaga, C.; Eren, M.; Amaral, A.P.; Shang, M.; Lux, E.; Khan, S.S.; Shah, S.J.; Omura, Y.; Pamir, N.; et al. Role of PAI-1 in hepatic steatosis and dyslipidemia. *Sci. Rep.* **2021**, *11*, 430. [CrossRef] [PubMed]
36. Song, C.; Burgess, S.; Eicher, J.D.; O'Donnell, C.J.; Johnson, A.D. Causal Effect of Plasminogen Activator Inhibitor Type 1 on Coronary Heart Disease. *J. Am. Heart Assoc.* **2017**, *6*, e004918. [CrossRef] [PubMed]
37. Testa, R.; Bonfigli, A.R.; Sirolla, C.; Marra, M.; Boemi, M.; Mari, D.; Sacchi, E.; Dolci, A.; Catalano, A.; Procopio, A.; et al. C-reactive protein is directly related to plasminogen activator inhibitor type 1 (PAI-1) levels in diabetic subjects with the 4G allele at position -675 of the PAI-1 gene. *Nutr. Metab. Cardiovasc. Dis.* **2008**, *18*, 220–226. [CrossRef] [PubMed]
38. Holmes, D.T.; Frohlich, J.; Buhr, K.A. The concept of precision extended to the atherogenic index of plasma. *Clin. Biochem.* **2008**, *41*, 631–635. [CrossRef] [PubMed]
39. Xie, F.; Zhou, H.; Wang, Y. Atherogenic index of plasma is a novel and strong predictor associated with fatty liver: A cross-sectional study in the Chinese Han population. *Lipids Health Dis.* **2019**, *18*, 170. [CrossRef] [PubMed]
40. Potze, W.; Siddiqui, M.S.; Sanyal, A.J. Vascular Disease in Patients with Nonalcoholic Fatty Liver Disease. *Semin. Thromb. Hemost.* **2015**, *41*, 488–493. [CrossRef]
41. Alessi, M.C.; Bastelica, D.; Mavri, A.; Morange, P.; Berthet, B.; Grino, M.; Juhan-Vague, I. Plasma PAI-1 levels are more strongly related to liver steatosis than to adipose tissue accumulation. *Arterioscler. Thromb. Vasc. Biol.* **2003**, *23*, 1262–1268. [CrossRef]

biomedicines

MDPI

Review

Methylglyoxal and Advanced Glycation End Products (AGEs): Targets for the Prevention and Treatment of Diabetes-Associated Bladder Dysfunction?

Akila Lara Oliveira, Mariana Gonçalves de Oliveira , Fabíola Zakia Mónica and Edson Antunes *

Department of Translational Medicine, Pharmacology Area, Faculty of Medical Sciences, University of Campinas (UNICAMP), Campinas 13084-971, SP, Brazil; a192906@dac.unicamp.br (A.L.O.); marigo@unicamp.br (M.G.d.O.); fzm@unicamp.br (F.Z.M.)

* Correspondence: eantunes@unicamp.br; Tel.: +55-19-9-9601-1516

Abstract: Methylglyoxal (MGO) is a highly reactive α-dicarbonyl compound formed endogenously from 3-carbon glycolytic intermediates. Methylglyoxal accumulated in plasma and urine of hyperglycemic and diabetic individuals acts as a potent peptide glycation molecule, giving rise to advanced glycation end products (AGEs) like arginine-derived hydroimidazolone (MG-H1) and carboxyethyl-lysine (CEL). Methylglyoxal-derived AGEs exert their effects mostly via activation of RAGE, a cell surface receptor that initiates multiple intracellular signaling pathways, favoring a pro-oxidant environment through NADPH oxidase activation and generation of high levels of reactive oxygen species (ROS). Diabetic bladder dysfunction is a bothersome urological complication in patients with poorly controlled diabetes mellitus and may comprise overactive bladder, urge incontinence, poor emptying, dribbling, incomplete emptying of the bladder, and urinary retention. Preclinical models of type 1 and type 2 diabetes have further confirmed the relationship between diabetes and voiding dysfunction. Interestingly, healthy mice supplemented with MGO for prolonged periods exhibit in vivo and in vitro bladder dysfunction, which is accompanied by increased AGE formation and RAGE expression, as well as by ROS overproduction in bladder tissues. Drugs reported to scavenge MGO and to inactivate AGEs like metformin, polyphenols, and alagebrium (ALT-711) have shown favorable outcomes on bladder dysfunction in diabetic obese leptin-deficient and MGO-exposed mice. Therefore, MGO, AGEs, and RAGE levels may be critically involved in the pathogenesis of bladder dysfunction in diabetic individuals. However, there are no clinical trials designed to test drugs that selectively inhibit the MGO–AGEs–RAGE signaling, aiming to reduce the manifestations of diabetes-associated bladder dysfunction. This review summarizes the current literature on the role of MGO–AGEs–RAGE–ROS axis in diabetes-associated bladder dysfunction. Drugs that directly inactivate MGO and ameliorate bladder dysfunction are also reviewed here.

Keywords: dicarbonyl stress; RAGE; oxidative stress; polyphenols; metformin; alagebrium

Citation: Oliveira, A.L.; de Oliveira, M.G.; Mónica, F.Z.; Antunes, E. Methylglyoxal and Advanced Glycation End Products (AGEs): Targets for the Prevention and Treatment of Diabetes-Associated Bladder Dysfunction?. *Biomedicines* **2024**, *12*, 939. https://doi.org/10.3390/biomedicines12050939

Academic Editors: Teresa Vezza and Zaida Abad-Jiménez

Received: 19 March 2024
Revised: 11 April 2024
Accepted: 17 April 2024
Published: 23 April 2024

1. Introduction

Methylglyoxal (MGO) is a highly reactive α-dicarbonyl compound endogenously generated during the glycolytic pathway [1]. Hyperglycemia in diabetic and obese patients markedly elevates the plasma and urine levels of MGO as a consequence of the glycolytic overload [2]. The abnormal accumulation of MGO has been referred to as dicarbonyl stress, which may be implicated in many diseases [3]. Methylglyoxal promotes post-translational modification of peptides or proteins, ultimately leading to the formation of advanced glycation end products (AGEs), the most studied of which include arginine-derived hydroimidazolone (MG-H1) and carboxyethyl-lysine (CEL) [1]. MGO also covalently modifies DNA, leading to nucleic acid AGE formation, consisting mainly of guanine adducts. AGEs bind their cell membrane-anchored ligand receptor, termed RAGE [4], triggering multiple intracellular signaling pathways, including the activation of NADPH oxidase that leads to increased production of reactive oxygen species (ROS), thus contributing to generate a pro-oxidant environment. Diabetic bladder dysfunction (DBD) is a highly prevalent condition that may affect the detrusor, nerve fiber terminals, urothelium, and urethra, and manifests as storage problems such as OAB and urge incontinence, and voiding problems such as decreased sensation and increased capacity [5,6]. DBD may progress from detrusor overactivity at initial stages to detrusor underactivity at advanced stages of this disease, a condition defined by the International Continence Society (ICS) as contraction of reduced strength and/or duration, resulting in prolonged bladder emptying and/or a failure to achieve complete bladder emptying within a normal timespan [5,7–9]. The underactive bladder comprises mostly voiding phase symptoms such as slow stream, intermittency, hesitancy, feeling of incomplete emptying of the bladder, and urinary retention [10]. Increased capacity and decreased sensation, together with recurrent urinary tract infections, may also be present in DBD [11,12]. Preclinical models of type 1 (streptozotocin, Akita mice) and type 2 diabetes (high-fat diets, ob/ob and db/db mice) have provided further evidence confirming the relationship between diabetes and obesity with voiding dysfunction. However, little is known about the importance of MGO generation and, hence, AGEs–RAGE activation in the pathophysiology of diabetic-associated bladder dysfunction [13]. Interestingly, in mice treated orally with MGO for prolonged periods, voiding spot assays in conscious mice and urodynamic evaluation in anesthetized mice revealed significant increases in total void volume, volume per void, micturition frequency, and nonvoiding contractions number, along with enhanced in vitro bladder contractility [14]. In addition, elevated levels of MGO, AGEs, RAGE, and ROS were found in bladder tissues from mice chronically treated with MGO, pointing out that they could be important markers of DBD pathophysiology [15]. Similar data were obtained in bladder tissues of diabetic obese ob/ob mice [16]. The antihyperglycemic drug metformin [17–19] and polyphenols like resveratrol and epigallocatechin-3-gallate [20] can directly scavenge MGO, explaining, at least in part, their capacity to ameliorate diabetes-associated bladder dysfunction. However, no clinical trials exist aiming to test inhibitors of the MGO–AGEs–RAGE signaling as potential drugs to prevent and treat manifestations of diabetes-associated bladder dysfunction. Therefore, the design and development of new drugs that inhibit the MGO–AGEs–RAGE axis may become an interesting approach for the prevention and treatment of bladder dysfunction in diabetic conditions. The present review aimed to provide an updated overview on bladder dysfunction in diabetic and obesity conditions in animals and humans, emphasizing the MGO–AGEs–RAGE signaling pathway as a potential mechanism implicated in the pathophysiology of this disorder, focusing on bladder overactivity. Drugs that inactivate MGO or inhibit AGEs formation in parallel to reducing diabetic-associated bladder dysfunction are also reviewed here.

2. Lower Urinary Tract Symptoms (LUTS) and Overactive Bladder (OAB) Syndrome

Urinary bladder function is regulated by a complex interaction of efferent and afferent fibers from the autonomic nervous system and somatic innervation [21]. An imbalance between these systems leads to lower urinary tract symptoms (LUTS), which comprise storage, voiding, and post-micturition symptoms [5]. Storage symptoms consist of altered bladder sensation, increased daytime frequency, nocturia, and urgency incontinence, whereas voiding symptoms consists of hesitancy, intermittency, weak or irregular stream, straining, and terminal dribble. Post-micturition symptoms include dribbling and sensation of incomplete voiding. The storage symptoms are generally more bothersome than voiding or post-micturition symptoms, as observed in both men and women. Overactive bladder (OAB) syndrome is a subgroup of storage symptoms consisting mainly of urinary urgency. In men, LUTS typically occur in association with bladder outlet obstruction (BOO) secondary to benign prostatic hyperplasia (BPH), despite that it may occur independently of BOO or prostatic diseases, whereas in women, the most frequent LUTS is stress urinary incontinence [22]. Epidemiological studies have shown OAB to be a widely prevalent condition in men and women [5,23], with an incidence of 16.6% in a sample from Europe [24],

16.9% in women and 16.0% in men in a sample from the USA [25], and an overall prevalence of OAB of 18.9% in a South American population [26,27]. Epidemiological studies applying the ICS definition of OAB across multiple countries found a prevalence of 11–13% [5]. LUTS negatively impacts the social quality of life and sexual health of patients [28,29].

3. Association of Metabolic Syndrome and Diabetes with Urinary Bladder Dysfunction

Metabolic syndrome, the medical term for a combination of cardiometabolic risk factors such as central obesity, hyperglycemia, hypertension, and dyslipidemia, is critically involved in the onset of many cardiovascular diseases, being the leading cause of death worldwide [30]. Outside the cardiovascular system, metabolic syndrome associated with increased body mass index (BMI) represents an important risk factor for LUTS/OAB and urinary incontinence [31–44], despite some studies showing no positive association between metabolic syndrome and LUTS in men and women [45,46]. An association between metabolic syndrome and interstitial cystitis/bladder pain syndrome (IC/BPS) has been reported in women [47]. Elevated body mass index and diabetes also increase the risk of urinary tract infections and pyelonephritis [48,49]. Metabolic syndrome is also associated with LUTS secondary to benign prostatic hyperplasia [50–52]. Surgical and nonsurgical weight loss leads to improvements in stress urinary incontinence [53], even though a definite conclusion has not been achieved [54].

4. Bladder Dysfunction in Type 1 and Type 2 Diabetes in Patients and Animals

Diabetes mellitus is a chronic metabolic disease characterized by high blood sugar levels (hyperglycemia) as a result of abnormal insulin production and/or insulin function. The most common and bothersome urological complication of diabetes mellitus is DBD (or diabetic cystopathy), which affects more than 80% of individuals diagnosed with diabetes [55–59]. The pathophysiology of DBD is multifactorial and may involve alterations at all levels of the urinary tract, including the detrusor, urethra, urothelium, and innervation [60]. Clinical DBD manifestations consist of storage bladder problems such as OAB and urge incontinence, and voiding problems like poor emptying with resultant elevated post-void residual urine [7,12]. Increased capacity and decreased sensation together with recurrent urinary tract infections may also be present in DBD [11]. Preclinical models of type 1 (T1DM) and type 2 diabetes (T2DM) have provided further evidence confirming the relationship between metabolic diseases and bladder dysfunction [61–67].

T1DM can be mimicked by injection of streptozotocin (STZ) in rodents, a cytotoxic glucose analogue that destroys pancreatic β-cells due to its high affinity for glucose transporter 2 (GLUT2) [68,69]. Analysis of bladders from STZ-induced diabetes in male and female rodents (rats and mice) revealed increased bladder mass [70–75], which is suggested to represent a physical adaptation to increased urine production [76–78]. Despite that high glucose levels and diabetic polyuria have been proposed as pathophysiological mechanisms explaining bladder enlargement in the STZ model [79], a recent study comprising different models of diabetes in rodents, including T1DM, did not confirm such a proposition [80]. Insulin administration can prevent, or even reverse, most of the morphological, functional, and molecular bladder alterations in the STZ model [79,81–83]. Moreover, increases in both volume and frequency of micturition [70,73], as well as in urinary frequency, capacity, and number of nonvoiding contraction (NVCs) [66,84], have been reported in STZ-induced diabetes, as revealed by urodynamic studies. Spontaneous voiding spot assays [85,86] also revealed significant increases in voiding frequency, total voided volume, and mean volume per micturition in STZ-injected mice [87], which are paralleled by in vitro detrusor overactivity [66,88]. However, after prolonged hyperglycemia and insulin resistance in response to STZ, bladders may progress to an underactive detrusor and an inability to produce an effective voiding [64] through mechanisms mediated by the activation of NLRP-3 inflammasome [89]. Therefore, in STZ-induced diabetes, a temporal effect of diabetes on bladder activity has been established, that is, an early phase of compensatory followed by a later phase of decompensated bladder function [7,63,77,78]. In female Akita mice (T1DM model), diabetic bladder dysfunction also progresses from overactivity to underactivity [90]. At the molecular level, the impairment of the nitric oxide—soluble guanylate cyclase (sGC)—cyclic GMP signaling [82,91] and NLRP3 inflammasome activation in urothelial cells [89] have been proposed as a critical mechanism contributing to bladder dysfunction. Nevertheless, conflicting data on different parameters of bladder activity in animal models of STZ-induced diabetes models have been obtained, which may possibly rely on both animal species and strain used, in addition to the disease time course [77,78,92,93]. Experimental T1DM in rats and rabbits can also be induced using alloxan, a hydrophilic unstable compound that shares a structure similar to glucose [94]. Increases in bladder weight, detrusor smooth muscle cells, capacity, and urinary output, along with irregular bladder contractions, were observed in alloxan-induced diabetic rats [95–98]. In rabbits made diabetic by alloxan, an increase in bladder weight [99] and a reduction in the vitro bladder contractions to carbachol were reported [100].

The main T2DM models that result in hyperinsulinemia and insulin resistance rely on allowing animals free access to diets highly enriched in fats [101,102]. In addition to producing the classical obesity-associated vascular dysfunction, male mice fed high-fat diets progress to an overactive bladder phenotype, as evidenced mainly by filling cystometry in anesthetized and awake rats and mice [103–105]. The resulting increased body weight, hyperglycemia, and insulin resistance by prolonged high-fat diet intake in mice is also accompanied by in vitro bladder overactivity as a consequence of high extracellular calcium influx through L-type voltage-operated calcium [106–109]. The importance of calcium channels to bladder dysfunction has also been confirmed in diabetic db/db mice [110]. High-fat diet-fed obese mice also display impaired urethral smooth muscle relaxations [111,112] and prostate hypercontractility [105,113], which are suggested to contribute to the resulting bladder overactivity. Impaired striated urethral muscle contractions were reported in Zucker obese rats [114]. Contrasting to these studies, no evidence of bladder dysfunction was observed in obese mice fed a high-fat diet for 16 weeks, as assessed by void spot assays [115]. The temporal effects (up to 42 weeks) of different diets consisting of fructose, cholesterol, and lard, at varying proportions and combinations, on 24 h urinary behavior and conscious cystometry were investigated in rats [116]. Compared with the control group, the total voided volume was lower in all experimental diets, and animals receiving 32.5% lard diet alone exhibited decreases in bladder capacity, mean voided volume, and inter-micturition intervals that were indicative of an overactive bladder phenotype [116].

Leptin is a satiety hormone that is synthesized by adipocytes, the levels of which increase with the adipose tissue mass [117]. Mice genetically deficient in leptin (ob/ob) or in the leptin receptor (db/db) are hyperphagic, obese, hyperinsulinemic, and hyperglycemic, and have been widely used as a T2DM model [118]. Similarly, in the STZ- or diet-induced obesity models, ob/ob male mice exhibit bladder dysfunction characterized by increases in urine volume and in vitro bladder smooth muscle contractions [119]. Increases in total void volume and volume per void with no alterations of spot number were observed in five-week-old male and female ob/ob mice, as evaluated by void spot assays [16]. Four- and six-month-old ob/ob mice exhibited some degree of bladder dysfunction such as increases in total urine volume and number of primary void spots, although that depended on animal sex and animal age [115]. In male db/db mice, increases in bladder weight, voiding frequency, and capacity together with elevated in vitro contractions were described [110]. Increases in detrusor smooth muscle area, urothelium area, and collagen content were also reported in male and female db/db mice at 12, 24, or 52 weeks of age, which was suggested to reflect a compensatory response to the increased urine output [120]. Double-knockout hepatic-specific insulin receptor substrate 1 and 2 (IRS1 and IRS2) female mice that develop T2DM exhibit bladder overactivity, high frequency of nonvoiding contractions, decreased voided volume, and dispersed urine spot patterns [121].

5. Methylglyoxal–Advanced Glycation End Products (AGEs)–RAGE Signaling Pathway

The abnormal accumulation of highly reactive dicarbonyl compounds as a consequence of glycolytic overload has been referred to as dicarbonyl stress [1,3]. 1,2-Dicarbonyl compounds include glucosone, 3-deoxyglucosone, methylglyoxal (MGO), and glyoxal, but MGO is one of the most studied, given that it exerts a critical role in diabetes-associated cardiovascular complications, such as diabetic nephropathy, endothelial dysfunction, postinfarct remodeling, and impairment of insulin signaling [122–126]. Methylglyoxal, chemically referred as acetylformaldehyde, 2-ketopropionaldehyde, pyruvaldehyde, or 2-oxopropanal, is a highly reactive dicarbonyl compound formed endogenously from 3-carbon glycolytic intermediates of glycolysis (dihydroxyacetone phosphate and glyceraldehyde-3-phosphate), although it can also be generated as a byproduct of protein, lipid, and ketones [127,128]. In addition to the endogenous production in mammalian cells, MGO may be present at marked levels in many food products and beverages, as well as in microorganisms [129]. In healthy conditions, glyoxalases (Glo) are the most important enzymatic detoxification system that converts MGO into its end-product D-lactate [1]. Glyoxalases comprise two major enzymes, namely, Glo1 (lactoylglutathione methylglyoxal lyase) and Glo2 (hydroxyacylglutathione hydrolase), with Glo1 described as a rate-limiting enzyme [130–132]. Interestingly, the increased levels of glucose and MGO are normalized in Glo-1 transgenic rats after induction of diabetes by intravenous injection of STZ [123,133].

The endogenous process by which endogenous MGO promotes post-translational modification of peptides or proteins, ultimately leading to generation of AGEs, is referred to as glycation [2]. The main MGO-derived AGEs in mammalian metabolism are arginine-derived hydroimidazolone (MG-H1) and carboxyethyl-lysine (CEL) [134], but other AGEs may be generated depending on the dicarbonyl species formed [135]. The mechanism of MG-H1 generation involves the replacement of the hydrophilic positively charged arginine residue by an uncharged, hydrophobic MG-H1 residue, producing misfolding and activation of the unfolded protein response [1,136]. Incubation of human plasma from heathy donors with different concentrations of MGO (10 and 100 μM) for 24 h induced a time- and dose-dependent increase in MG-H1 levels, as detected within the first 6 h [137].

Once generated, AGEs bind their cell membrane-anchored ligand receptor (termed RAGE), which is a member of the immunoglobulin superfamily of cell surface receptors able to recognize endogenous ligands [4]. Structurally, RAGE consists of three immunoglobulin domains, that is, (i) an extracellular part consisting of one V type and two C types (C1 and C2), (ii) a transmembrane spanning helix, and (iii) a short, highly charged intracellular cytoplasmic "C" terminal tail that is primarily associated with the downstream signaling pathways [138]. The extracellular domain devoid of cytoplasmic and transmembrane domains is called soluble RAGE (sRAGE), which comprise two forms, namely, cleaved RAGE (cRAGE) and endogenous secretory RAGE (esRAGE or RAGEv1) [139]. cRAGE is generated at the cell surface by the proteolytic cleavage of RAGE at the boundary between its extracellular and transmembrane portions, whereas esRAGE results from alternative splicing of RAGE pre-mRNA. In advanced chronic kidney disease (CKD) patients, an inverse association between risk of mortality and cRAGE/esRAGE ratio was reported [140]. RAGE plays an important role in the innate immune response and as a mediator of proinflammatory processes, triggering multiple intracellular signaling pathways, including the generation of proinflammatory mediators such as IL-1β, VCAM-1, and TNF-α via the transcription factor NF-κB [138] and phosphorylation of JNK and p38MAPK [141]. Furthermore, many of the RAGE actions have been attributed to the activation of NADPH oxidase [142], which leads to excess formation of ROS, thus contributing to generate a pro-oxidant environment [143–145]. Figure 1 illustrates the MGO–AGEs–RAGE signaling and glyoxalase system (Glo1 and Glo2 enzymes).

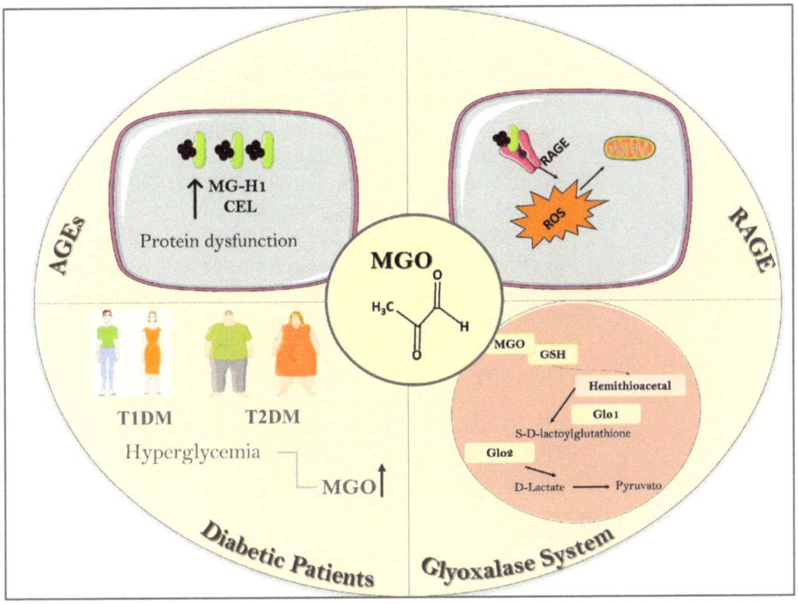

Figure 1. Methylglyoxal (MGO) is a highly reactive α-dicarbonyl compound generated endogenously during the glycolytic pathway. Hyperglycemia in type 1 (T1DM) and type 2 diabetic (T2DM) individuals markedly elevates plasma and urinary levels of MGO as a consequence of glycolytic overload. The abnormal accumulation of MGO (dicarbonyl stress) has been implicated in many diseases. Methylglyoxal promotes post-translational modification of peptides or proteins, leading to the formation of advanced glycation end products (AGEs), including hydroimidazolone derived from arginine (MG-H1) and carboxyethyl-lysine (CEL). AGEs bind to their receptor ligand (termed RAGE) anchored in cell membranes, triggering multiple intracellular signaling pathways, leading to increased reactive oxygen species (ROS) production. Under healthy conditions, glyoxalases (Glo) are the most important enzymatic detoxification system converting MGO into its final product D-lactate. Glyoxalases comprise two main enzymes, namely, Glo1 (lactoylglutathione methylglyoxal lyase) and Glo2 (hydroxyacylglutathione hydrolase), with Glo1 described as a rate-limiting enzyme in detoxification. This image was produced with the assistance of Servier Medical Art (Servier; https://smart.servier.com/ accessed on 18 March 2024), licensed under a Creative Commons Attribution 4.0 Unported License.

Plasma levels of MGO are markedly elevated in conditions of hyperglycemia associated with diabetes mellitus in men and women, as usually detected by liquid chromatography–mass spectrometry, enzyme-linked immunosorbent, or electrochemical biosensor assays [146–153]. The patients included in these studies comprised men and women aged 54–61 years old, insulin- and noninsulin users, with accompanying diseases such as chronic renal failure, diabetic nephropathy, and coronary heart disease. Obese patients also have increased MGO levels in plasma, which can be even higher if the obese patients have diabetes [125,154]. The urine levels of MGO in diabetic patients are also higher than those in nondiabetic individuals [155]. In addition, in healthy volunteers, a rapid increase (49 min) in plasma levels of MGO was observed after oral glucose tolerance test (OGTT) [152]. Likewise, fasted healthy mice intraperitoneally injected with a glucose solution displayed a rapid elevation in plasma levels of MGO, as detected at 30 min after glucose administration [137]. Interestingly, lower levels of Glo1 activity in red blood cells paralleled the increased plasma MGO levels in T2DM patients displaying acute coronary syndrome [156].

6. MGO–AGEs–RAGE Axis as a Key Player of Bladder Dysfunction in Animals and Humans

There is a large amount of data on MGO in different organs at physiological and pathological conditions [2], but surprisingly few studies have explored the role of the MGO–AGEs–RAGE signaling pathway in the pathophysiology of the lower urinary tract system. The existing literature in this field has been restricted to bladder pain via the release of high mobility group box 1 protein (HMGB1) [157,158] and bladder cancer [159,160], which are not the focus of the present review.

In T2DM patients diagnosed with moderate/severe LUTS, serum levels of AGEs are positively correlated with symptoms and overactive bladder, suggesting that levels of AGEs may be early markers of diabetes-associated LUTS [13]. In addition, an immunohistochemical study in human bladders showed positive sites for carboxymethyl-lysine and pentosidine in the connective tissue between muscle bundles and muscle fibers, suggesting that extracellular matrix is the main site of action for AGE accumulation [161]. The MG-H1 free adduct has been described as the most responsive AGE associated with chronic kidney disease status, with higher levels in diabetic compared with nondiabetic individuals [162]. The MG-H1 residue contents of plasma protein are also elevated in male spontaneously diabetic Torii (SDT) rats at the age of 16 weeks [163].

The model of chronic overload intake of MGO at doses of 50 to 75 mg/kg for 6 to 12 weeks, as supplemented in the drinking water of the animals or injected intraperitoneally in rats and mice, has been shown to mimic some cardiovascular complications of diabetes in the absence of hyperglycemia such as endothelial dysfunction, microvascular damage, atherogenesis [164–167], cardiac dysfunction [168,169], and renal damage [170–172] (Table 1). However, the direct contribution of MGO to bladder dysfunction remains poorly investigated. This model of exogenous animal supplementation with MGO clearly differs from the classical diabetic animals in that MGO is not generated from the endogenous glucose metabolism, and, consequently, does not itself affect the glucose levels and insulin sensitivity [164,173–175]. Intake of MGO to healthy mice for 7 days (500 to 2000 mg/kg) significantly increased the urine levels of this dicarbonyl molecule [176]. Serum levels achieved by a 12-week intake of 0.5% MGO to healthy mice [174,175,177] reached comparable levels to those found in plasma of diabetic/obese individuals [125,154]. Levels of MGO levels were also increased in both plasma and urine after a 6-month MGO administration to mice at the doses of 200 mg/kg [176] and 500 mg/kg [178]. Likewise, in high-fat fed mice, levels of plasma and urine levels of MGO were significantly higher than animals kept on low-fat diet [179]. Diabetic obese ob/ob mice also displayed elevated serum MGO compared with normoglycemic animals [16] (Table 1).

Table 1. Main findings produced by methylglyoxal (MGO) treatment in rodents.

Reference Number	Dose	Route of Administration	Animal and Strain	Sex	Treatment with MGO
[164]	50–75 mg/kg	Intraperitoneal	Wistar rat	Male	Microvascular damage Microvessel degeneration
[165]	50–75 mg/Kg	Drinking water	Spontaneously diabetic (GK) rats	Male	Endothelial dysfunction
[166]	50 mmol/L	Drinking water	C57Bl6 $ApoE^{-/-}$	Male	Atherosclerosis
[170]	50–75 mg/kg	Drinking water	Goto-Kakizaki (GK), nonobese type 2 diabetic rats	ND	Renal disease
[171]	17.25 mg/kg	Intraperitoneal	Sprague Dawley (SD) rats	ND	Renal disease
[172]	600 mg/kg/day	Oral	NMRI mice	Male	Diabetic nephropathy
[176]	500–2000 mg/kg	Drinking water	RAGE−/Glo1 ++ mice	Male Female	Renal dysfunction
[178]	500 mg/kg	Drinking water	RAGE-KO mice	Male Female	Increased airway resistance/decreased maximal inspiratory flow
[173,175]	0.5%	Drinking water	C57BL/6Junib mice	Male	Aggravation of allergic airway disease and acute lung injury

ND, nondetermined; T2DM, type 2 diabetes mellitus; KO, knockout.

Bladders from male mice treated orally with MGO for 4 weeks revealed tissue disorganization, partial loss of the urothelium, and mucosal edema along with marked cell infiltration [14]. Urodynamic evaluation (cystometric assays) in these male animals showed marked increases in micturition frequency and number of nonvoiding contractions (NVCs) with no alterations in bladder capacity [14]. Cystometric assays in male mice treated orally with MGO for an extended period (12 weeks) showed significant increases in the frequency of NVCs, bladder capacity, inter-micturition pressure, and residual volume [177]. In female mice treated with MGO for 12 weeks, cystometric assays confirmed urodynamic alterations such as increases in NVCs frequency, bladder capacity, inter-micturition pressure, and residual volume [15]. Using the model of spontaneous void spot assay (VSA) on filter paper, male mice treated with MGO for 12 weeks revealed an increased volume per void with no changes in the spot number as compared with the untreated group. In the female group, this treatment increased the spot number (mainly the number of microvolume spots) but, rather, reduced the volume per void [177,180]. During the MGO treatment, no alterations in the water consumption are observed in any group [177]. In this VSA assay, the term thigmotaxis refers to the wall-seeking behavior, that is, the tendency of mice to urinate next to the walls of the cage, which is interpreted as a rational response related to the fear of predation [181,182]. In healthy conditions, mice of both sexes will urinate at the corner of the filter paper, and when the animal loses this outlet control, urinating in the center of the filter, this may indicate bladder dysfunction. In the 12-week MGO treatment, whist the male mice had 95% of the voided spots in the corners of the filter paper, the voided spots in the female group were also detected in the center of the filter, indicating an altered outlet behavior in favor of an overactive bladder phenotype. The in vitro contractions to electrical-field stimulation (EFS; neurogenic contractions) as well as those induced by selective muscarinic and purinergic P2X1 receptor activation (using carbachol and α, β-methylene ATP as receptor agonists, respectively) were also evaluated in bladders of male and female mice treated with MGO for 12 weeks (Table 2). In intact bladder preparations of male mice, higher contractions to EFS, carbachol, and α,β-methylene ATP were observed after MGO treatment [177]. In the female group, higher contractile responses to EFS and α,β-methylene ATP (but not to carbachol) were also observed in intact bladder preparations from animals treated with MGO for 12 weeks [15]. An increased carbachol-induce response by MGO treatment in the female mice is solely observed when the urothelium is removed from the preparations. Moreover, the higher EFS-induced contractions in the MGO group were normalized by prior tissue incubation with the selective TRPA1 blocker HC-030031, suggesting that MGO exposure via TRPA1 activation leads to enhancement of purinergic over cholinergic neurotransmission in the bladder [180] (Figure 2). Table 2 summarizes the main in vivo and in vitro bladder alterations observed in male and female mice treated with MGO for 12 weeks.

Table 2. In vivo and in vitro bladder alterations in male and female mice treated with methylglyoxal (MGO) 12 weeks [15,177,180].

	Parameter	Male	Female
Urodynamic evaluation	Number of nonvoiding contractions (NVCs)	↑	—
	Frequency of voiding	↓	↑
	Bladder capacity	↑	↓
Bladder smooth Muscle contractility in vitro (presence of urothelium)	Neurogenic contractions (electrical-Field Stimulation, EFS)	↑	↑
	Muscarinic-mediated contractions (carbachol)	↑	—
	Purinergic-mediated contractions (α,β-methylene ATP)	↑	↑

Table 2. *Cont.*

	Parameter	Male	Female
Void spot analysis	Total void volume	↑	—
	Volume per void	↑	↓
	Urine spot number	—	↑
	Urine spot in center	—	↑
	Urine spot in corner	—	—
Histology	Collagen content	↑	↑

Arrows indicate ↑ increased; ↓ decreased or — unaltered parameters.

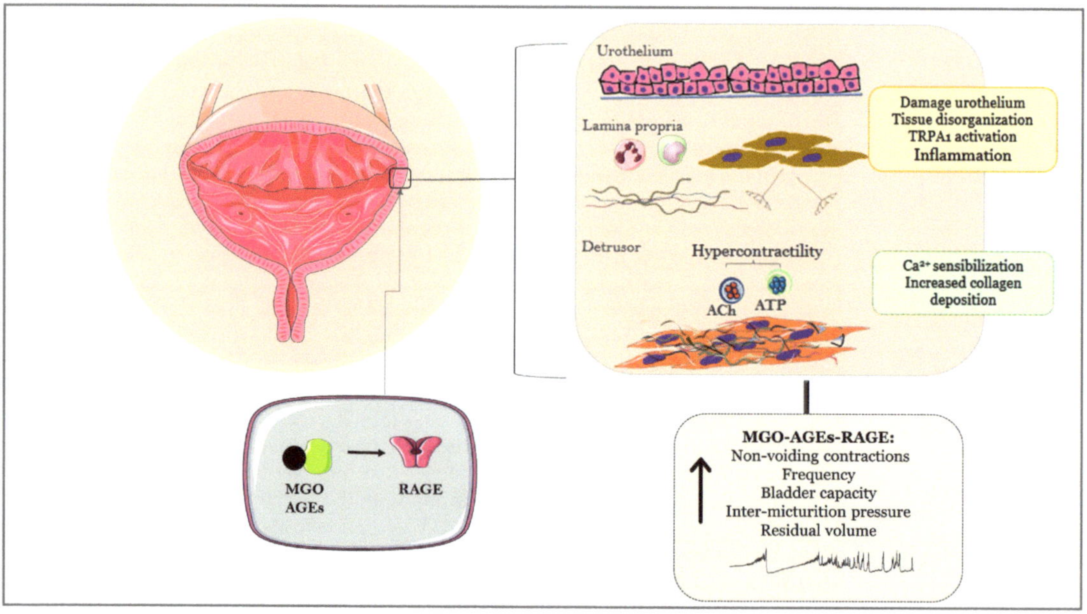

Figure 2. Bladder alterations at the level of urothelium, lamina propria, and detrusor smooth muscle in mice treated with methylglyoxal (MGO) for 4 and 12 weeks. Activation of the MGO–AGEs–RAGE axis leads to urothelial damage, tissue disorganization, edema, and inflammatory cellular infiltration, along with sensitivity alterations due to TRPA1 channel activation. The in vitro detrusor contractile responses to electrical-field stimulation (EFS), α,β-methylene ATP (purinergic P2X1 receptor agonist), and carbachol (nonselective muscarinic agonist) due to increased Ca^{2+} sensitization machinery are higher in MGO-treated mice. Higher collagen deposition is seen in bladders of MGO-treated mice. Urodynamic changes, including increases in nonvoiding contractions (NVCs), frequency, bladder capacity, inter-micturition pressure, and residual volume, may also be observed in MGO groups. Drugs capable of scavenging MGO and protecting bladder cells from oxidative insult, such as the polyphenols resveratrol and epigallocatechin-3-gallate, and the antihyperglycemic metformin exert reduce AGEs levels and oxidative stress in bladder tissues. This image was produced with the assistance of Servier Medical Art (Servier; https://smart.servier.com/ accessed on 18 March 2024), licensed under a Creative Commons Attribution 4.0 Unported License.

In MGO-treated mice, elevated levels of AGEs and RAGE in bladder tissues were also observed [15,177]. Likewise, hyperglycemic diabetic leptin-deficient male and female mice (ob/ob) exhibit bladder dysfunction, as evidenced by the increases in total void volume and volume per void (void spot assay) in addition to high collagen content in the bladders [16]. These bladder alterations were associated with high levels of total AGEs, MG-H1 and RAGE found in bladder tissues, which is consistent with the findings that the AGE breaker alagebrium (ALT-711) at 1 mg/kg during 8 weeks in the drinking water nearly reversed all the molecular and functional alterations in ob/ob mice [16] (Table 3).

Table 3. Protective effects of alagebrium (ALT-711) on the levels of total AGEs, MG-H1, RAGE and collagen in bladder tissues of obese diabetic ob/ob mice [16].

Parameter	ob/ob	ob/ob + ALT-711
Blood glucose	↑	—
AGEs in bladder	↑	↓
MG-H1 content in bladder	↑	↓
RAGE content in bladder	↑	↓
Collagen content in the bladders	↑	↓
Volume per void	↑	↓
Number of voids	↓	↑
Void size	↑	↓

Arrows indicate ↑ increased; ↓ decreased or — unaltered parameters.

7. Drugs Presenting Potential to Downregulate AGEs Formation and Oxidative Stress in Bladder Tissues

It is well established that NADPH oxidase and increased levels of superoxide anion (O_2^-) and hydrogen peroxide (H_2O_2) play a critical role in diabetic complications [55,183–192]. Oxidative stress at excessive levels also plays an important role in pathophysiology of bladder outlet obstruction [193], cyclophosphamide-induced cystitis [194], benign prostatic hyperplasia [195] and STZ-induced bladder dysfunction [196]. Hydrogen peroxide (H_2O_2) is reported to activate bladder afferent signaling inducing detrusor overactivity [197]. In human and dog bladders in vitro, H_2O_2 itself produced contractions and potentiated the contractions induced by electrical-field stimulation, an effect attenuated by the natural NADPH oxidase inhibitor apocynin [198]. Given obesity-associated bladder dysfunction correlates with increased oxidative stress and that MGO treatment leads to excess ROS production, it is plausible that drugs that inactivate MGO [20,199] or that protect bladder cells from the oxidative insult [200,201] offer an interesting approach to reduce the deleterious effects of AGEs in the bladder (Figure 3). Therefore, we summarized below some drugs reported to ameliorate bladder dysfunction in animals including some polyphenols and metformin whose protective mechanisms may be related to their ability to downregulate AGEs formation and oxidative stress in bladder tissues.

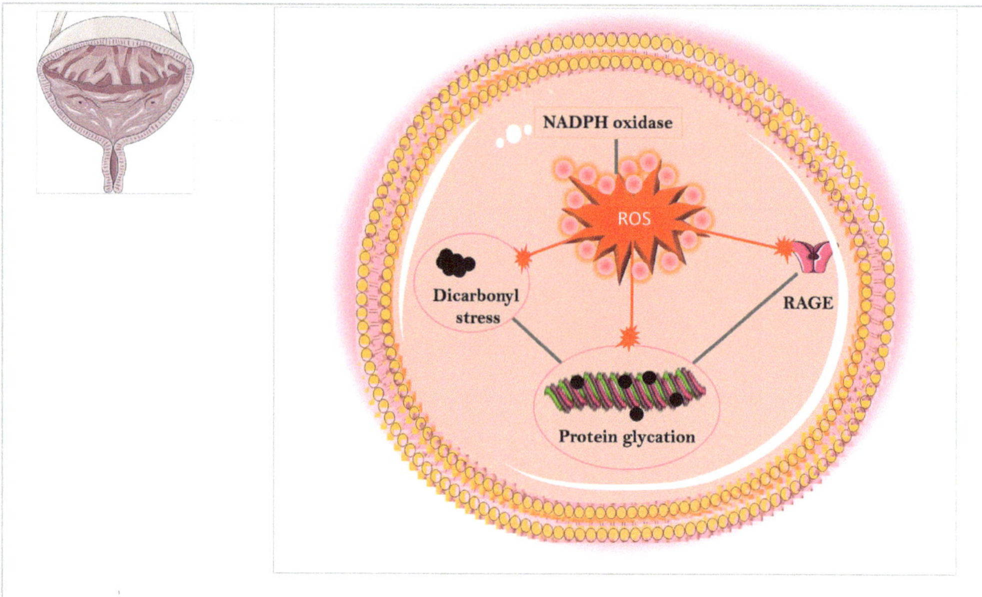

Figure 3. The abnormal accumulation of methylglyoxal (MGO) enhances the dicarbonyl stress leading to protein glycation and excessive RAGE-mediated ROS production in the urinary bladder. This image was produced with the assistance of Servier Medical Art (Servier; https://smart.servier.com/ accessed on 18 March 2024), licensed under a Creative Commons Attribution 4.0 Unported License.

7.1. Polyphenols: Resveratrol and Epigallocatechin-3-Gallate

Resveratrol is a polyphenol present in numerous plant-based foods that increases lipolysis and reduces lipogenesis in adipocytes, being suggested as a therapeutic alternative to treat obesity-related diseases [202,203]. Two-week therapy with resveratrol (100 mg/kg/day, given by gavage) in high-fat-diet-fed obese mice reduced the in vivo urodynamic changes, the in vitro bladder overactivity, and the ROS production in bladder tissues [104]. Resveratrol treatment also increased the nitric oxide levels and restored the impaired urethral relaxations in obese mice, an effect mimicked by the antioxidant enzyme SOD [112]. Likewise, the in vitro urethral hyperactivity was restored by resveratrol in obese mice [112]. In the bladders of STZ-induced diabetic rats, daily oral treatment with resveratrol (10 mg/kg) reduced the histological abnormalities and inhibited the expression and localization of markers of oxidative stress and DNA oxidative damage [204]. Intragastric administration of resveratrol (20 mg/kg/day) reduced bladder hypertrophy, tissue damage, inflammatory cell infiltration, and levels of inflammatory cytokines in the bladders of STZ-induced diabetic rats [205]. In the chronic prostatitis model in rats, oral administration of resveratrol (10 mg/kg) for 10 days reduced the resulting overactive bladder and fibrosis by reducing the protein expressions SCF, c-Kit, and p-AKT [206,207]. At the molecular level, resveratrol exhibited a high inhibition rate on the fluorescent formation of AGEs mainly due to scavenging free radicals and capturing MGO [20]. Epigallocatechin-3-gallate is another polyphenol compound present in green tea that has also favorable effects on bladder overactivity, as evidenced in ovariectomized rats fed standard chow [201] and high-fat, high-sugar diet [208]. Treatment with epigallocatechin-3-gallate reduced the expressions of transforming growth factor-β (TGF-β) and type I collagen, as well as the apoptosis and oxidative stress in the bladders [208,209]. In a bladder outlet obstruction (BOO) model in rats, intraperitoneal injection of epigallocatechin-3-gallate (4.5 mg/kg/day) reduced the histologic changes and submucosal endoplasmic reticulum (ER) stress-related apoptosis, recovering the bladder compliance and inter-contractile intervals [210]. At 6 mM, epigallocatechin-3-gallate was also shown to exert anti-AGEs activity through its capacity to strongly trap and inactivate MGO [211]. In diabetic db/db mice, 16-week oral administration of (+)-catechin (15, 30, and 60 mg/kg) directly trapped MGO, hence downregulating the downstream signal transduction and inflammatory response induced by AGE–RAGE interaction in the kidney [212]. Therefore, although the mechanisms behind the uroprotective actions of polyphenols on diabetes-associated bladder dysfunction deserve further investigation, they could involve the capacity of these molecules to directly trap MGO, further inhibiting MGO-induced glycation, AGEs formation, and RAGE activation.

7.2. Metformin

Metformin is a first-line pharmacological treatment for T2DM patients as monotherapy or in combination with sulfonylureas or dipeptidyl peptidase 4 inhibitors [213]. The orally administered doses of metformin (as immediate-release or extended-release formulations) usually vary from 0.5 to 2.5 g daily, being safety and effective for long-term glycemic control. Metformin is associated with low risk of hypoglycemia and documented cardiovascular benefits [214]. Metformin increases tissue sensitivity to insulin and decreases the levels of glycated hemoglobin by mechanisms involving the activation of adenosine monophosphate-activated protein kinase (AMPK) and non-AMPK pathways [215], but its exact mechanism of action remains largely incomplete [213]. Recently, metformin was shown to increase intestinal glucose uptake, influencing hepatic glucose production through a gut–liver crosstalk [216]. Metformin is among the molecules reported to strongly react with MGO [17], forming an imidazolinone metabolite [19]. In addition, in the plasma of T2DM patients, metformin, through its guanidine group, was shown to bind to MGO, reducing this dicarbonyl concentration [18], hence reducing AGEs formation, which paralleled a significant increase in Glo1 activity [217]. A two-week treatment of high-fat-diet-fed mice with metformin (300 mg/kg) reversed the bladder overactivity, as evidenced by in vivo and in vitro studies [106]. Metformin also normalized the enhanced serum levels of MGO and fluorescent AGEs in mice treated chronically with MGO [177]. In bladders of MGO-treated mice, metformin treatment reduced Glo1 expression and activity, urothelium thickness, and collagen content, as well as the in vitro and in vivo micturition dysfunction [177]. It is, therefore, plausible to suggest that the beneficial effects of metformin in obesity-associated bladder dysfunction rely at least in part on its MGO capturing property. Of interest, oral administration of metformin (150 mg/kg, gavage) reduced both bladder remodeling and dysfunction in models of partial bladder outlet obstruction in rats [218], erectile dysfunction in obese mice [219], and diabetic nephropathy in STZ-induced diabetes [220].

8. Concluding Remarks and Future Therapeutics

Diabetic bladder dysfunction is a highly prevalent condition manifesting as storage (such as OAB and urge incontinence) and voiding problems (poor emptying with resultant elevated capacity). Increased capacity and decreased sensation together with recurrent urinary tract infections may also be present in DBD. Preclinical models of T1DM and T2DM in rodents have provided further evidence confirming the relationship between diabetes and bladder dysfunction. Hyperglycemia in diabetic/obese patients significantly elevates the levels of α-dicarbonyl compounds, including MGO, in plasma and urine as a consequence of the glycolytic overload. MGO promotes post-translational modification of peptides and proteins, ultimately leading to the formation of AGEs such as MG-H1. AGEs bind their cell membrane-anchored ligand receptor RAGE, triggering multiple intracellular signaling pathways, among which ROS production at excessive levels plays a critical role. However, little is known about the importance of MGO generation and AGEs–RAGE activation in the pathophysiology of diabetic-associated bladder dysfunction. Voiding spot assays and cystometrical evaluation in mice treated chronically with MGO have revealed significant increases in total void volume, volume per void, micturition frequency, and nonvoiding contractions number, along with enhanced in vitro bladder contractility. Moreover, levels of MGO, AGEs, RAGE, and ROS are all elevated in the bladder tissues obtained from

MGO-treated animals and diabetic ob/ob mice. The antihyperglycemic drug metformin and the polyphenols resveratrol and epigallocatechin-3-gallate can directly scavenger MGO, exerting uroprotective actions. Therefore, we propose here that evaluation of MGO, AGEs, and RAGE levels may constitute important biomarkers of DBD pathophysiology. The design and development of new drugs that inhibit the MGO–AGEs–RAGE axis may become an interesting approach for the prevention and treatment of bladder dysfunction in diabetic conditions.

Author Contributions: Conceptualization and original draft preparation, E.A.; review and editing, A.L.O., M.G.d.O., F.Z.M. and E.A.; supervision, E.A.; funding acquisition, F.Z.M. and E.A. All authors have read and agreed to the published version of the manuscript.

Funding: This work was supported by "Fundação de Amparo à Pesquisa do Estado de São Paulo" (FAPESP), grant number 2017/15175-1.

Conflicts of Interest: The authors declare no conflicts of interest.

References

1. Rabbani, N.; Xue, M.; Thornalley, P.J. Dicarbonyl stress, protein glycation and the unfolded protein response. *Glycoconj. J.* **2021**, *38*, 331–340. [CrossRef] [PubMed]
2. Schalkwijk, C.G.; Stehouwer, C.D.A. Methylglyoxal, a Highly Reactive Dicarbonyl Compound, in Diabetes, Its Vascular Complications, and Other Age-Related Diseases. *Physiol. Rev.* **2020**, *100*, 407–461. [CrossRef]
3. Stratmann, B. Dicarbonyl Stress in Diabetic Vascular Disease. *Int. J. Mol. Sci.* **2022**, *23*, 6186. [CrossRef] [PubMed]
4. Kim, H.J.; Jeong, M.S.; Jang, S.B. Molecular characteristics of RAGE and advances in small-molecule inhibitors. *Int. J. Mol. Sci.* **2021**, *22*, 6904. [CrossRef]
5. Abrams, P.; Cardozo, L.; Fall, M.; Griffiths, D.; Rosier, P.; Ulmsten, U.; Van Kerrebroeck, P.; Victor, A.; Wein, A.; Standardisation Sub-Committee of the International Continence Society. The standardisation of terminology in lower urinary tract function: Report from the standardisation sub-committee of the International Continence Society. *Urology* **2003**, *61*, 37–49. [CrossRef] [PubMed]
6. D'Ancona, C.; Haylen, B.; Oelke, M.; Abranches-Monteiro, L.; Arnold, E.; Goldman, H.; Hamid, R.; Homma, Y.; Marcelissen, T.; Rademakers, K.; et al. Standardisation Steering Committee ICS and the ICS Working Group on Terminology for Male Lower Urinary Tract & Pelvic Floor Symptoms and Dysfunction. The International Continence Society (ICS) report on the terminology for adult male lower urinary tract and pelvic floor symptoms and dysfunction. *Neurourol. Urodyn.* **2019**, *38*, 433–477. [CrossRef] [PubMed]
7. Liu, G.; Daleeneshgari, F. Diabetic bladder dysfunction. *Chin. Med. J.* **2014**, *127*, 1357–1364. [PubMed]
8. Chapple, C.R.; Osman, N.I.; Birder, L.; van Koeveringe, G.A.; Oelke, M.; Nitti, V.W.; Drake, M.J.; Yamaguchi, O.; Abrams, P.; Smith, P.P. The underactive bladder: A new clinical concept? *Eur. Urol.* **2015**, *68*, 351–353. [CrossRef]
9. Cohn, J.A.; Brown, E.T.; Kaufman, M.R.; Dmochowski, R.R.; Reynolds, W.S. Underactive bladder in women: Is there any evidence? *Curr. Opin. Urol.* **2016**, *26*, 309–314. [CrossRef]
10. Wang, J.; Ren, L.; Liu, X.; Liu, J.; Ling, Q. Underactive bladder and detrusor underactivity: New advances and prospectives. *Int. J. Mol. Sci.* **2023**, *24*, 15517. [CrossRef]
11. Yoshimura, N.; Chancellor, M.B.; Andersson, K.E.; Christ, G.J. Recent advances in understanding the biology of diabetes-associated bladder complications and novel therapy. *BJU Int.* **2005**, *95*, 733–738. [CrossRef] [PubMed]
12. Daneshgari, F.; Liu, G.; Hanna-Mitchell, A.T. Path of translational discovery of urological complications of obesity and diabetes. *Am. J. Physiol. Renal Physiol.* **2017**, *312*, F887–F896. [CrossRef] [PubMed]
13. Galì, A.; Mucciardi, G.; Butticè, S.; Subba, E.; D'Amico, C.; Lembo, F.; Magno, C. Correlation between advanced glycation end-products, lower urinary tract symptoms and bladder dysfunctions in patients with type 2 diabetes mellitus. *Low. Urin. Tract. Symptoms* **2017**, *9*, 15–20. [CrossRef] [PubMed]
14. De Oliveira, M.G.; Medeiros, M.L.; Tavares, E.B.G.; Mónica, F.Z.; Antunes, E. Methylglyoxal, a reactive glucose metabolite, induces bladder overactivity in addition to inflammation in mice. *Front. Physiol.* **2020**, *11*, 290. [CrossRef] [PubMed]
15. Oliveira, A.L.; Medeiros, M.L.; de Oliveira, M.G.; Teixeira, C.J.; Mónica, F.Z.; Antunes, E. Enhanced RAGE expression and excess reactive-oxygen species production mediates Rho Kinase-dependent detrusor overactivity after methylglyoxal exposure. *Front. Physiol.* **2022**, *13*, 860342. [CrossRef] [PubMed]
16. Oliveira, A.L.; Medeiros, M.L.; Ghezzi, A.C.; Dos Santos, G.A.; Mello, G.C.; Mónica, F.Z.; Antunes, E. Evidence that methylglyoxal and receptor for advanced glycation end products are implicated in bladder dysfunction of obese diabetic *ob/ob* mice. *Am. J. Physiol. Renal Physiol.* **2023**, *325*, F436–F447. [CrossRef] [PubMed]
17. Ruggiero-Lopez, D.; Lecomte, M.; Moinet, G.; Patereau, G.; Lagarde, M.; Wiernsperger, N. Reaction of metformin with dicarbonyl compounds. Possible implication in the inhibition of advanced glycation end product formation. *Biochem. Pharmacol.* **1999**, *58*, 1765–1773. [CrossRef] [PubMed]

18. Beisswenger, P.J.; Howell, S.K.; Touchette, A.D.; Lal, S.; Szwergold, B.S. Metformin reduces systemic methylglyoxal levels in type 2 diabetes. *Diabetes* **1999**, *48*, 198–202. [CrossRef] [PubMed]
19. Kinsky, O.R.; Hargraves, T.L.; Anumol, T.; Jacobsen, N.E.; Dai, J.; Snyder, S.A.; Monks, T.J.; Lau, S.S. Metformin scavenges methylglyoxal to form a novel imidazolinone metabolite in humans. *Chem. Res. Toxicol.* **2016**, *29*, 227–234. [CrossRef] [PubMed]
20. Liu, H.; Huo, X.; Wang, S.; Yin, Z. The inhibitory effects of natural antioxidants on protein glycation as well as aggregation induced by methylglyoxal and underlying mechanisms. *Colloids Surf. B Biointerfaces* **2022**, *212*, 112360. [CrossRef]
21. Roy, H.A.; Green, A.L. The central autonomic network and regulation of bladder function. *Front. Neurosci.* **2019**, *13*, 535. [CrossRef] [PubMed]
22. Cox, L.; Rovner, E.S. Lower urinary tract symptoms in women: Epidemiology, diagnosis, and management. *Curr. Opin. Urol.* **2016**, *26*, 328–333. [CrossRef] [PubMed]
23. Irwin, D.E.; Milsom, I.; Hunskaar, S.; Reilly, K.; Kopp, Z.; Herschorn, S.; Coyne, K.; Kelleher, C.; Hampel, C.; Artibani, W.; et al. Population-based survey of urinary incontinence, overactive bladder, and other lower urinary tract symptoms in five countries: Results of the EPIC study. *Eur. Urol.* **2006**, *50*, 1306–1315. [CrossRef] [PubMed]
24. Milsom, I.; Abrams, P.; Cardozo, L.; Roberts, R.G.; Thüroff, J.; Wein, A.J. How widespread are the symptoms of an overactive bladder and how are they managed? A population-based prevalence study. *BJU Int.* **2001**, *87*, 760–766. [CrossRef] [PubMed]
25. Stewart, W.F.; Van Rooyen, J.B.; Cundiff, G.W.; Abrams, P.; Herzog, A.R.; Corey, R.; Hunt, T.L.; Wein, A.J. Prevalence and burden of overactive bladder in the United States. *World J. Urol.* **2003**, *20*, 327–336. [CrossRef]
26. Teloken, C.; Caraver, F.; Weber, F.A.; Teloken, P.E.; Moraes, J.F.; Sogari, P.R.; Graziottin, T.M. Overactive bladder: Prevalence and implications in Brazil. *Eur. Urol.* **2006**, *49*, 1087–1092. [CrossRef] [PubMed]
27. Moreira, E.D., Jr.; Neves, R.C.; Neto, A.F.; Duarte, F.G.; Moreira, T.L.; Lobo, C.F.; Glasser, D.B. A population-based survey of lower urinary tract symptoms (LUTS) and symptom-specific bother: Results from the Brazilian LUTS epidemiology study (BLUES). *World J. Urol.* **2013**, *31*, 1451–1458. [CrossRef]
28. Coyne, K.S.; Kvasz, M.; Ireland, A.M.; Milsom, I.; Kopp, Z.S.; Chapple, C.R. Urinary incontinence and its relationship to mental health and health-related quality of life in men and women in Sweden, the United Kingdom, and the United States. *Eur. Urol.* **2012**, *61*, 88–95. [CrossRef] [PubMed]
29. Gomes, C.M.; Averbeck, M.A.; Koyama, M.; Soler, R. Impact of OAB symptoms on work, quality of life and treatment-seeking behavior in Brazil. *Curr. Med. Res. Opin.* **2020**, *36*, 1403–1415. [CrossRef] [PubMed]
30. Jia, G.; Bai, H.; Mather, B.; Hill, M.A.; Jia, G.; Sowers, J.R. Diabetic vasculopathy: Molecular mechanisms and clinical insights. *Int. J. Mol. Sci.* **2024**, *25*, 804. [CrossRef] [PubMed]
31. Rohrmann, S.; Smit, E.; Giovannucci, E.; Platz, E.A. Association between markers of the metabolic syndrome and lower urinary tract symptoms in the Third National Health and Nutrition Examination Survey (NHANES III). *Int. J. Obes.* **2005**, *29*, 310–316. [CrossRef] [PubMed]
32. Hunskaar, S. A systematic review of overweight and obesity as risk factors and targets for clinical intervention for urinary incontinence in women. *Neurourol. Urodyn.* **2008**, *27*, 749–757. [CrossRef] [PubMed]
33. Tai, H.C.; Chung, S.D.; Ho, C.H.; Tai, T.Y.; Yang, W.S.; Tseng, C.H.; Wu, H.P.; Yu, H.J. Metabolic syndrome components worsen lower urinary tract symptoms in women with type 2 diabetes. *J. Clin. Endocrinol. Metab.* **2010**, *95*, 1143–1150. [CrossRef] [PubMed]
34. Uzun, H.; Zorba, O.U. Metabolic syndrome in female patients with overactive bladder. *Urology* **2012**, *79*, 72–75. [CrossRef] [PubMed]
35. Bang, W.J.; Lee, J.Y.; Koo, K.C.; Hah, Y.S.; Lee, D.H.; Cho, K.S. Is type-2 diabetes mellitus associated with overactive bladder symptoms in men with lower urinary tract symptoms? *Urology* **2014**, *84*, 670–674. [CrossRef] [PubMed]
36. Bunn, F.; Kirby, M.; Pinkney, E.; Cardozo, L.; Chapple, C.; Chester, K.; Cruz, F.; Haab, F.; Kelleher, C.; Milsom, I.; et al. Is there a link between overactive bladder and the metabolic syndrome in women? A systematic review of observational studies. *Int. J. Clin. Pract.* **2015**, *69*, 199–217. [CrossRef] [PubMed]
37. Zacche, M.M.; Giarenis, I.; Thiagamoorthy, G.; Robinson, D.; Cardozo, L. Is there an association between aspects of the metabolic syndrome and overactive bladder? A prospective cohort study in women with lower urinary tract symptoms. *Eur. J. Obstet. Gynecol. Reprod. Biol.* **2017**, *217*, 1–5. [CrossRef] [PubMed]
38. Lai, H.H.; Helmuth, M.E.; Smith, A.R.; Wiseman, J.B.; Gillespie, B.W.; Kirkali, Z. Symptoms of lower urinary tract dysfunction research network (LURN). relationship between central Obesity, general obesity, overactive bladder syndrome and urinary incontinence among male and female patients seeking care for their lower urinary tract symptoms. *Urology* **2019**, *123*, 34–43. [CrossRef] [PubMed]
39. Kim, J.K.; Lee, Y.G.; Han, K.; Han, J.H. Obesity, metabolic health, and urological disorders in adults: A nationwide population-based study. *Sci. Rep.* **2021**, *11*, 8687. [CrossRef] [PubMed]
40. Pereira, T.A.; D'ancona, C.A.L.; Cândido, E.C.; Achermann, A.P.P.; Chaim, E.A. Prevalence of LUTS and urodynamics results in obese women. *Neurourol. Urodyn.* **2022**, *41*, 468–474. [CrossRef] [PubMed]
41. Heo, J.E.; Kim, D.G.; Yoo, J.W.; Lee, K.S. Metabolic syndrome-related factors as possible targets for lower urinary tract symptoms in Korean males. *Aging Male* **2023**, *26*, 6–12. [CrossRef] [PubMed]
42. Chilaka, C.; Toozs-Hobson, P.; Chilaka, V. Pelvic floor dysfunction and obesity. *Best Pract. Res. Clin. Obstet. Gynaecol.* **2023**, *90*, 102389. [CrossRef] [PubMed]

43. Li, S.; Zou, J.; Wang, Z.; Wang, M.; Yuan, Y.; Lv, H. Correlation Between Insulin resistance and urinary incontinence in female patients with type 2 Diabetes Mellitus. *Int. Urogynecol. J.* **2024**, *35*, 431–440. [CrossRef] [PubMed]
44. Alsannan, B.; Laganà, A.S.; Alhermi, J.; Almansoor, S.; Ayed, A.; Venezia, R.; Etrusco, A. Prevalence of overactive bladder among overweight and obese women: A prospective cross-sectional cohort study. *Eur. J. Obstet. Gynecol. Reprod. Biol.* **2024**, *295*, 59–64. [CrossRef] [PubMed]
45. Park, H.K.; Lee, H.W.; Lee, K.S.; Byun, S.S.; Jeong, S.J.; Hong, S.K.; Lee, S.E.; Park, J.H.; Lee, S.B.; Kim, K.W.; et al. Relationship between lower urinary tract symptoms and metabolic syndrome in a community-based elderly population. *Urology* **2008**, *72*, 556. [CrossRef] [PubMed]
46. Eom, C.S.; Park, J.H.; Cho, B.L.; Choi, H.C.; Oh, M.J.; Kwon, H.T. Metabolic syndrome and accompanying hyperinsulinemia have favorable effects on lower urinary tract symptoms in a generally healthy screened population. *J. Urol.* **2011**, *186*, 175–179. [CrossRef] [PubMed]
47. Peng, L.; Di, X.P.; He, S.X.; Zeng, X.; Shen, H.; Zhu, H.L.; Luo, D.Y. Metabolic syndrome in women with and without interstitial cystitis/bladder pain syndrome. *Int. Urogynecol. J.* **2021**, *32*, 1299–1306. [CrossRef] [PubMed]
48. Pugliese, G.; Liccardi, A.; Graziadio, C.; Barrea, L.; Muscogiuri, G.; Colao, A. Obesity and infectious diseases: Pathophysiology and epidemiology of a double pandemic condition. *Int. J. Obes.* **2022**, *46*, 449–465. [CrossRef]
49. Pari, B.; Gallucci, M.; Ghigo, A.; Brizzi, M.F. Insight on Infections in diabetic setting. *Biomedicines* **2023**, *11*, 971. [CrossRef] [PubMed]
50. Kirby, M.G.; Wagg, A.; Cardozo, L.; Chapple, C.; Castro-Diaz, D.; de Ridder, D.; Espuna-Pons, M.; Haab, F.; Kelleher, C.; Kölbl, H.; et al. Overactive bladder: Is there a link to the metabolic syndrome in men? *Neurourol. Urodyn.* **2010**, *29*, 1360–1364. [CrossRef] [PubMed]
51. Calogero, A.E.; Burgio, G.; Condorelli, R.A.; Cannarella, R.; La Vignera, S. Epidemiology and risk factors of lower urinary tract symptoms/benign prostatic hyperplasia and erectile dysfunction. *Aging Male* **2019**, *22*, 12–19. [CrossRef] [PubMed]
52. Omran, A.; Leca, B.M.; Oštarijaš, E.; Graham, N.; Da Silva, A.S.; Zaïr, Z.M.; Miras, A.D.; le Roux, C.W.; Vincent, R.P.; Cardozo, L.; et al. Metabolic syndrome is associated with prostate enlargement: A systematic review, meta-analysis, and meta-regression on patients with lower urinary tract symptom factors. *Ther. Adv. Endocrinol. Metab.* **2021**, *12*, 20420188211066210. [CrossRef] [PubMed]
53. Subak, L.L.; Richter, H.E.; Hunskaar, S. Obesity and urinary incontinence: Epidemiology and clinical research update. *J. Urol.* **2009**, *182*, S2–S7. [CrossRef]
54. Bauer, S.R.; Harrison, S.L.; Cawthon, P.M.; Senders, A.; Kenfield, S.A.; Suskind, A.M.; McCulloch, C.E.; Covinsky, K.; Marshall, L.M. Longitudinal changes in adiposity and lower urinary tract symptoms among older men. *J. Gerontol. Ser. A* **2022**, *77*, 2102–2109. [CrossRef] [PubMed]
55. Daneshgari, F.; Liu, G.; Birder, L.; Hanna-Mitchell, A.T.; Chacko, S. Diabetic bladder dysfunction: Current translational knowledge. *J. Urol.* **2009**, *182*, S18–S26. [CrossRef] [PubMed]
56. Wang, C.C.; Chancellor, M.B.; Lin, J.M.; Hsieh, J.H.; Yu, H.J. Type 2 diabetes but not metabolic syndrome is associated with an increased risk of lower urinary tract symptoms and erectile dysfunction in men aged <45 years. *BJU Int.* **2010**, *105*, 1136–1140. [CrossRef] [PubMed]
57. Moul, S.; McVary, K.T. Lower urinary tract symptoms, obesity and the metabolic syndrome. *Curr. Opin. Urol.* **2010**, *20*, 7–12. [CrossRef] [PubMed]
58. Arrellano-Valdez, F.; Urrutia-Osorio, M.; Arroyo, C.; Soto-Vega, E. A comprehensive review of urologic complications in patients with diabetes. *Springerplus* **2014**, *3*, 549. [CrossRef] [PubMed]
59. Blair, Y.; Wessells, H.; Pop-Busui, R.; Ang, L.; Sarma, A.V. Urologic complications in diabetes. *J. Diabetes Complicat.* **2022**, *36*, 108288. [CrossRef] [PubMed]
60. Yuan, Z.; Tang, Z.; He, C.; Tang, W. Diabetic cystopathy: A review. *J. Diabetes* **2015**, *7*, 442–447. [CrossRef] [PubMed]
61. Gasbarro, G.; Lin, D.L.; Vurbic, D.; Quisno, A.; Kinley, B.; Daneshgari, F.; Damaser, M.S. Voiding function in obese and type 2 diabetic female rats. *Am. J. Physiol. Renal Physiol.* **2010**, *298*, F72–F77. [CrossRef] [PubMed]
62. Fry, C.H.; Daneshgari, F.; Thor, K.; Drake, M.; Eccles, R.; Kanai, A.J.; Birder, L.A. Animal models and their use in understanding lower urinary tract dysfunction. *Neurourol. Urodyn.* **2010**, *29*, 603–608. [CrossRef]
63. Liu, G.; Li, M.; Vasanji, A.; Daneshgari, F. Temporal diabetes and diuresis-induced alteration of nerves and vasculature of the urinary bladder in the rat. *BJU Int.* **2011**, *107*, 1988–1993. [CrossRef] [PubMed]
64. Nirmal, J.; Tyagi, P.; Chuang, Y.C.; Lee, W.C.; Yoshimura, N.; Huang, C.C.; Rajaganapathy, B.; Chancellor, M.B. Functional and molecular characterization of hyposensitive underactive bladder tissue and urine in streptozotocin-induced diabetic rat. *PLoS ONE* **2014**, *9*, e102644. [CrossRef] [PubMed]
65. He, Q.; Babcook, M.A.; Shukla, S.; Shankar, E.; Wang, Z.; Liu, G.; Erokwu, B.O.; Flask, C.A.; Lu, L.; Daneshgari, F.; et al. Obesity-initiated metabolic syndrome promotes urinary voiding dysfunction in a mouse model. *Prostate* **2016**, *76*, 964–976. [CrossRef] [PubMed]
66. De Oliveira, M.G.; Nascimento, D.M.; Alexandre, E.C.; Bonilla-Becerra, S.M.; Zapparoli, A.; Mónica, F.Z.; Antunes, E. Menthol ameliorates voiding dysfunction in types I and II diabetic mouse model. *Neurourol. Urodyn.* **2018**, *37*, 2510–2518. [CrossRef] [PubMed]

67. Blaha, I.; López-Oliva, M.E.; Martínez, M.P.; Recio, P.; Agis-Torres, Á.; Martínez, A.C.; Benedito, S.; García-Sacristán, A.; Prieto, D.; Fernandes, V.S.; et al. Bladder dysfunction in an obese Zucker rat: The role of TRPA1 channels, oxidative Stress, and hydrogen sulfide. *Oxid. Med. Cell Longev.* **2019**, *2019*, 5641645. [CrossRef] [PubMed]
68. Eleazu, C.O.; Eleazu, K.C.; Chukwuma, S.; Essien, U.N. Review of the mechanism of cell death resulting from streptozotocin challenge in experimental animals, its practical use and potential risk to humans. *J. Diabetes Metab. Disord.* **2013**, *12*, 60. [CrossRef] [PubMed]
69. Pandey, S.; Chmelir, T.; Chottova, D.M. Animal models in diabetic research-history, presence, and future perspectives. *Biomedicines* **2023**, *11*, 2852. [CrossRef] [PubMed]
70. Andersson, P.O.; Malmgren, A.; Uvelius, B. Cystometrical and in vitro evaluation of urinary bladder function in rats with STZ-induced diabetes. *J. Urol.* **1988**, *139*, 1359–1362. [CrossRef]
71. Eika, B.; Levin, R.M.; Longhurst, P.A. Comparison of urinary bladder function in rats with hereditary diabetes insipidus, streptozotocin-induced diabetes mellitus, and nondiabetic osmotic diuresis. *J. Urol.* **1994**, *151*, 496–502. [CrossRef] [PubMed]
72. Tammela, T.L.; Leggett, R.E.; Levin, R.M.; Longhurst, P.A. Temporal changes in micturition and bladder contractility after sucrose diuresis and streptozotocin-induced diabetes mellitus in rats. *J. Urol.* **1995**, *153*, 2014–2021. [CrossRef] [PubMed]
73. Turner, W.H.; Brading, A.F. Smooth muscle of the bladder in the normal and the diseased state: Pathophysiology, diagnosis and treatment. *Pharmacol. Ther.* **1997**, *75*, 77–110. [CrossRef] [PubMed]
74. Pitre, D.A.; Ma, T.; Wallace, L.J.; Bauer, J.A. Time-dependent urinary bladder remodeling in the streptozotocin-induced diabetic rat model. *Acta Diabetol.* **2002**, *39*, 23–27. [CrossRef] [PubMed]
75. Arioglu, I.E.; Ellenbroek, J.H.; Michel, M.C. A systematic review of urinary bladder hypertrophy in experimental diabetes: Part I. Streptozotocin-induced rat models. *Neurourol. Urodyn.* **2018**, *37*, 1212–1219. [CrossRef] [PubMed]
76. Liu, G.; Daneshgari, F. Alterations in neurogenically mediated contractile responses of urinary bladder in rats with diabetes. *Am. J. Physiol. Renal Physiol.* **2005**, *288*, F1220–F1226. [CrossRef] [PubMed]
77. Daneshgari, F.; Huang, X.; Liu, G.; Bena, J.; Saffore, L.; Powell, C.T. Temporal differences in bladder dysfunction caused by diabetes, diuresis, and treated diabetes in mice. *Am. J. Physiol. Regul. Integr. Comp. Physiol.* **2006**, *290*, R1728–R1735. [CrossRef] [PubMed]
78. Daneshgari, F.; Liu, G.; Imrey, P.B. Time dependent changes in diabetic cystopathy in rats include compensated and decompensated bladder function. *J. Urol.* **2006**, *176*, 380–386. [CrossRef] [PubMed]
79. Xiao, N.; Huang, Y.; Kavran, M.; Elrashidy, R.A.; Liu, G. Short-term diabetes- and diuresis-induced alterations of the bladder are mostly reversible in rats. *Int. J. Urol.* **2015**, *22*, 410–415. [CrossRef] [PubMed]
80. Yesilyurt, Z.E.; Matthes, J.; Hintermann, E.; Castañeda, T.R.; Elvert, R.; Beltran-Ornelas, J.H.; Silva-Velasco, D.L.; Xia, N.; Kannt, A.; Christen, U.; et al. Analysis of 16 studies in nine rodent models does not support the hypothesis that diabetic polyuria is a main reason of urinary bladder enlargement. *Front. Physiol.* **2022**, *13*, 923555. [CrossRef] [PubMed]
81. Cao, N.; Alexandre, E.C.; Gotoh, D.; Kurobe, M.; Mizoguchi, S.; Gu, B.; Yoshimura, N. Urethral dysfunction and alterations of nitric oxide mechanisms in streptozotocin-induced diabetic rats with or without low-dose insulin treatment. *Life Sci.* **2020**, *249*, 117537. [CrossRef] [PubMed]
82. Gotoh, D.; Cao, N.; Alexandre, E.C.; Saito, T.; Morizawa, Y.; Hori, S.; Miyake, M.; Torimoto, K.; Fujimoto, K.; Yoshimura, N. Effects of low-dose insulin or a soluble guanylate cyclase activator on lower urinary tract dysfunction in streptozotocin-induced diabetic rats. *Life Sci.* **2021**, *286*, 120001. [CrossRef] [PubMed]
83. Erdogan, B.R.; Liu, G.; Arioglu-Inan, E.; Michel, M.C. Established and emerging treatments for diabetes-associated lower urinary tract dysfunction. *Naunyn-Schmiedeberg's Arch. Pharmacol.* **2022**, *395*, 887–906. [CrossRef] [PubMed]
84. Leiria, L.O.; Mónica, F.Z.; Carvalho, F.D.; Claudino, M.A.; Franco-Penteado, C.F.; Schenka, A.; Grant, A.D.; De Nucci, G.; Antunes, E. Functional, morphological and molecular characterization of bladder dysfunction in streptozotocin-induced diabetic mice: Evidence of a role for L-type voltage-operated Ca^{2+} channels. *Br. J. Pharmacol.* **2011**, *163*, 1276–1288. [CrossRef]
85. Yu, W.; Ackert-Bicknell, C.; Larigakis, J.D.; MacIver, B.; Steers, W.D.; Churchill, G.A.; Hill, W.G.; Zeidel, M.L. Spontaneous voiding by mice reveals strain-specific lower urinary tract function to be a quantitative genetic trait. *Am. J. Physiol. Renal Physiol.* **2014**, *306*, F1296–F1307. [CrossRef] [PubMed]
86. Bjorling, D.E.; Wang, Z.; Vezina, C.M.; Ricke, W.A.; Keil, K.P.; Yu, W.; Guo, L.; Zeidel, M.L.; Hill, W.G. Evaluation of voiding assays in mice: Impact of genetic strains and sex. *Am. J. Physiol. Renal Physiol.* **2015**, *308*, F1369–F1378. [CrossRef] [PubMed]
87. Mossa, A.H.; Galan, A.; Cammisotto, P.G.; Velasquez Flores, M.; Shamout, S.; Barcelona, P.; Saragovi, H.U.; Campeau, L. Antagonism of proNGF or its receptor $p75^{NTR}$ reverses remodelling and improves bladder function in a mouse model of diabetic voiding dysfunction. *Diabetologia* **2020**, *63*, 1932–1946. [CrossRef] [PubMed]
88. Yoshizawa, T.; Hayashi, Y.; Yoshida, A.; Yoshida, S.; Ito, Y.; Yamaguchi, K.; Yamada, S.; Takahashi, S. Concomitant alteration in number and affinity of P_2X and muscarinic receptors are associated with bladder dysfunction in early stage of diabetic rats. *Int. Urol. Nephrol.* **2018**, *50*, 451–458. [CrossRef] [PubMed]
89. Hughes, F.M., Jr.; Allkanjari, A.; Odom, M.R.; Jin, H.; Purves, J.T. Diabetic bladder dysfunction progresses from an overactive to an underactive phenotype in a type-1 diabetic mouse model (Akita female mouse) and is dependent on NLRP3. *Life Sci.* **2022**, *299*, 120528. [CrossRef] [PubMed]

90. Hughes, F.M., Jr.; Allkanjari, A.; Odom, M.R.; Mulcrone, J.E.; Jin, H.; Purves, J.T. Male Akita mice develop signs of bladder underactivity independent of NLRP3 as a result of a decrease in neurotransmitter release from efferent neurons. *Am. J. Physiol. Renal Physiol.* **2023**, *325*, F61–F72. [CrossRef]
91. Mónica, F.Z.; Antunes, E. Stimulators and activators of soluble guanylate cyclase for urogenital disorders. *Nat. Rev. Urol.* **2018**, *15*, 42–54. [CrossRef] [PubMed]
92. Torimoto, K.; Fraser, M.O.; Hirao, Y.; De Groat, W.C.; Chancellor, M.B.; Yoshimura, N. Urethral dysfunction in diabetic rats. *J. Urol.* **2004**, *171*, 1959–1964. [CrossRef] [PubMed]
93. Melman, A.; Zotova, E.; Kim, M.; Arezzo, J.; Davies, K.; DiSanto, M.; Tar, M. Longitudinal studies of time-dependent changes in both bladder and erectile function after streptozotocin-induced diabetes in Fischer 344 male rats. *BJU Int.* **2009**, *104*, 1292–1300. [CrossRef] [PubMed]
94. Radenković, M.; Stojanović, M.; Prostran, M. Experimental diabetes induced by alloxan and streptozotocin: The current state of the art. *J. Pharmacol. Toxicol. Methods* **2016**, *78*, 13–31. [CrossRef] [PubMed]
95. Uvelius, B. Detrusor smooth muscle in rats with alloxan-induced diabetes. *J. Urol.* **1986**, *136*, 949–952. [CrossRef] [PubMed]
96. Paro, M.; Italiano, G.; Travagli, R.A.; Petrelli, L.; Zanoni, R.; Prosdocimi, M.; Fiori, M.G. Cystometric changes in alloxan diabetic rats: Evidence for functional and structural correlates of diabetic autonomic neuropathy. *J. Auton. Nerv. Syst.* **1990**, *30*, 1–11. [CrossRef]
97. Rodrigues, A.A., Jr.; Suaid, H.J.; Tucci, S., Jr.; Fazan, V.P.; Foss, M.C.; Cologna, A.J.; Martins, A.C. Long term evaluation of functional and morphological bladder alterations on alloxan-induced diabetes and aging: Experimental study in rats. *Acta Cir. Bras.* **2008**, *23* (Suppl. S1), 53–58; discussion 58. [CrossRef] [PubMed]
98. Rocha, J.N. Functional and biochemical characteristics of urinary bladder muscarinic receptors in long-term alloxan diabetic rats. *Einstein* **2015**, *13*, 404–409. [CrossRef] [PubMed]
99. Chang, S.; Hypolite, J.A.; DiSanto, M.E.; Changolkar, A.; Wein, A.J.; Chacko, S. Increased basal phosphorylation of detrusor smooth muscle myosin in alloxan-induced diabetic rabbit is mediated by upregulation of Rho-kinase beta and CPI-17. *Am. J. Physiol. Renal Physiol.* **2006**, *290*, F650–F656. [CrossRef]
100. Ichiyanagi, N.; Tsujii, T.; Masuda, H.; Kihara, K.; Goto, M.; Azuma, H. Changed responsiveness of the detrusor in rabbits with alloxan induced hyperglycemia: Possible role of 5-hydroxytryptamine for diabetic bladder dysfunction. *J. Urol.* **2002**, *168*, 303–307. [CrossRef] [PubMed]
101. Aravani, D.; Kassi, E.; Chatzigeorgiou, A.; Vakrou, S. Cardiometabolic syndrome: An update on available mouse models. *Thromb. Haemost.* **2021**, *121*, 703–715. [CrossRef] [PubMed]
102. Winzell, M.S.; Ahrén, B. The high-fat diet-fed mouse: A model for studying mechanisms and treatment of impaired glucose tolerance and type 2 diabetes. *Diabetes* **2004**, *53* (Suppl. S3), S215–S219. [CrossRef] [PubMed]
103. Rahman, N.U.; Phonsombat, S.; Bochinski, D.; Carrion, R.E.; Nunes, L.; Lue, T.F. An animal model to study lower urinary tract symptoms and erectile dysfunction: The hyperlipidaemic rat. *BJU Int.* **2007**, *100*, 658–663. [CrossRef] [PubMed]
104. Alexandre, E.C.; Calmasini, F.B.; de Oliveira, M.G.; Silva, F.H.; da Silva, C.P.V.; André, D.M.; Leonardo, F.C.; Delbin, M.A.; Antunes, E. Chronic treatment with resveratrol improves overactive bladder in obese mice via antioxidant activity. *Eur. J. Pharmacol.* **2016**, *788*, 29–36. [CrossRef] [PubMed]
105. Calmasini, F.B.; de Oliveira, M.G.; Alexandre, E.C.; da Silva, F.H.; da Silva, C.P.V.; Candido, T.Z.; Antunes, E.; Mónica, F.Z. Long-term treatment with the beta-3 adrenoceptor agonist, mirabegron ameliorates detrusor overactivity and restores cyclic adenosine monophosphate (cAMP) levels in obese mice. *Neurourol. Urodyn.* **2017**, *36*, 1511–1518. [CrossRef] [PubMed]
106. Leiria, L.O.; Sollon, C.; Calixto, M.C.; Lintomen, L.; Mónica, F.Z.; Anhê, G.F.; De Nucci, G.; Zanesco, A.; Grant, A.D.; Antunes, E. Role of PKC and CaV1.2 in detrusor overactivity in a model of obesity associated with insulin resistance in mice. *PLoS ONE* **2012**, *7*, e48507. [CrossRef] [PubMed]
107. Leiria, L.O.; Sollon, C.; Báu, F.R.; Mónica, F.Z.; D'Ancona, C.L.; De Nucci, G.; Grant, A.D.; Anhê, G.F.; Antunes, E. Insulin relaxes bladder via PI3K/AKT/eNOS pathway activation in mucosa: Unfolded protein response-dependent insulin resistance as a cause of obesity-associated overactive bladder. *J. Physiol.* **2013**, *591*, 2259–2273. [CrossRef] [PubMed]
108. Leiria, L.O.; Silva, F.H.; Davel, A.P.; Alexandre, E.C.; Calixto, M.C.; De Nucci, G.; Mónica, F.Z.; Antunes, E. The soluble guanylyl cyclase activator BAY 60-2770 ameliorates overactive bladder in obese mice. *J. Urol.* **2014**, *191*, 539–547. [CrossRef] [PubMed]
109. Malysz, J.; Petkov, G.V. Urinary bladder smooth muscle ion channels: Expression, function, and regulation in health and disease. *Am. J. Physiol. Renal Physiol.* **2020**, *319*, F257–F283. [CrossRef] [PubMed]
110. Jiang, X.; Luttrell, I.; Chitaley, K.; Yang, C.C. T- and L-type voltage-gated calcium channels: Their role in diabetic bladder dysfunction. *Neurourol. Urodyn.* **2014**, *33*, 147–152. [CrossRef] [PubMed]
111. Alexandre, E.C.; Leiria, L.O.; Silva, F.H.; Mendes-Silvério, C.B.; Calmasini, F.B.; Davel, A.P.; Mónica, F.Z.; De Nucci, G.; Antunes, E. Soluble guanylyl cyclase (sGC) degradation and impairment of nitric oxide-mediated responses in urethra from obese mice: Reversal by the sGC activator BAY 60-2770. *J. Pharmacol. Exp. Ther.* **2014**, *349*, 2–9. [CrossRef] [PubMed]
112. Alexandre, E.C.; Calmasini, F.B.; Sponton, A.C.D.S.; de Oliveira, M.G.; André, D.M.; Silva, F.H.; Delbin, M.A.; Mónica, F.Z.; Antunes, E. Influence of the periprostatic adipose tissue in obesity-associated mouse urethral dysfunction and oxidative stress: Effect of resveratrol treatment. *Eur. J. Pharmacol.* **2018**, *836*, 25–33. [CrossRef] [PubMed]

113. Calmasini, F.B.; McCarthy, C.G.; Wenceslau, C.F.; Priviero, F.B.M.; Antunes, E.; Webb, R.C. Toll-like receptor 9 regulates metabolic profile and contributes to obesity-induced benign prostatic hyperplasia in mice. *Pharmacol. Rep.* **2020**, *72*, 179–187. [CrossRef] [PubMed]
114. Lee, Y.C.; Lin, G.; Wang, G.; Reed, A.; Lu, Z.; Wang, L.; Banie, L.; Lue, T.F. Impaired contractility of the circular striated urethral sphincter muscle may contribute to stress urinary incontinence in female zucker fatty rats. *Neurourol. Urodyn.* **2017**, *36*, 1503–1510. [CrossRef] [PubMed]
115. Kim, A.K.; Hamadani, C.; Zeidel, M.L.; Hill, W.G. Urological complications of obesity and diabetes in males and females of three mouse models: Temporal manifestations. *Am. J. Physiol. Renal Physiol.* **2020**, *318*, F160–F174. [CrossRef] [PubMed]
116. Wu, L.; Wang, M.; Maher, S.; Fu, P.; Cai, D.; Wang, B.; Gupta, S.; Hijaz, A.; Daneshgari, F.; Liu, G. Effects of different diets used to induce obesity/metabolic syndrome on bladder function in rats. *Am. J. Physiol. Regul. Integr. Comp. Physiol.* **2023**, *324*, R70–R81. [CrossRef] [PubMed]
117. Oswal, A.; Yeo, G. Leptin and the control of body weight: A review of its diverse central targets, signaling mechanisms, and role in the pathogenesis of obesity. *Obesity* **2010**, *18*, 221–229. [CrossRef] [PubMed]
118. Ellenbroek, J.H.; Arioglu, E.; Michel, M.C. A systematic review of urinary bladder hypertrophy in experimental diabetes: Part 2. Comparison of animal models and functional consequences. *Neurourol. Urodyn.* **2018**, *37*, 2346–2360. [CrossRef] [PubMed]
119. Nobe, K.; Yamazak, T.; Tsumita, N.; Hashimoto, T.; Honda, K. Glucose-dependent enhancement of diabetic bladder contraction is associated with a rho kinase-regulated protein kinase C pathway. *J. Pharmacol. Exp. Ther.* **2009**, *328*, 940–950. [CrossRef] [PubMed]
120. Wu, L.; Zhang, X.; Xiao, N.; Huang, Y.; Kavran, M.; Elrashidy, R.A.; Wang, M.; Daneshgari, F.; Liu, G. Functional and morphological alterations of the urinary bladder in type 2 diabetic FVB (db/db) mice. *J. Diabetes Complicat.* **2016**, *30*, 778–785. [CrossRef] [PubMed]
121. Wang, Z.; Cheng, Z.; Cristofaro, V.; Li, J.; Xiao, X.; Gomez, P.; Ge, R.; Gong, E.; Strle, K.; Sullivan, M.P.; et al. Inhibition of TNF-α improves the bladder dysfunction that is associated with type 2 diabetes. *Diabetes* **2012**, *61*, 2134–2145. [CrossRef] [PubMed]
122. Riboulet, C.A.; Pierron, I.; Durand, J.; Murdaca, J.; Giudicelli, E.; Van, O. Methylglyoxal impairs the insulin signaling pathways independently of the formation of intracellular reactive oxygen species. *Diabetes* **2006**, *55*, 1289–1299. [CrossRef] [PubMed]
123. Brouwers, O.; Niessen, P.M.; Ferreira, I.; Miyata, T.; Scheffer, P.G.; Teerlink, T.; Schrauwen, P.; Brownlee, M.; Stehouwer, C.D.; Schalkwijk, C.G. Overexpression of glyoxalase-I reduces hyperglycemia-induced levels of advanced glycation end products and oxidative stress in diabetic rats. *J. Biol. Chem.* **2011**, *286*, 1374–1380. [CrossRef] [PubMed]
124. Shamsaldeen, Y.A.; Mackenzie, L.S.; Lione, L.A.; Benham, C.D. Methylglyoxal, a metabolite increased in diabetes is associated with insulin resistance, vascular dysfunction and neuropathies. *Curr. Drug Metab.* **2016**, *17*, 359–367. [CrossRef] [PubMed]
125. De la Cruz-Ares, S.; Cardelo, M.P.; Gutiérrez-Mariscal, F.M.; Torres-Peña, J.D.; García-Rios, A.; Katsiki, N.; Malagón, M.M.; López-Miranda, J.; Pérez-Martínez, P.; Yubero-Serrano, E.M. Endothelial dysfunction and advanced glycation end products in patients with newly diagnosed versus established diabetes: From the CORDIOPREV Study. *Nutrients* **2020**, *12*, 238. [CrossRef] [PubMed]
126. Heber, S.; Haller, P.M.; Kiss, A.; Jäger, B.; Huber, K.; Fischer, M.J.M. Association of plasma methylglyoxal increase after myocardial infarction and the left ventricular ejection fraction. *Biomedicines* **2024**, *10*, 605. [CrossRef] [PubMed]
127. Thornalley, P.J.; Langborg, A.; Minhas, H.S. Formation of glyoxal, methylglyoxal and 3-deoxyglucosone in the glycation of proteins by glucose. *Biochem. J.* **1999**, *344 Pt 1*, 109–116. [CrossRef] [PubMed]
128. Kalapos, M.P.; de Bari, L. Hidden biochemical fossils reveal an evolutionary trajectory for glycolysis in the prebiotic era. *FEBS Lett.* **2022**, *596*, 1955–1968. [CrossRef] [PubMed]
129. Lai, S.W.T.; Lopez Gonzalez, E.J.; Zoukari, T.; Ki, P.; Shuck, S.C. Methylglyoxal and Its Adducts: Induction, Repair, and Association with Disease. *Chem. Res. Toxicol.* **2022**, *35*, 1720–1746. [CrossRef] [PubMed]
130. Maessen, D.E.; Stehouwer, C.D.; Schalkwijk, C.G. The role of methylglyoxal and the glyoxalase system in diabetes and other age-related diseases. *Clin. Sci.* **2015**, *28*, 839–861. [CrossRef] [PubMed]
131. Hanssen, N.M.J.; Stehouwer, C.D.A.; Schalkwijk, C.G. Methylglyoxal stress, the glyoxalase system, and diabetic chronic kidney disease. *Curr. Opin. Nephrol. Hypertens.* **2019**, *28*, 26–33. [CrossRef] [PubMed]
132. Miranda, E.R.; Haus, J.M. Glyoxalase I is a novel target for the prevention of metabolic derangement. *Pharmacol. Ther.* **2023**, *250*, 108524. [CrossRef] [PubMed]
133. Brouwers, O.; Niessen, P.M.; Haenen, G.; Miyata, T.; Brownlee, M.; Stehouwer, C.D.; De Mey, J.G.; Schalkwijk, C.G. Hyperglycaemia-induced impairment of endothelium-dependent vasorelaxation in rat mesenteric arteries is mediated by intracellular methylglyoxal levels in a pathway dependent on oxidative stress. *Diabetologia* **2010**, *53*, 989–1000. [CrossRef] [PubMed]
134. Thornalley, P.J.; Waris, S.; Fleming, T.; Santarius, T.; Larkin, S.J.; Winklhofer-Roob, B.M.; Stratton, M.R.; Rabbani, N. Imidazopurinones are markers of physiological genomic damage linked to DNA instability and glyoxalase 1-associated tumour multidrug resistance. *Nucleic Acids Res.* **2010**, *38*, 5432–5442. [CrossRef] [PubMed]
135. Ban, I.; Sugawa, H.; Nagai, R. Protein Modification with Ribose Generates N^{δ}-(5-hydro-5-methyl-4-imidazolone-2-yl)-ornithine. *Int. J. Mol. Sci.* **2022**, *23*, 1224. [CrossRef] [PubMed]
136. Irshad, Z.; Xue, M.; Ashour, A.; Larkin, J.R.; Thornalley, P.J.; Rabbani, N. Activation of the unfolded protein response in high glucose treated endothelial cells is mediated by methylglyoxal. *Sci. Rep.* **2019**, *9*, 7889. [CrossRef]

137. Zhang, X.; Scheijen, J.L.; Stehouwer, C.D.A.; Wouters, K.; Schalkwijk, C. Increased methylglyoxal formation in plasma and tissues during a glucose tolerance test is derived from exogenous glucose. *Clin. Sci.* **2023**, *20*, CS20220753. [CrossRef] [PubMed]
138. Erusalimsky, J.D. The use of the soluble receptor for advanced glycation-end products (sRAGE) as a potential biomarker of disease risk and adverse outcomes. *Redox Biol.* **2021**, *42*, 101958. [CrossRef]
139. Jangde, N.; Ray, R.; Rai, V. RAGE and its ligands: From pathogenesis to therapeutics. *Crit. Rev. Biochem. Mol. Biol.* **2020**, *55*, 555–575. [CrossRef] [PubMed]
140. Dozio, E.; Vettoretti, S.; Caldiroli, L.; Nerini-Molteni, S.; Tacchini, L.; Ambrogi, F.; Messa, P.; Corsi Romanelli, M.M. Advanced glycation end products (AGE) and soluble forms of AGE receptor: Emerging role as mortality risk factors in CKD. *Biomedicines* **2020**, *8*, 638. [CrossRef]
141. Ganbaatar, B.; Fukuda, D.; Shinohara, M.; Yagi, S.; Kusunose, K.; Yamada, H.; Soeki, T.; Hirata, K.I.; Sata, M. Empagliflozin ameliorates endothelial dysfunction and suppresses atherogenesis in diabetic apolipoprotein E-deficient mice. *Eur. J. Pharmacol.* **2020**, *875*, 173040. [CrossRef] [PubMed]
142. Korac, B.; Kalezic, A.; Pekovic-Vaughan, V.; Korac, A.; Jankovic, A. Redox changes in obesity, metabolic syndrome, and diabetes. *Redox Biol.* **2021**, *42*, 101887. [CrossRef] [PubMed]
143. Seryogina, E.S.; Kamynina, A.V.; Koroev, D.O.; Volpina, O.M.; Vinokurov, A.Y.; Abramov, A.Y. RAGE induces physiological activation of NADPH oxidase in neurons and astrocytes and neuroprotection. *FEBS J.* **2024**. [CrossRef]
144. Dobrucki, I.T.; Miskalis, A.; Nelappana, M.; Applegate, C.; Wozniak, M.; Czerwinski, A.; Kalinowski, L.; Dobrucki, L.W. Receptor for advanced glycation end-products: Biological significance and imaging applications. *Wiley Interdiscip. Rev. Nanomed. Nanobiotechnol.* **2023**, *5*, e1935. [CrossRef] [PubMed]
145. Reddy, V.P. Oxidative Stress in Health and Disease. *Biomedicines* **2023**, *11*, 2925. [CrossRef] [PubMed]
146. Odani, H.; Shinzato, T.; Matsumoto, Y.; Usami, J.; Maeda, K. Increase in three alpha, beta-dicarbonyl compound levels in human uremic plasma: Specific in vivo determination of intermediates in advanced Maillard reaction. *Biochem. Biophys. Res. Commun.* **1999**, *256*, 89–93. [CrossRef] [PubMed]
147. Lapolla, A.; Flamini, R.; Dalla, V.A.; Senesi, A.; Reitano, R.; Fedele, D.; Basso, E.; Seraglia, R.; Traldi, P. Glyoxal and methylglyoxal levels in diabetic patients: Quantitative determination by a new GC/MS method. *Clin. Chem. Lab. Med.* **2003**, *41*, 1166–1173. [CrossRef] [PubMed]
148. Ogawa, S.; Nakayama, K.; Nakayama, M.; Mori, T.; Matsushima, M.; Okamura, M.; Senda, M.; Nako, K.; Miyata, T.; Ito, S. Methylglyoxal is a predictor in type 2 diabetic patients of intima-media thickening and elevation of blood pressure. *Hypertension* **2010**, *56*, 471–476. [CrossRef] [PubMed]
149. Lu, J.; Randell, E.; Han, Y.; Adeli, K.; Krahn, J.; Meng, Q.H. Increased plasma methylglyoxal level, inflammation, and vascular endothelial dysfunction in diabetic nephropathy. *Clin. Biochem.* **2011**, *44*, 307–311. [CrossRef] [PubMed]
150. Terawaki, H.; Nakao, M.; Nakayama, K.; Nakayama, M.; Kimura, A.; Takane, K.; Mitome, J.; Hamaguchi, A.; Ogura, M.; Yokoyama, K.; et al. Peritoneal clearance and transport of methylglyoxal. *Nephrol. Dial. Transplant.* **2011**, *26*, 753–754. [CrossRef] [PubMed]
151. Senda, M.; Ogawa, S.; Nako, K.; Okamura, M.; Sakamoto, T.; Ito, S. The strong relation between post-hemodialysis blood methylglyoxal levels and post-hemodialysis blood glucose concentration rise. *Clin. Exp. Nephrol.* **2015**, *9*, 527–533. [CrossRef] [PubMed]
152. Ramachandra, B.L.; Vedantham, S.; Krishnan, U.M.; Rayappan, J.B.B. Methylglyoxal—An emerging biomarker for diabetes mellitus diagnosis and its detection methods. *Biosens. Bioelectron.* **2019**, *133*, 107–124. [CrossRef] [PubMed]
153. Gutierrez-Mariscal, F.M.; Cardelo, M.P.; de la Cruz, S.; Alcala-Diaz, J.F.; Roncero-Ramos, I.; Guler, I.; Vals-Delgado, C.; López-Moreno, A.; Luque, R.M.; Delgado-Lista, J.; et al. Reduction in circulating advanced glycation end products by mediterranean diet is associated with increased likelihood of yype 2 diabetes remission in patients with coronary heart disease: From the Cordioprev Study. *Mol. Nutr. Food Res.* **2021**, *65*, e1901290. [CrossRef] [PubMed]
154. Maessen, D.E.; Hanssen, N.M.; Lips, M.A.; Scheijen, J.L.; Willems, D.K.; Pijl, H.; Stehouwer, C.D.; Schalkwijk, C.G. Energy restriction and Roux-en-Y gastric bypass reduce postprandial α-dicarbonyl stress in obese women with type 2 diabetes. *Diabetologia* **2016**, *59*, 2013–2017. [CrossRef] [PubMed]
155. Pastor-Belda, M.; Fernández-García, A.J.; Campillo, N.; Pérez-Cárceles, M.D.; Motas, M.; Hernández-Córdoba, M.; Viñas, P. Glyoxal and methylglyoxal as urinary markers of diabetes. Determination using a dispersive liquid-liquid microextraction procedure combined with gas chromatography-mass spectrometry. *J. Chromatogr. A* **2017**, *1509*, 43–49. [CrossRef] [PubMed]
156. Bora, S.; Shankarrao Adole, P.; Vishwanath Vinod, K.; Ananthkrishna Pillai, A.; Ahmed, S. The genetic polymorphisms and activity of glyoxalase 1 as a risk factor for acute coronary syndrome in South Indians with type 2 diabetes mellitus. *Gene* **2023**, *10*, 147701. [CrossRef] [PubMed]
157. Tanaka, J.; Yamaguchi, K.; Ishikura, H.; Tsubota, M.; Sekiguchi, F.; Seki, Y.; Tsujiuchi, T.; Murai, A.; Umemura, T.; Kawabata, A. Bladder pain relief by HMGB1 neutralization and soluble thrombomodulin in mice with cyclophosphamide-induced cystitis. *Neuropharmacology* **2014**, *79*, 112–118. [CrossRef]
158. Hiramoto, S.; Tsubota, M.; Yamaguchi, K.; Okazaki, K.; Sakaegi, A.; Toriyama, Y.; Tanaka, J.; Sekiguchi, F.; Ishikura, H.; Wake, H.; et al. Cystitis-related bladder pain involves ATP-dependent HMGB1 release from macrophages and its downstream $H_2S/Ca_v3.2$ signaling in mice. *Cells* **2020**, *9*, 1748. [CrossRef]

159. Khorramdelazad, H.; Bagheri, V.; Hassanshahi, G.; Karami, H.; Moogooei, M.; Zeinali, M.; Abedinzadeh, M. S100A12 and RAGE expression in human bladder transitional cell carcinoma: A role for the ligand/RAGE axis in tumor progression? *Asian Pac. J. Cancer Prev.* **2015**, *16*, 2725–2729. [CrossRef] [PubMed]
160. Karkin, K.; İzol, V.; Kaplan, M.; Değer, M.; Akdoğan, N.; Tansuğ, M.Z. Demonstration of advanced glycation end product (AGE) expression in bladder cancer tissue in type-2 diabetic and non-diabetic patients and the relationship between AGE accumulation and endoplasmic reticulum stress with bladder cancer. *Int. J. Clin. Pract.* **2021**, *75*, e14526. [CrossRef] [PubMed]
161. Matsumoto, K.; Fujiwara, Y.; Nagai, R.; Yoshida, M. Immunohistochemical detection of advanced glycation end products in human bladder with specific monoclonal antibody. *Int. J. Urol.* **2009**, *16*, 402–405. [CrossRef] [PubMed]
162. Rabbani, N.; Thornalley, P.J. Emerging glycation-based therapeutics-glyoxalase 1 inducers and glyoxalase 1 inhibitors. *Int. J. Mol. Sci.* **2022**, *23*, 2453. [CrossRef] [PubMed]
163. Chen, S.J.; Aikawa, C.; Yoshida, R.; Matsui, T. Methylglyoxal-derived hydroimidazolone residue of plasma protein can behave as a predictor of prediabetes in Spontaneously Diabetic Torii rats. *Physiol. Rep.* **2015**, *3*, e12477. [CrossRef] [PubMed]
164. Berlanga, J.; Cibrian, D.; Guillén, I.; Freyre, F.; Alba, J.S.; Lopez-Saura, P.; Merino, N.; Aldama, A.; Quintela, A.M.; Triana, M.E.; et al. Methylglyoxal administration induces diabetes-like microvascular changes and perturbs the healing process of cutaneous wounds. *Clin. Sci.* **2005**, *109*, 83–95. [CrossRef]
165. Sena, C.M.; Matafome, P.; Crisóstomo, J.; Rodrigues, L.; Fernandes, R.; Pereira, P.; Seiça, R.M. Methylglyoxal promotes oxidative stress and endothelial dysfunction. *Pharmacol. Res.* **2012**, *65*, 497–506. [CrossRef] [PubMed]
166. Tikellis, C.; Pickering, R.J.; Tsorotes, D.; Huet, O.; Cooper, M.E.; Jandeleit-Dahm, K.; Thomas, M.C. Dicarbonyl stress in the absence of hyperglycemia increases endothelial inflammation and atherogenesis similar to that observed in diabetes. *Diabetes* **2014**, *63*, 3915–3925. [CrossRef] [PubMed]
167. Hanssen, N.M.J.; Tikellis, C.; Pickering, R.J.; Dragoljevic, D.; Lee, M.K.S.; Block, T.; Scheijen, J.L.; Wouters, K.; Miyata, T.; Cooper, M.E.; et al. Pyridoxamine prevents increased atherosclerosis by intermittent methylglyoxal spikes in the aortic arches of ApoE$^{-/-}$ mice. *Biomed. Pharmacother.* **2023**, *158*, 114211. [CrossRef] [PubMed]
168. Crisóstomo, J.; Matafome, P.; Santos-Silva, D.; Rodrigues, L.; Sena, C.M.; Pereira, P.; Seiça, R. Methylglyoxal chronic administration promotes diabetes-like cardiac ischaemia disease in Wistar normal rats. *Nutr. Metab. Cardiovasc. Dis.* **2013**, *23*, 1223–1230. [CrossRef]
169. Peyret, H.; Konecki, C.; Terryn, C.; Dubuisson, F.; Millart, H.; Feliu, C.; Djerada, Z. Methylglyoxal induces cardiac dysfunction through mechanisms involving altered intracellular calcium handling in the rat heart. *Chem. Biol. Interact.* **2024**, *394*, 110949. [CrossRef] [PubMed]
170. Rodrigues, L.; Matafome, P.; Crisóstomo, J.; Santos-Silva, D.; Sena, C.; Pereira, P.; Seiça, R. Advanced glycation end products and diabetic nephropathy: A comparative study using diabetic and normal rats with methylglyoxal-induced glycation. *J. Physiol. Biochem.* **2014**, *70*, 173–184. [CrossRef] [PubMed]
171. Jung, E.; Kang, W.S.; Jo, K.; Kim, J. Ethyl pyruvate prevents renal damage induced by methylglyoxal-derived advanced glycation end products. *J. Diabetes Res.* **2019**, *2019*, 4058280. [CrossRef]
172. Mojadami, S.; Ahangarpour, A.; Mard, S.A.; Khorsandi, L. Diabetic nephropathy induced by methylglyoxal: Gallic acid regulates kidney microRNAs and glyoxalase1-Nrf2 in male mice. *Arch. Physiol. Biochem.* **2023**, *29*, 655–662. [CrossRef]
173. Medeiros, M.L.; de Oliveira, M.G.; Tavares, E.G.; Mello, G.C.; Anhê, G.F.; Mónica, F.Z.; Antunes, E. Long-term methylglyoxal intake aggravates murine Th2-mediated airway eosinophil infiltration. *Int. Immunopharmacol.* **2020**, *81*, 106254. [CrossRef] [PubMed]
174. Medeiros, M.L.; Oliveira, A.L.; de Oliveira, M.G.; Mónica, F.Z.; Antunes, E. Methylglyoxal exacerbates lipopolysaccharide-induced acute lung injury via RAGE-induced ROS generation: Protective effects of metformin. *J. Inflamm. Res.* **2021**, *14*, 6477–6489. [CrossRef] [PubMed]
175. Medeiros, M.L.; Oliveira, A.L.; Mello, G.C.; Antunes, E. Metformin counteracts the deleterious effects of methylglyoxal on ovalbumin-induced airway eosinophilic inflammation and remodeling. *Int. J. Mol. Sci.* **2023**, *24*, 9549. [CrossRef] [PubMed]
176. Zunkel, K.; Simm, A.; Bartling, B. Long-term intake of the reactive metabolite methylglyoxal is not toxic in mice. *Food Chem. Toxicol.* **2020**, *141*, 111333. [CrossRef] [PubMed]
177. Oliveira, A.L.; de Oliveira, M.G.; Medeiros, M.L.; Mónica, F.; Antunes, E. Metformin abrogates the voiding dysfunction induced by prolonged methylglyoxal intake. *Eur. J. Pharmacol.* **2021**, *910*, 174502. [CrossRef] [PubMed]
178. Al-Robaiy, S.; Navarrete, S.A.; Simm, A. RAGE-dependent effect of exogenous methylglyoxal intake on lung biomechanics in mice. *Nutrients* **2022**, *15*, 23. [CrossRef] [PubMed]
179. Tang, Y.; Zhao, Y.; Wang, P.; Sang, S. Simultaneous determination of multiple reactive carbonyl species in high fat diet-induced metabolic disordered mice and the inhibitory effects of rosemary on carbonyl stress. *J. Agric. Food Chem.* **2021**, *69*, 1123–1131. [CrossRef]
180. Oliveira, A.L.; Medeiros, M.L.; Gomes, E.T.; Mello, G.C.; Costa, S.K.; Mónica, F.Z.; Antunes, E. TRPA1 channel mediates methylglyoxal-induced mouse bladder dysfunction. *Front. Physiol.* **2023**, *14*, 1308077. [CrossRef] [PubMed]
181. Hill, W.G.; Zeidel, M.L.; Bjorling, D.E.; Vezina, C.M. Void spot assay: Recommendations on the use of a simple micturition assay for mice. *Am. J. Physiol. Renal Physiol.* **2018**, *315*, F1422–F1429. [CrossRef] [PubMed]

182. Wegner, K.A.; Abler, L.L.; Oakes, S.R.; Mehta, G.S.; Ritter, K.E.; Hill, W.G.; Zwaans, B.M.; Lamb, L.E.; Wang, Z.; Bjorling, D.E.; et al. Void spot assay procedural optimization and software for rapid and objective quantification of rodent voiding function, including overlapping urine spots. *Am. J. Physiol. Renal Physiol.* **2018**, *315*, F1067–F1080. [CrossRef] [PubMed]
183. Laddha, A.P.; Kulkarni, Y.A. NADPH oxidase: A membrane-bound enzyme and its inhibitors in diabetic complications. *Eur. J. Pharmacol.* **2020**, *881*, 173206. [CrossRef] [PubMed]
184. Poladia, D.P.; Bauer, J.A. Early cell-specific changes in nitric oxide synthases, reactive nitrogen species formation, and ubiquitinylation during diabetes-related bladder remodeling. *Diabetes Metab. Res. Rev.* **2003**, *19*, 313–319. [CrossRef] [PubMed]
185. Beshay, E.; Carrier, S. Oxidative stress plays a role in diabetes-induced bladder dysfunction in a rat model. *Urology* **2004**, *64*, 1062–1067. [CrossRef] [PubMed]
186. Kanika, N.D.; Chang, J.; Tong, Y.; Tiplitsky, S.; Li, J.; Yohannes, E.; Tar, M.; Chance, M.; Christ, G.J.; Melman, A.; et al. Oxidative stress status accompanying diabetic bladder cystopathy results in the activation of protein degradation pathways. *BJU Int.* **2011**, *107*, 1676–1684. [CrossRef]
187. Tomechko, S.E.; Liu, G.; Tao, M.; Schlatzer, D.; Powell, C.T.; Gupta, S.; Chance, M.R.; Daneshgari, F. Tissue specific dysregulated protein subnetworks in type 2 diabetic bladder urothelium and detrusor muscle. *Mol. Cell Proteom.* **2015**, *14*, 635–645. [CrossRef] [PubMed]
188. Wang, Y.; Deng, G.G.; Davies, K.P. Novel insights into development of diabetic bladder disorder provided by metabolomic analysis of the rat nondiabetic and diabetic detrusor and urothelial layer. *Am. J. Physiol. Endocrinol. Metab.* **2016**, *311*, E471–E479. [CrossRef] [PubMed]
189. Andersson, K.E. Oxidative stress and its possible relation to lower urinary tract functional pathology. *BJU Int.* **2018**, *121*, 527–533. [CrossRef] [PubMed]
190. Andersson, K.E. Oxidative stress and its relation to lower urinary tract symptoms. *Int. Neurourol. J.* **2022**, *26*, 261–267. [CrossRef]
191. Song, Q.X.; Sun, Y.; Deng, K.; Mei, J.Y.; Chermansky, C.J.; Damaser, M.S. Potential role of oxidative stress in the pathogenesis of diabetic bladder dysfunction. *Nat. Rev. Urol.* **2022**, *19*, 581–596. [CrossRef] [PubMed]
192. Khosla, L.; Gong, S.; Weiss, J.P.; Birder, L.A. Oxidative stress biomarkers in age-related lower urinary tract disorders: A systematic review. *Int. Neurourol. J.* **2022**, *26*, 3–19. [CrossRef]
193. Miyata, Y.; Matsuo, T.; Mitsunari, K.; Asai, A.; Ohba, K.; Sakai, H. A review of oxidative stress and urinary dysfunction caused by bladder outlet obstruction and treatments using antioxidants. *Antioxidants* **2019**, *8*, 132. [CrossRef] [PubMed]
194. De Oliveira, M.; Monica, F.Z.; Passos, G.R.; Victorio, J.A.; Davel, A.P.; Oliveira, A.L.L.; Parada, C.A.; D'Ancona, C.A.L.; Hill, W.G.; Antunes, E. Selective pharmacological inhibition of NOX2 by GSK2795039 improves bladder dysfunction in cyclophosphamide-induced cystitis in mice. *Antioxidants* **2022**, *12*, 92. [CrossRef] [PubMed]
195. Calmasini, F.B.; de Oliveira, M.G.; Alexandre, E.C.; Silva, F.H.; Tavares, E.B.G.; André, D.M.; Zapparoli, A.; Antunes, E. Obesity-induced mouse benign prostatic hyperplasia (BPH) is improved by treatment with resveratrol: Implication of oxidative stress, insulin sensitivity and neuronal growth factor. *J. Nutr. Biochem.* **2018**, *55*, 53–58. [CrossRef] [PubMed]
196. Elrashidy, R.A.; Kavran, M.; Asker, M.E.; Mohamed, H.E.; Daneshgari, F.; Liu, G. Smooth muscle-specific deletion of MnSOD exacerbates diabetes-induced bladder dysfunction in mice. *Am. J. Physiol. Renal Physiol.* **2019**, *317*, F906–F912. [CrossRef] [PubMed]
197. Masuda, H.; Kihara, K.; Saito, K.; Matsuoka, Y.; Yoshida, S.; Chancellor, M.B.; de Groat, W.C.; Yoshimura, N. Reactive oxygen species mediate detrusor overactivity via sensitization of afferent pathway in the bladder of anaesthetized rats. *BJU Int.* **2008**, *101*, 775–780. [CrossRef]
198. Frara, N.; Giaddui, D.; Braverman, A.S.; Jawawdeh, K.; Wu, C.; Ruggieri, M.R.; Barbe, M.F. Mechanisms involved in nicotinamide adenine dinucleotide phosphate (NADPH) oxidase (Nox)-derived reactive oxygen species (ROS) modulation of muscle function in human and dog bladders. *PLoS ONE* **2023**, *18*, e0287212. [CrossRef] [PubMed]
199. Rowan, S.; Bejarano, E.; Taylor, A. Mechanistic targeting of advanced glycation end-products in age-related diseases. *Biochim. Biophys. Acta Mol. Basis Dis.* **2018**, *1864*, 3631–3643. [CrossRef] [PubMed]
200. Coyle, C.H.; Philips, B.J.; Morrisroe, S.N.; Chancellor, M.B.; Yoshimura, N. Antioxidant effects of green tea and its polyphenols on bladder cells. *Life Sci.* **2008**, *83*, 12–18. [CrossRef] [PubMed]
201. Nomiya, M.; Andersson, K.E.; Yamaguchi, O. Chronic bladder ischemia and oxidative stress: New pharmacotherapeutic targets for lower urinary tract symptoms. *Int. J. Urol.* **2015**, *22*, 40–46. [CrossRef] [PubMed]
202. Szkudelska, K.; Szkudelski, T. Resveratrol, obesity and diabetes. *Eur. J. Pharmacol.* **2010**, *635*, 1–8. [CrossRef] [PubMed]
203. Tabatabaei-Malazy, O.; Lavari, N.; Abdollahi, M. Natural products in the clinical management of metabolic syndrome. In *Handbook of Experimental Pharmacology*; Springer: Berlin/Heidelberg, Germany, 2024. [CrossRef]
204. Tsounapi, P.; Honda, M.; Hikita, K.; Sofikitis, N.; Takenaka, A. Oxidative stress alterations in the bladder of a short-period type 2 diabetes rat model: Antioxidant treatment can be beneficial for the bladder. *In Vivo* **2019**, *33*, 1819–1826. [CrossRef] [PubMed]
205. Xu, F.; Du, H.; Hou, J.; Li, N. Anti-inflammation properties of resveratrol in the detrusor smooth muscle of the diabetic rat. *Int. Urol. Nephrol.* **2022**, *54*, 2833–2843. [CrossRef] [PubMed]
206. He, Y.; Zeng, H.; Yu, Y.; Zhang, J.; Liu, Q.; Yang, B. Resveratrol improved detrusor fibrosis induced by mast cells during progression of chronic prostatitis in rats. *Eur. J. Pharmacol.* **2017**, *815*, 495–500. [CrossRef]
207. Yu, Y.; Jiang, J.; He, Y.; Wang, W.; Shen, C.; Yang, B. Resveratrol improves urinary dysfunction in rats with chronic prostatitis and suppresses the activity of the stem cell factor/c-Kit signaling pathway. *Mol. Med. Rep.* **2017**, *16*, 1395–1400. [CrossRef] [PubMed]

208. Lee, Y.L.; Lin, K.L.; Wu, B.N.; Chuang, S.M.; Wu, W.J.; Lee, Y.C.; Ho, W.T.; Juan, Y.S. Epigallocatechin-3-gallate alleviates bladder overactivity in a rat model with metabolic syndrome and ovarian hormone deficiency through mitochondria apoptosis pathways. *Sci. Rep.* **2018**, *8*, 5358. [CrossRef] [PubMed]
209. Juan, Y.S.; Chuang, S.M.; Lee, Y.L.; Long, C.Y.; Wu, T.H.; Chang, W.C.; Levin, R.M.; Liu, K.M.; Huang, C.H. Green tea catechins decrease oxidative stress in surgical menopause-induced overactive bladder in a rat model. *BJU Int.* **2012**, *110*, E236–E244. [CrossRef] [PubMed]
210. Hsieh, J.T.; Kuo, K.L.; Liu, S.H.; Shi, C.S.; Chang, H.C.; Lin, W.C.; Chou, C.T.; Hsu, C.H.; Liao, S.M.; Wang, Z.H.; et al. Epigallocatechin Gallate Attenuates Partial Bladder Outlet obstruction-induced bladder injury via suppression of endoplasmic reticulum stress-related apoptosis-in vivo study. *Urology* **2016**, *91*, 242.e1–242.e9. [CrossRef] [PubMed]
211. Sang, S.; Shao, X.; Bai, N.; Lo, C.Y.; Yang, C.S.; Ho, C.T. Tea polyphenol (-)-epigallocatechin-3-gallate: A new trapping agent of reactive dicarbonyl species. *Chem. Res. Toxicol.* **2007**, *20*, 1862–1870. [CrossRef]
212. Zhu, D.; Wang, L.; Zhou, Q.; Yan, S.; Li, Z.; Sheng, J.; Zhang, W. (+)-Catechin ameliorates diabetic nephropathy by trapping methylglyoxal in type 2 diabetic mice. *Mol. Nutr. Food Res.* **2014**, *58*, 2249–2260. [CrossRef] [PubMed]
213. Foretz, M.; Guigas, B.; Viollet, B. Metformin: Update on mechanisms of action and repurposing potential. *Nat. Rev. Endocrinol.* **2023**, *19*, 460–476. [CrossRef] [PubMed]
214. Schernthaner, G.; Schernthaner, G.H. The right place for metformin today. *Diabetes Res. Clin. Pract.* **2020**, *159*, 107946. [CrossRef] [PubMed]
215. Mbara, K.C.; Mofo, P.E.; Driver, C.; Nzuza, S.; Mkhombo, N.T.; Gcwensa, S.K.; Mcobothi, E.N.; Owira, P.M. Metformin turns 62 in pharmacotherapy: Emergence of non-glycaemic effects and potential novel therapeutic applications. *Eur. J. Pharmacol.* **2021**, *898*, 173934. [CrossRef]
216. Tobar, N.; Rocha, G.Z.; Santos, A.; Guadagnini, D.; Assalin, H.B.; Camargo, J.A.; Gonçalves, A.E.S.S.; Pallis, F.R.; Oliveira, A.G.; Rocco, S.A.; et al. Metformin acts in the gut and induces gut-liver crosstalk. *Proc. Natl. Acad. Sci. USA* **2023**, *120*, e2211933120. [CrossRef] [PubMed]
217. Kender, Z.; Fleming, T.; Kopf, S.; Torzsa, P.; Grolmusz, V.; Herzig, S.; Schleicher, E.; Rácz, K.; Reismann, P.; Nawroth, P.P. Effect of metformin on methylglyoxal metabolism in patients with type 2 diabetes. *Exp. Clin. Endocrinol. Diabetes* **2014**, *122*, 316–319. [CrossRef] [PubMed]
218. Chen, L.; Lv, L.; Zhang, L.; Gao, Z.; Liu, Y.; Wang, S.; Zhou, N.; Xia, Y.; Cui, J.; Jiang, X.; et al. Metformin ameliorates bladder dysfunction in a rat model of partial bladder outlet obstruction. *Am. J. Physiol. Renal Physiol.* **2021**, *320*, F838–F858. [CrossRef] [PubMed]
219. Silva, F.H.; Alexandre, E.C.; Calmasini, F.B.; Calixto, M.C.; Antunes, E. Treatment with metformin improves erectile dysfunction in a murine model of obesity associated with insulin resistance. *Urology* **2015**, *86*, 423.e1–423.e6. [CrossRef] [PubMed]
220. Zhang, S.; Xu, H.; Yu, X.; Wu, Y.; Sui, D. Metformin ameliorates diabetic nephropathy in a rat model of low-dose streptozotocin-induced diabetes. *Exp. Ther. Med.* **2017**, *14*, 383–390. [CrossRef] [PubMed]

biomedicines

Review

Thermogenic Fat as a New Obesity Management Tool: From Pharmaceutical Reagents to Cell Therapies

Ying Cheng [1,2], Shiqing Liang [2], Shuhan Zhang [2] and Xiaoyan Hui [2,*]

[1] Zhongshan Hospital (Xiamen), Fudan University, Xiamen 361015, China; cheng.ying@zsxmhospital.com
[2] School of Biomedical Sciences, The Chinese University of Hong Kong, Hong Kong 999077, China; liangshiqingll@outlook.com (S.L.); 1155160276@link.cuhk.edu.hk (S.Z.)
* Correspondence: hannahhui@cuhk.edu.hk; Tel.: +852-39435101

Abstract: Obesity is a complex medical condition caused by a positive imbalance between calorie intake and calorie consumption. Brown adipose tissue (BAT), along with the newly discovered "brown-like" adipocytes (called beige cells), functions as a promising therapeutic tool to ameliorate obesity and metabolic disorders by burning out extra nutrients in the form of heat. Many studies in animal models and humans have proved the feasibility of this concept. In this review, we aim to summarize the endeavors over the last decade to achieve a higher number/activity of these heat-generating adipocytes. In particular, pharmacological compounds, especially agonists to the β3 adrenergic receptor (β3-AR), are reviewed in terms of their feasibility and efficacy in elevating BAT function and improving metabolic parameters in human subjects. Alternatively, allograft transplantation of BAT and the transplantation of functional brown or beige adipocytes from mesenchymal stromal cells or human induced pluripotent stem cells (hiPSCs) make it possible to increase the number of these beneficial adipocytes in patients. However, practical and ethical issues still need to be considered before the therapy can eventually be applied in the clinical setting. This review provides insights and guidance on brown- and beige-cell-based strategies for the management of obesity and its associated metabolic comorbidities.

Keywords: brown adipose tissue; beige adipocyte; β3 adrenergic receptor; brown adipose tissue transplantation; human induced pluripotent stem cell; obesity

Citation: Cheng, Y.; Liang, S.; Zhang, S.; Hui, X. Thermogenic Fat as a New Obesity Management Tool: From Pharmaceutical Reagents to Cell Therapies. *Biomedicines* **2024**, *12*, 1474. https://doi.org/10.3390/biomedicines12071474

Academic Editors: Jae-Hyung Park, Teresa Vezza and Zaida Abad-Jiménez

Received: 12 May 2024
Revised: 14 June 2024
Accepted: 19 June 2024
Published: 4 July 2024

1. Introduction

Obesity has reached epidemic levels worldwide and has approximately tripled since 1975 [1]. It is estimated that nearly 40% of adults are overweight or obese, and this incidence will increase in the coming decades [1]. In addition, the number of children with obesity is steeply increasing, along with the fact that children of overweight and obese people have a higher chance of remaining obese into adulthood and developing non-communicable chronic diseases at a younger age, including diabetes and cardiovascular diseases [2].

Obesity is fundamentally caused by a positive energy balance, consequently leading to excessive lipid storage in the adipose tissue [1]. Under obese conditions, the functions of multiple organs are impaired, which results in disturbed whole-body energy homeostasis and thus gives rise to a vicious cycle forming the relapsing character of the disease. Therefore, obesity substantially increases the risk of numerous diseases, such as type 2 diabetes mellitus (T2DM), hypertension, cardiovascular diseases, metabolic-associated fatty liver disease (MAFLD), and certain types of cancer, all of which contribute to a decline in life quality and expectancy [3,4].

Due to the multiple means of biological adaptations (e.g., hormones and neurochemical defense adaptations), diet and exercise alone are often inefficient in maintaining long-term weight loss [5]. Luckily, brown adipose tissue (BAT) and "brown-like" beige adipocytes, which dissipate extra energy in the form of heat, have recently been detected in adult humans. The activation of these thermogenic fats has been suggested as an alternative strategy for treating obesity, diabetes, and cardiovascular diseases, which has been comprehensively reviewed elsewhere [6]. This review summarizes and discusses the current progress in applying thermogenic fat for therapeutic purposes, including the in vitro differentiation/mobilization of the cells and transplantation therapy.

2. Brown/Beige Adipose Tissue Biology

Adipose tissue constitutes approximately 20–25% of the total body weight in healthy subjects and is classified into three distinct types. White adipose tissue (WAT), which predominates in the body, serves as the energy storage reservoir as well as a thermal insulator and a cushion for internal organs. WAT is also an abundant source of hormones and bioactive factors [7]. White adipocytes are characterized by the presence of a single lipid droplet (unilocular) that pushes the nucleus and other organelles aside. WAT accumulates in individuals who are overweight and obese, especially in the abdominal region, and the amount of WAT is closely correlated with insulin resistance, diabetes, and other metabolic diseases [8]. In particular, the location of WAT is also essential for metabolic health. Epidemiological studies have shown that WAT in the visceral cavity is more harmful when it is deposited in the lower trunk of the body.

In contrast to WAT, BAT can dissipate energy through non-shivering thermogenesis [9]. Compared to white adipocytes, brown adipocytes are smaller in size (15–50 μm in diameter). Multiple lipid droplets (multilocular) and abundant mitochondria exist in each brown adipocyte, and the iron-rich mitochondria give them a brownish color. In addition to the morphological differences, brown adipocytes express a unique mitochondrial resident protein called uncoupling protein-1 (UCP1). As a proton leakage channel, UCP1 uncouples oxidative phosphorylation from ATP production, thus resulting in energy waste and heat generation [9].

Interestingly, unlike the classical BAT found in babies, many of these glucose-responsive adipocytes, which are called beige or brite (brown-in-white) cells, are inducible and convertible within adult human WAT [7,10]. Upon specific stimuli (such as cold acclimation), these otherwise quiescent beige cells are remodeled to demonstrate brown-like characteristics [10–12]. Therefore, beige adipocytes possess the properties of both white and brown adipocytes, acting as an energy reservoir in the quiescent condition while dissipating energy upon activation.

Both classical brown adipocytes and beige adipocytes have multilocular lipid droplets and a high mitochondria content. They are positive for UCP1 expression, giving them the general term of thermogenic adipocytes. However, beige adipocytes are not simply brown adipocytes in WAT [10,13]. Despite their similar features, classical brown and beige adipocytes are developmentally derived from distinct precursors and mobilized by both overlapping and different mechanisms [7]. Specifically, brown adipocytes originate from a *Myf5*$^+$ myogenic precursor, a special progenitor for skeletal muscle [13,14]. In contrast, beige adipocytes lack the historical expression of *Myf5*$^+$. This suggests that brown adipocytes behave in more myocytic ways. Additionally, the UCP1 of beige adipocytes is only induced after appropriate stimuli; otherwise, the expression level is as low as that in white adipocytes [10,15]. Finally, they differ in other gene expression patterns despite both *UCP1* and other thermogenic genes being exhibited. Specific markers, such as *TMEM26*, *TBX1*, and *CD137*, are unique to beige adipocytes. On the other hand, brown adipocytes preferentially express *ZIC1* and *LHX8* [16]. Adipose tissue reservoirs in the supraclavicular and paraspinal regions do not consist of "pure" classical brown adipocytes. Instead, the brown-to-beige ratio is elevated when moving deeper. Despite the distinct developmental lineages and mechanisms of activation, whether brown and beige adipocytes have different functions remains unknown. Wu et al. suggested that fully stimulated brown and beige adipocytes contain comparable levels of UCP1 [10]. However, it will be interesting to investigate and compare the functions of these two adipocytes, such as their lipid-lowering, glucose uptake, anti-inflammation, and cytokine secretion patterns.

UCP1 plays a key role in adaptive thermogenesis through uncoupling the mitochondrial respiratory chain from ATP production. During a cold challenge, highly innervated sympathetic neurons are evoked and release catecholamine, mainly norepinephrine (NE). Followed by the provocation of β3-AR after NE binding, adenylyl cyclase (AC) is activated and converts ATP to cyclic adenosine monophosphate (cAMP). The downstream protein kinase A (PKA) is evoked to phosphorylate triacylglycerol lipase, resulting in lipid catabolism. Afterward, a large amount of free fatty acid is liberated and acts as a fuel to induce UCP1 expression. UCP1 uncouples the mitochondrial respiratory chain, speeding up the substrate oxidation rate to produce heat and lower ATP generation in brown fat [9].

However, UCP1 KO animal models still maintain cold resistance, indicating the presence of UCP1-independent thermogenesis. In beige adipocytes, specifically in UCP1 KO mice, pathways related to Ca^{2+} cycling were upregulated upon cold exposure, including sarco/endoplasmic reticulum Ca^{2+}-ATPase2b (SERCA2b) [17]. This Ca^{2+} cycling uncoupled ATP production and resulted in heat production [17]. Furthermore, in thermogenic adipocytes, the creatine phosphorylation and de-phosphorylation cycle stimulates mitochondrial respiration, a process that occurs independently of UCP1 [18]. Josef et al. identified a futile triglyceride/fatty acid cycle as a novel mechanism in UCP1-independent thermogenesis [19]. This cycle involves concurrent triglyceride lipolysis and fatty acid re-esterification within the lipid droplets of brown adipocytes, thus accelerating ATP consumption and heat generation through the interaction with the mitochondrial electron transport chain (ETC) [19].

BAT was previously thought to be present only in human newborns. However, in the past decade, [18F]-fluorodeoxyglucose positron emission tomography/computed tomography (18^F-PET/CT) unequivocally revealed the presence of functional BAT that extends from the anterior neck to the thorax in healthy adults [20,21]. The amount of BAT was inversely associated with age and BMI in either sex, although a higher mass and activity of BAT were detected in women compared to men [20]. Furthermore, a large-scale retrospective study recently revealed that the amount of BAT was closely correlated with cardiometabolic health [22]. There is also evidence that activation of BAT antagonized multiple types of cancer progression in preclinical animal models and in one lymphoma patient [23].

The basal and maximum respiration of endogenous mature brown adipocytes were approximately 5–10 times higher compared to the stromal vascular fraction (SVF) cells or mature white adipocytes [24]. Therefore, these thermogenic adipocytes (beige and brown adipocytes) are active combustors of glucose and lipid. It is noteworthy that brown and beige cells also conferred a series of metabolic benefits beyond the direct glucose- and lipid-lowering effects, including anti-inflammation, insulin-sensitization, and anti-atherosclerosis (Figure 1) [25–27]. In line with these mechanistic studies, transplantation studies have demonstrated the essential role of BAT in improving adiposity, glucose metabolism, and insulin resistance in mice (reviewed in detail below). In clinical studies, the activation of brown/beige adipocytes by cold stimulation ameliorated glucose metabolism and insulin sensitivity in both healthy subjects and patients with T2DM (Figure 1) [28–30]. This evidence collectively highlights the therapeutic potential of thermogenic adipocytes in treating obesity and its related comorbidities.

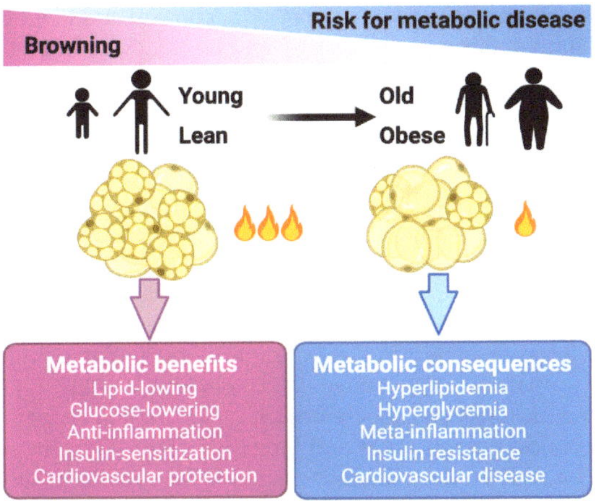

Figure 1. Brown and beige fat cells hold therapeutic potential for treating obesity and metabolic diseases. Brown and beige fat cells (adipocytes) generate heat to burn off extra amounts of glucose and lipids. Activation of these thermogenic adipocytes also confers benefits beyond lowering glucose and lipids. The activity of brown and beige adipose declines in obese and aging people.

3. Pharmaceutical Approaches for Activation of Brown and Beige Adipocytes

Recent research has provided insights into small molecules with the potential to activate BAT or beige adipocytes. Dietary components, including capsaicin, resveratrol, curcumin, green tea, menthol, and fish-derived omega-3 fatty acids, have shown signs of stimulating BAT and beige adipocyte activities, which at least partially explains their anti-obesity and anti-metabolic syndrome benefits. This information has been well summarized [31–33]. Therefore, the current review will focus on compounds beyond these categories that include either FDA-approved drugs or pharmaceutical reagents, for their effects on enhancing the thermogenic activity and energy expenditure rate in adipose tissues.

3.1. FDA-Approved Drugs Repurposed for Metabolic Diseases

3.1.1. β3 Adrenergic Receptor (β3-AR) Agonist

β3 adrenergic receptor (β3-AR) is expressed in rodent brown adipocytes. It plays a vital role in engaging lipolysis and thermogenesis [34]. In humans with type 2 diabetes, short-term cold exposure activates brown/beige fat and improves insulin sensitivity [30]. However, cold exposure is clinically impractical owing to the inconvenience of the regimen, the discomfort, and the complex and indirect pathways elicited by the cold. Cold mimetics, i.e., nonspecific sympathomimetic drugs, also pose cardiovascular risks because of the widespread expression of sympathetic receptors, which limits their clinical application [35]. By this logic, β3-AR is a superior target since its expression is restricted in adipocytes and the urinary bladder. Although it is still under debate as to whether β3-AR is the predominant isoform expressed in human BAT [36], extensive efforts have been made to establish β3-AR agonism as a promising treatment regimen for metabolic disorders. Mirabegron is a selective β3-AR agonist with satisfactory bioavailability [37,38]. Sold under the brand name Myrbetriq®, it was initially approved by the U.S. Food and Drug Administration in 2012 for the treatment of overactive bladder (OBA) at daily doses of 25 or 50 mg. In recent years, a number of clinical trials have been carried out to explore the repurposing of this drug for treating obesity and metabolic diseases at a super-therapeutic dosage [11]. In particular, one study that explored an effective and safe dosage for individual use revealed that a 100 mg daily intake increased the skin temperature and energy expenditure in participants without any cardiovascular side effects, while increased blood pressure and heart rate were observed at 150 and 200 mg doses, respectively [39]. However, the exact β-AR isoforms that predominate in human brown and beige adipocytes are under debate [36,40]. Different β-AR isoforms may function in distinct brown and beige subtypes. Adipose-selective or even subtype-specific delivery of agonists might be of interest in future studies.

3.1.2. Glucagon-like peptide-1 Receptor Agonists (GLP-1RA)

The glucagon-like peptide-1 receptor agonists (GLP-1RAs), including liraglutide and semaglutide, were initially used to manage T2DM due to their ability to amplify glucose-dependent insulin secretion [41,42]. They decrease plasma glucose concentrations, induce weight loss, have cardioprotective effects, improve insulin sensitivity, and reduce low-grade inflammation in humans [43]. In particular, a remarkable weight loss effect was observed clinically [44–46]. GLP-1RA acts directly on the brain and gastrointestinal tract to suppress appetite and delay gastric emptying. Whether GLP-1R is present in adipose tissue is less clear. It has been reported that liraglutide stimulates BAT thermogenesis and browning via hypothalamic AMP-activated protein kinase (AMPK) [47]. Liraglutide demonstrated a therapeutic effect on mitochondrial dysfunction in human adipocytes in vitro, promoting mitochondrial respiration and biogenesis [48]. Liraglutide also suppressed obesity and induced a brown-fat-like phenotype via the soluble-guanylyl-cyclase-mediated pathway in vivo and in vitro [49]. A recent study found that liraglutide promoted brown remodeling of visceral WAT by regulating miRNAs [50]. More importantly, in a longitudinal study of 25 patients with obese T2DM who were treated with exenatide or liraglutide for 1 year, both of these two GLP-1RAs elevated the energy expenditure rates of the subjects [47]. These observations suggest that long-acting GLP-1RAs control body weight at least partially through increasing energy consumption, in addition to their effect on regulating food intake. Repurposing the GLP-1RAs for obesity is quite popular in clinical scenarios. However, there is an upper limitation on weight loss in obesity, even without considering relapse after treatment cessation. Considerably obese patients also show resistance or poor response to the drugs. Furthermore, there are contraindications, such as pancreatitis and thyroid medulla carcinoma.

3.1.3. Thiazolidinediones (TZDs)

Thiazolidinediones (TZDs) are potent agonists for peroxisome proliferator-activated receptor gamma (PPARγ) and exert strong stimulatory effects on fatty acid storage via adipogenesis and fatty acid flux into adipocytes in adipose tissue [51]. TZDs have shown the potential to enhance the brown features within WAT, leading to a reduction in obesity-related disorders [52,53]. In vivo or in vitro exposure of white adipocytes to TZDs activated the browning process, characterized by enhanced levels of mitochondrial biogenesis, oxygen consumption, and lipid oxidation [52,54]. Meanwhile, TZDs and other PPARγ ligands led to the suppression of multiple "bad" adipokines, such as resistin, α1-acid glycoprotein, and haptoglobin, which probably contributed to the anti-obesity effect of these drugs [55–57]. Unfortunately, TZDs have largely been withdrawn from the market due to their risks of inducing hepatotoxicity, myocardial infarction, bladder cancer, and heart failure [58]. Therefore, next-generation PPARγ agonists are warranted and are being tested for BAT and beige adipocyte activation.

3.1.4. Others

Thanks to the development of high-throughput screening platforms, a broader scope of pharmacological reagents has been found with the potential for repurposing to treat obesity and metabolic dysfunction. A group of medicines has been demonstrated to activate UCP1 expression levels, including indirubin, rutin, and myricetin [47]. Indirubin is made from indigo naturalis and has a long history in traditional Chinese medicine. It has been mainly used for the prevention and treatment of cancers because of its suppressive effects on oncogenic gene expression [59]. It is now known that indirubin improves body weight, lipid accumulation, and glucose homeostasis in high-fat diet (HFD)-fed mice via enhancing thermogenesis in BAT and WAT [60]. Melatonin is recognized as a hormone that regulates sleep and circadian rhythms; it also shows a potential thermogenic effect. Oral melatonin-treated Zucker rats showed induced browning of inguinal WAT, with around double the expression level of UCP1 and PGC-1α [61]. Meanwhile, melatonin increased the inguinal temperature and made the rats more sensitive to acute cold exposure without physical activity modification [61]. The cellular screen approach is an alternative system with a directly visible fluorescence readout by which new candidates were identified. Sutent, for example, was found to be capable of upregulating UCP1 in BAT and elevating calorie consumption to protect against obesity [62].

In addition to unbiased screening-based studies, hypothesis-driven studies have also revealed some promising candidates. UCP1 is mechanistically regulated by signaling pathways such as Janus kinase (JAK) and PPARα. Tofacitinib and R604, two JAK inhibitors, induced brown-like metabolic properties, including elevated UCP1 expression and enhanced mitochondrial activity, in human adipocytes [63]. The activation of PPARα by fenofibrate, a PPARα agonist, triggered *UCP1* and other thermogenic gene expressions in BAT [64]. Along with insulin resistance, obesity and WAT dysfunction in obese mice were reversed [64]. Bexarotene, sold under the brand Targretin, is an antineoplastic agent used for the treatment of cutaneous T-cell lymphoma. A study showed that it enhanced the conversion of myoblastoma to brown adipocytes and promoted thermogenesis [65].

However, it should be noted that except for Myrbetriq, the current pro-browning evidence for the FDA-approved drugs mentioned earlier is based only on mouse or cell line studies. Clinical studies are needed to prove the effect of these drugs in enhancing energy expenditure and adipose tissue browning. In addition, repurposing raises concerns about the possible side effects of these drugs on obese/diabetic patients, posing a major obstacle. Table 1 summarizes the drugs examined in the studies described above.

Table 1. FDA-approved drugs showing the potential to promote adipose browning.

Name	Original Purpose	Reference
MIRABEGRON	Overactive bladder	[11,39]
GLP-1RA	Type 2 diabetes	[47–50]
TZDs	Type 2 diabetes	[52–54]
INDIRUBIN	Chronic myeloid leukemia	[60]
MELATONIN	Insomnia	[61]
SUTENT	Gastrointestinal stromal tumors, renal cell carcinoma	[62]
TOFACITINIB	Rheumatoid arthritis, psoriatic arthritis, ankylosing spondylitis, juvenile idiopathic arthritis, ulcerative colitis	[63]
FENOFIBRATE	Hypertriglyceridemia	[64]
BEXAROTENE	Cutaneous T-cell lymphoma	[65]

3.2. Small-Molecule Compounds under Preclinical Trial Studies for Obesity and Metabolic Diseases

3.2.1. Five Catalogs under Preclinical Trial Studies

As mentioned, PPARγ is a promising anti-obesity target, with multiple PPARγ agonists currently under development. Formononetin is a natural compound isolated from Astragalus membranaceus that shows putative PPARγ agonism activity. Formononetin protected mice from diet-induced obesity and facilitated a higher level of energy expenditure through increasing UCP1 following binding to PPARγ [66]. Mu Q et al. treated 3T3-L1 mature adipocytes with ginsenoside, the major active pharmacological component of ginseng. They found that 10 μM of ginsenoside Rb1 elevated basal and insulin-stimulated glucose uptake [67]. Moreover, ginsenoside Rb1 treatment also promoted adipocyte browning through the PPARγ signaling pathway, and its effect was abolished by the PPARγ antagonist [67]. Physical activity increases human energy expenditure and lowers the risk of obesity and T2DM [68]. β-aminoisobutyric acid (BAIBA) is a small-molecule myokine secreted from *PGC-1α*-expressing myocytes. In mice and humans, circulating BAIBA levels rose following exercise, ameliorating thermogenesis in WAT via PPAR [69].

Retinoid X receptor (RXR) is the heterodimeric partner for PPARγ. Both RXR and retinoid acid receptor (RAR) are endogenously activated by retinoid acid (RA). In 1995,

Alvarez and Puigserver demonstrated that RA serves as a transcriptional coactivator contributing to UCP1 expression both in vitro and in vivo [70,71]. Fenretinide protected against weight gain and insulin resistance in obese mice [72]. Inhibition of preadipocyte differentiation was proposed as a possible mechanism [72]. However, fenretinide's role as a thermogenesis inducer has not been directly tested yet. Benzoic acid and methoprene, working via RAR and RXR, respectively, significantly induced the mRNA expression of *UCP1* in BAT [73].

PR domain-containing 16 (PRDM16) is a 140 kDa zinc-finger PR (PRD1–BF1–RIZ1 homologous) domain-containing protein that was first identified by Nishikata et al. in 2003 [74]. Earlier studies found that PRDM16 was selectively expressed in brown adipocytes and had the capability to drive the brown phenotype and induce mitochondrial respiration [75]. It is also a coregulator of and interacts with PGC-1α/β or CtBPs to promote brown adipocyte-specific genes while suppressing white adipocyte gene expression [76]. Recently, rutaecarpine was identified as inducing the browning process in WAT after screening 500 natural compounds. With KEGG pathway analysis from RNA sequencing, it was proposed that the AMPK signaling pathway played a vital role as the downstream PRDM16 was activated [77]. Likewise, RGFP966, a selective class I histone deacetylase (HDAC3) inhibitor, was found to potentially drive the thermogenic pattern in brown and beige adipocytes in vitro, such as PGC-1α, UCP1, and FGF21 [78]. This impact of RGFP966 was proved to act through PRDM16 since PRDM16 knock-down blunted the effect. PRDM4 is another PRDM family member showing a positive function in stimulating browning in white adipose tissue. Butein is a biologically active flavonoid with an anti-cancer effect. It was identified as a potent means of inducing the expression of UCP1, increasing energy expenditure, and stimulating the generation of thermogenic adipocytes, thus highlighting a PRDM4-dependent pathway [79].

Bile acids (BAs) are natural products of the liver that participate in cholesterol catabolism and lipid absorption to reverse diet-induced obesity. Recent studies have shown that BAs are capable of enhancing energy expenditure in BAT and oxygen consumption in brown adipocytes [80]. The rising level of cAMP directly regulates these impacts after BAs bind to the G-protein-coupled receptor TGR5. Additionally, the genes involved in energy metabolism and uncoupling, iodothyronine deiodinase type 2 (*DIO2*), peroxisome proliferator-activated receptor γ coactivator-1α (*PGC-1α*), and *UCP1*, were found to be remarkably elevated with an increasing concentration of plasma BAs [81]. However, other hormonal functions of BAs via the farnesoid X receptor might have been overlooked, which makes them a less viable candidate.

During BA synthesis, liver X receptors (LXRs) are essential nuclear receptors with two isoforms (LXRα and LXRβ). New evidence has emerged that LXRs and PPARγ work either together to promote lipogenesis and differentiation in adipocytes or in an antagonistic manner to induce insulin resistance and suppress adiponectin signaling [82–84]. Moreover, LXR agonists, such as TO901317 and GW3965, inhibit adaptive thermogenesis via down-regulating *DIO2* and *UCP1*, respectively [85,86]. On the contrary, as a natural compound derived from *Rheum palmatum* L., rhein evokes UCP1 expression by antagonizing LXRs in BAT, thus maintaining the energy balance in mice [87]. Further clinical investigations are necessary in light of its multi-target character [88].

3.2.2. Other Chemicals Showing Efficacy in BAT, Beige Adipocytes, or Both

Several other compound molecules acting through distinct mechanisms have been observed to activate BAT in preclinical models and serve as candidates for the clinical treatment of obesity and associated diseases. We now summarize these molecules according to their reported target cells/organs.

WWL113 and capsiate are two molecules that mainly contribute to BAT activation. WWL113 inhibits the lipolysis enzyme of mouse carboxylesterase 3 (Ces3) and the human orthologue CES1 [89]. A recent study proved that CES1 activity was doubled in obese T2DM patients compared with lean individuals, and obese T2DM patients were noted to produce excessive fatty acids that were deposited in ectopic tissues [89]. WWL113 treatment resulted in a predominant augmentation in mRNA levels of *UCP1* and thermogenic-related genes, including *PGC-1α* and *DIO2*, in BAT both in vivo and in vitro via the PPARα signaling pathway [90]. WWL113 exhibited a more robust response in the presence of adrenergic stimulation in that it increased the energy expenditure of mice without affecting their locomotor activity, food intake, or heart rate [90]. These data support the potential of WWL113 in stimulating BAT activity. Similarly, capsiate, extracted from red peppers, is a modified nonpungent capsaicin analog. After treatment with 10 mg/kg body weight for 2 weeks, mice exhibited elevated UCP1 levels in their BAT and raised oxygen consumption. In another study, capsiate activated the sympathetic nervous system in mice by promoting adrenaline secretion. These results suggest that capsiate is a promising means to achieve BAT activation [91,92].

Other small molecules reportedly show selective effects on beige adipocyte induction, including WIN18446, linifanib, KY19334, and TG003. Aldehyde dehydrogenase 1 family member A1 (ALDH1a1) belongs to the aldehyde dehydrogenase family and is the second enzyme in the oxidative pathway of alcohol metabolism. WIN18446 is an ALDH1a1-specific inhibitor. It has been demonstrated to suppress weight gain and decrease the amount of visceral adipose tissue in obese mice [93]. Another independent study determined the possible underlying mechanism, indicating that ALDH1a1 ablation induced a BAT-remodeled transcriptional process in WAT, which enhanced mitochondrial respiration and adaptive browning via retinaldehyde and acetate [94,95]. In a study using a UCP1 reporter cell, linifanib was found to increase the expression of UCP1 via inhibiting STAT3 phosphorylation [96]. The Wnt/b-catenin signaling pathway is a negative regulator in adipogenic differentiation. In mice fed a high-fat diet, KY19334, a Wnt/b-catenin pathway activator via CXXC5–disheveled interaction, repressed adipogenesis and ameliorated insulin resistance and adipocyte hypertrophy, resulting in lower weight gain and potentiating browning [97]. PGC-1α is a substrate for Clk1 kinases, and inhibition of PGC-1α phosphorylation resulted in adipocyte beiging, indicating that Clk1 suppression may stimulate beige-like adipocytes. TG003, a Clk1 inhibitor, increased the number of mitochondria and induced fewer and smaller lipid droplets in adipocytes [98].

Considering the similar functions of brown and beige adipocytes, it is not surprising that several molecules show efficacy in activating both BAT and beige adipocytes. Nuclear factor kappa B (NF-κB) activation drives the activation of noncanonical IκB kinases IKKε and TANK-binding kinase 1 (TBK1) in adipose tissue and liver, subsequently counteracting energy storage [99]. It is intriguing that amlexanox, an inhibitor of these two kinases, increases thermogenesis and elevates energy expenditure; it has also been found to cause weight loss and ameliorate insulin resistance in mice with obesity [99]. Furthermore, the small-molecule compound BAY 41-8543 against soluble guanylyl cyclase (sGC) showed efficacy in protecting against diet-induced weight gain as a result of enhancing whole-body energy expenditure, enhancing the differentiation of brown adipocytes and inducing the thermogenesis of white adipocytes [100]. BIBO3304 is a selective antagonist of peripheral Y1R. It specifically induces thermogenesis in BAT and browning of WAT through the UCP1-dependent pathway [101]. BIBO3304 improves glucose homeostasis by driving Akt activity in BAT [101]. New data also provided the novel insight that a natural compound, harmine, is a thermogenic activator in both white and brown adipocytes, mediated by the RAC1/MEK/ERK pathway [102]. Studies have shown that berberine has a range of metabolic benefits, such as improving insulin resistance and hyperlipidemia. Zhang et al. found that in berberine-treated mice, fatty acids became the preferred fuel and generated more heat during cold exposure. Consistent with these processes, berberine also induced a thermogenic program in BAT and WAT via AMPK/PGC-1α [103]. However, the underlying mechanisms regarding the toxicity and side effects of these compounds in humans await further clarification.

Likewise, there is not enough current evidence to test these candidates on humans. The pharmacodynamics, pharmacokinetics, and toxicity of these compound candidates must be comprehensively studied before they can be further translated into clinical trials. Furthermore, although UCP1 was deemed the exclusive effector for thermoregulation, it is now clear that adipose thermogenesis is engaged by both UCP1-dependent and -independent mechanisms [17,18,104–106]. It is expected that new classes of pharmacological molecules targeting the non-canonical thermogenic pathways will be characterized and serve as novel drug leads for obesity and metabolic diseases. Table 2 summarizes the small-molecule compounds mentioned above.

Table 2. Small compounds showing browning potential.

Chemicals	Receptors/Downstream Target	Target	Models	Reference
FORMONONETIN	PPARγ	Beige adipocyte	3T3-L1	[66]
GINSENOSIDE	PPARα	Beige adipocyte	3T3-L1	[67]
BAIBA	PPARγ	BAT; beige adipocyte	Mouse; human	[69]
RETINOID ACID	RXR, RAR	BAT	Mouse	[70,71]
FENRETINIDE	RXR, RAR	Adipocyte differentiation and hypertrophy	3T3-L1	[72]
BENZOIC ACID	RAR	BAT	Mouse	[73]
METHOPRENE	RXR	BAT	Mouse	[73]
RUTAECARPINE	PRDM16	Beige adipocyte	Mouse	[77]
RGFP966	HDAC3	BAT; beige adipocyte	Mouse; human	[78]
BUTEIN	PPAR	Beige adipocyte	Mouse	[79]
BILE ACID	TGR5	BAT	Mouse; human	[80,81]
RHEIN	LXRs	BAT	Mouse	[87]
WWL113	CES1	BAT	Mouse; human	[90]
CAPSIATE	Sympathetic nervous system	BAT	Mouse	[91,92]
WIN18446	ALDH1a1	Beige adipocyte	Mouse	[93]
LINIFANIB	STAT3	Beige adipocyte	Mouse	[96]
KY19334	Wnt/b-catenin	Beige adipocyte	Mouse	[97]
TG003	Clk1	Beige adipocyte	3T3-L1	[98]
AMLEXANOX	TBK1	BAT; beige adipocyte	Mouse	[99]
BAY 41-8543	sGC	BAT; beige adipocyte	Mouse; human	[100]
BIBO3304	Y1R	BAT; beige adipocyte	Mouse; human	[101]
HARMINE	(DNA binding protein 4) CHD4	BAT; beige adipocyte	Mouse	[102]
BERBERINE	ERK/p38 MAPK	BAT; beige adipocyte	Human	[103]

4. BAT Transplantation to Counteract Metabolic Disorders

Although activating brown and beige adipocytes is a feasible strategy for managing metabolic diseases, one of the major clinical issues is that adult humans have only a limited amount of BAT [107]. Additionally, BAT mass further declines in aging and diabetic conditions [108], making the pharmaceutical method less feasible in older individuals and diabetic patients. Furthermore, heterogeneity has been uncovered within BAT, where distinct subtypes of brown adipocytes with different thermogenic activities are present [24]. Therefore, transplantation of BAT to increase its mass and activity is emerging as an alternative approach to reduce the comorbidities associated with obesity (Figure 2).

4.1. BAT Transplantation: Proof-of-Concept Studies

The first attempt at BAT transplantation was carried out in 1960, predominantly aiming to delineate the in vivo function of BAT in mice [109]. Since then, a series of studies have been conducted to evaluate the metabolic outcomes of BAT transplantation. BAT was collected and transplanted into the subcutaneous region, visceral cavity, or dorsal inter-scapular region of HFD-induced obese mice [110–112]. In one of these studies, BAT was subcutaneously transplanted into the dorsal inter-scapular region of recipient mice, which then exhibited reduced weight gain, decreased liver mass, improved glucose uptake, raised insulin responsiveness, and increased body temperature [110]. BAT transplantation increased the level of circulating adiponectin, whereas it reduced the levels of circulating free T3 and T4, which regulate thyroid hormone sensitivity in peripheral tissues. Higher expression levels of fatty acid oxidation genes and improved glucose uptake in primary BAT were observed after BAT transplantation [113]. Gunawardana et al. investigated the effect of BAT on glucose metabolism in a mouse model with type 1 diabetes [112]. Embryonic BAT was obtained and transplanted subcutaneously into streptozotocin (STZ)-treated recipient mice. Mice with BAT transplantation exhibited normoglycemia, reduced tissue inflammation, improved glucose tolerance, and reversed clinical diabetic symptoms such as polyuria, polydipsia, and polyphagia [112]. Furthermore, elevated serum levels of leptin, adiponectin, and insulin-like growth factor 1 (IGF-1) were observed after BAT transplantation. However, it remained unaddressed whether the increased levels of these adipokines came directly from the transplanted BAT or via an indirect mechanism to improve the function of the endogenous adipose tissues. Since IGF-1 is one possible candidate for activating the insulin receptor, the authors proposed that the insulin receptor was activated, leading to an improvement in glucose homeostasis in their model. Indeed, the IGF-1 mechanism was further investigated in a non-obese diabetic recipient [114]. The success rate was elevated to 57% in adult BAT transplantation with IGF-1 supplementation [115]. It is also worth noting that BAT transplantation was accompanied by a significant increase in the expression of β3-AR in WAT and mitochondrial-specific OXPHOS proteins in endogenous BAT [116], suggesting that BAT transplantation increased whole-body thermogenesis and reduced obesity and related diseases by activating endogenous BAT.

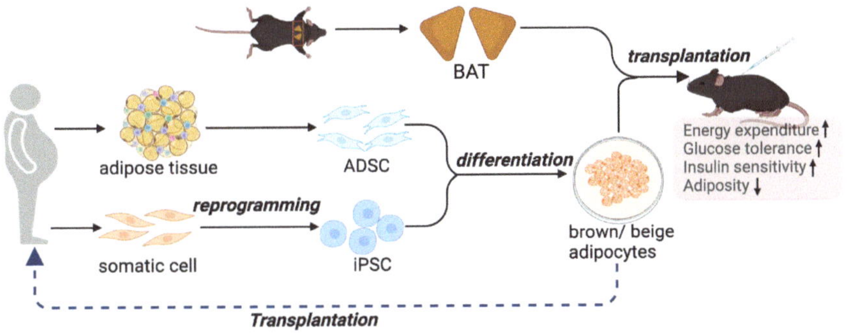

Figure 2. BAT, or thermogenic adipocyte transplantation, represents an alternative approach to counteract metabolic disorders. A proof-of-concept study demonstrates the therapeutic potential of thermogenic adipocytes and adipose tissue in protection against obesity and metabolic diseases. In the clinical setting, iPSCs or ADSCs are possible sources of precursors for brown and beige adipocytes.

4.2. Practicality and Ethical Concerns

Several concerns compromise the practicality of BAT transplantation from a human donor as a therapy. As in other organ donation cases, the high cost, availability of organ donors, and use of immunosuppressive medication are obstacles to be overcome before BAT transplantation can be deemed as a possible therapy. Furthermore, it should be noted that we currently lack unified standards for a "qualified BAT donor". A comprehensive and standardized guideline to quantitively evaluate the amount and activity of functional and healthy BAT in adult humans across different sexes and ethnic groups is urgently needed. Furthermore, unlike other solid organ transplantations, the anatomical location of functional BAT in adult humans is heterogeneous and scattered throughout the body. In most cases, BAT is combined with WAT. Therefore, further elucidation is necessary to identify the location of the adipose tissue depot to be biopsied. Likewise, the location of transplantation to the BAT donor in humans is not yet optimized.

What is equally important is evaluating the short- and long-term health and psychological risks of post-surgical procedures. Answers to these questions are key to addressing the ethical considerations in BAT transplantation because BAT transplantation involves the removal of a healthy organ from the donor. Therefore, whether the traditional first rule of medicine—primum non nocere (above all, do no harm)—will be violated needs to be considered.

5. Brown and Beige Adipocyte Transplantations as Cell Therapies for Metabolic Diseases

Owing to the limitations of BAT transplantation, therapies using the in vitro culture, expansion, and differentiation of brown and beige progenitor cells to functional thermogenic adipocytes are emerging as alternative approaches and more feasible sources of cells for transplantation (Figure 2). The transplanted adipocytes can be derived from human adipose-derived mesenchymal stromal cells (ADSCs) or human induced pluripotent stem cells (hiPSCs).

5.1. ADSCs for Brown and Beige Adipocyte Differentiation and Transplantation

Brown adipocytes were induced from ADSCs and then transplanted into HFD-fed obese mice [117]. Very-low-density lipoprotein (VLDL) and low-density lipoprotein (LDL) levels declined in the circulation of obese mice after transplantation, accompanied by an increasing HDL level [117]. Additionally, the concentrations of proinflammatory cytokines, including IL-6, tumor necrosis factor-alpha (TNF-α), and IL-1β, were diminished by inhibition of the ITGAM/NF-κB-mediated proinflammatory responses and polarization of M2 macrophages. This study indicated that the ADSC-derived brown adipocytes could promote lipid catabolism and alternative polarization of M2 macrophages to ameliorate adipose inflammation in obese animal models. Similarly, a human ADSC suspension, conditioned medium, and cell lysate were intramuscularly injected into obese mice, resulting in a remarkable improvement in insulin resistance accompanied by a decrease in oxidized LDL and IL-6 [118]. In addition, mice receiving the human ADSC suspension injection exhibited lower lipid content and macrophage infiltration in the liver and adipose tissue, respectively; enhanced glucose tolerance; and pancreatic islet hypertrophy [118]. Recently, the generation of beige adipocytes from human ADSCs in a serum-free medium was reported [119]. The derived beige adipocytes showed a similar molecular profile to primary beige adipocytes isolated from human tissue and exhibited uncoupled mitochondrial respiration and cAMP-induced lipolytic activity. Transplantation of these beige adipocytes increased whole-body energy expenditure and oxygen consumption, and reduced body weight in recipient mice.

5.2. hiPSC-Derived Brown and Beige Adipocytes as Sources of Transplantation

hiPSCs were first generated in 2007 [120,121]. hiPSCs are generated from various somatic cells and self-renew in vitro; therefore, they represent a more accessible and unlimited source of brown and beige precursor cells. The differentiation ability of hiPSCs lays the foundation for cell therapies and has the potential to counteract obesity and associated comorbidities [120]. Significant advances have been made in the area of hiPSC-derived brown and beige adipocytes, showing promise for obesity treatment.

With the aim of treating patients with lipodystrophy, Taura et al. were the first to demonstrate the adipogenic potential of hiPSCs [122]. Approximately 15% of the induced cells were lipid-containing. Key adipogenic markers were observed in the hiPSC-derived adipocytes, including *CEBP*α, *PPAR*γ*2*, and fatty-acid-binding protein 4 (*FABP4*). However, in this study, the phenotype of the generated adipocytes, i.e., whether they were white or brown, was not examined. In another study, hiPSC-derived fibroblasts were selected and induced into white adipocytes and brown adipocytes using the gene transfer method [123]. In this approach, the hiPSCs were driven into embryoid bodies, followed by generation into mesenchymal progenitor cells (MPCs). The MPCs were then programmed to express *PPAR*γ*2* alone or combined with *CEBP-*β and/or *PRDM16*. After culturing in an adipogenic cocktail (insulin, 3-isobutyl-1-methylxanthine, dexamethasone) and doxycycline for 14–16 days and further maintenance without doxycycline until 21 days, these cells differentiated into white or brown adipocytes with a high efficiency of 85–90% [123]. These adipocytes exhibited the properties of mature adipocytes, such as expression of the mature adipogenic markers *FABP4*, *PPARγ2*, and *CEBPα*. They also showed adipocyte functions, including lipid metabolism and glucose uptake. Notably, after these mature adipocytes were engrafted into the recipient mice for 4–6 weeks, ectopic fat pads were observed with the morphological and functional properties of white and brown adipose tissue. However, no evidence has yet proven that these programmed cells expressing mature adipocyte markers can mimic physiological adipogenesis in vivo.

Studies have reported the successful differentiation of brown adipocytes without gene transfer [124–126]. Nishio and colleagues reported for the first time a highly efficient (>90%) differentiation method to induce hiPSCs into functional brown adipocytes via a specific hemopoietin cocktail without gene transfer [124]. On the other hand, hiPSCs were differentiated into white adipocytes without a hemopoietin cocktail [124]. The resultant brown adipocytes were functional since they exhibited potent thermogenic activation and mitochondrial respiratory response under β-adrenergic receptor exposure, accompanied by increased lipid and glucose tolerance. Moreover, the immune-competent mice showed metabolic improvements after hiPSC-derived brown adipocyte transplantation. Another method was described to achieve the direct differentiation of hiPSCs into brown and white adipocyte progenitors without gene transfer [125]. Treatment with the TGFβ pathway inhibitor SB431542 and ascorbic acid and EGF promoted hiPSCs to brown adipocyte differentiation, thus highlighting the critical role of the TGFβ pathway in switching off hiPSC-brown adipogenesis [125]. Most recently, hiPSCs were selectively induced into functional brown adipocytes via the paraxial mesoderm progenitor stage at high efficiency, characterized by increased rates of uncoupled respiration, glycolysis, and lipolysis [127]. Afterward, they were transplanted into the inter-scapular region of mice and showed thermogenic activity. That is, the recipient mice had enhanced respiratory exchange rates, metabolic activity, and whole-body energy consumption. Moreover, transplanted brown adipocytes decreased circulating glucose levels in STZ-induced diabetic non-obese diabetic/severe combined immunodeficiency (NOD/SCID) mice.

Efforts are also underway to generate beige adipocytes from hiPSCs. Guenantin et al. reported a straightforward and efficient procedure to derive functional beige adipocytes from hiPSCs through the mesodermal and adipogenic progenitor state [128]. These hiPSC-derived beige adipocytes expressed the beige-specific markers, displayed higher expression of thermogenic genes (but not *UCP1*), increased mitochondrial content, and improved oxygen consumption after cAMP analog stimulation. hiPSC-derived beige adipocytes formed a well-organized and vascularized adipocyte tissue with a higher glucose uptake rate after transplantation. In contrast, Aaron Brown's team described the generation of human beige adipocytes from hiPSCs in a stepwise manner via forkhead box F1 (FOXF1+) splanchnic mesoderm, mural-like MSCs, and preadipocytes [129]. The mature hiPSC-derived beige adipocytes showed upregulated *UCP1* expression and uncoupled respiration. Cytokines were also secreted from hiPSC-derived beige adipocytes to improve insulin sensitivity and glucose uptake. At the molecular level, progenitors expressed beige/brite markers such as *CD137* and *TMEM26* but not the brown-specific marker *ZIC1*.

5.3. Practicality and Ethical Considerations for Brown and Beige Cell Therapies

Compared to BAT transplantation, the acquisition of brown and beige adipocytes in vitro is a more feasible means of transplantation. ADSCs can be harvested from the abdomen, thighs, flanks, and axilla using liposuction or direct excision techniques [130]. However, the in vitro expansion of ADSCs remains limited, making hiPSCs a superior and unlimited cellular source for brown and beige adipocyte differentiation, considering that hiPSCs can be derived from various skin, blood, and urine somatic cells [131–134]. However, knowledge about the developmental origins and pathways of human brown and beige adipocytes is still limited, especially regarding how the precursor cells of brown and beige adipocytes are developmentally derived. Consequently, compared to the in vitro differentiation from ADSCs, highly efficient differentiation protocols to generate authentic and functional thermogenic adipocytes from hiPSCs are still lacking and need to be developed.

Apart from the in vitro acquisition of functional brown and beige adipocytes, additional considerations and challenges remain. In particular, how long these cells stay alive/functional in the human body, especially in the "pathological" micro-environment of diabetic and aging patients, and how these transplanted cells react to endogenous stimuli and cause unforeseeable consequences require further investigation. The reactions between the physical body and differentiated cells determine how long the transplanted cells can stay alive and functional within the human body, thus determining the frequency of transplantation. If repeated transplantations are needed, additional financial considerations and safety issues must be considered. Whether the differentiated cells should be transplanted by local injection, intravenous infusion, or kept in macroencapsulation devices is another issue for discussion. In the case of allogeneic transplantation, the use of immunosuppressive medicine is critical to cell survival. Therefore, the appropriate immunosuppressive protocol requires further investigation before clinical use.

Meanwhile, using either ADSCs or hiPSCs involves ethical issues, and the donation should be governed by applicable laws and regulations to ensure the appropriate use of the cells. Table 3 summarizes all the cell transplantation therapies mentioned above.

Table 3. Cell transplantation therapy for improvement of metabolism.

Source	Recipient	Effects	Reference
MICE BAT	HFD-fed obese mice	Reduced weight gain; Decreased liver mass; Improved glucose uptake and insulin responsiveness; Increased body temperature.	[110]
MICE EMBRYONIC BAT	Type 1 diabetes mice	Normoglycemia; Reduced tissue inflammation; Improved glucose tolerance; Reversed clinical diabetic symptoms.	[112]
MICE BAT	Ob/Ob mice BAT-deficient obese mice	Increased circulating adiponectin; Improved glucose uptake.	[113]
MICE BAT	Hyperglycemic non-obese diabetic mice	Elevated serum level of IGF-1.	[115]
MICE BAT	Ob/Ob mice	Increased β3-AR expression in WAT; Activated endogenous BAT.	[116]
MICE ADSC-DERIVED BROWN ADIPOCYTES	HFD-fed obese mouse	Decreased levels of VLDL and LDL; Increased level of HDL.	[117]
HUMAN ADSCs	HFD-fed obese mouse	Improvement in insulin resistance; Decreased oxidized LDL and IL-6.	[118]
HUMAN ADSC-DERIVED BEIGE ADIPOCYTES	NOD/SCID mice	Increased whole-body energy expenditure and oxygen consumption; Reduced body weight.	[119]
HIPSC-DERIVED BROWN ADIPOCYTES	Immune-competent mouse	Metabolic improvements.	[125]
HIPSC-DERIVED BROWN ADIPOCYTES	Mouse	Enhanced respiratory exchange rates, metabolic activity, and whole-body energy consumption.	[127]
HIPSC-DERIVED BROWN ADIPOCYTES	NOD/SCID mouse	Decreased circulating glucose levels.	[127]
HIPSC-DERIVED BEIGE ADIPOCYTES	$FoxN1^{Nu}$ athymic mice	Higher glucose uptake rate.	[128]

6. Conclusions

The discovery of brown and beige adipocytes in human adults has inspired substantial efforts to develop new medicinal strategies targeting these thermogenic adipocytes. The original idea that these energy-dissipating adipocytes reverse the positive energy balance has been expanded in clinical and animal-based studies. Brown and beige fat has been directly implicated in glucose and lipid metabolism, the suppression of chronic inflammation, and hormone production [17,27,135,136]. Therefore, regimens to trigger the activity of brown and beige adipocytes are showing therapeutic potential beyond obesity and its related comorbidities, such as atherosclerosis, arterial hypertension, and polycystic ovary syndrome (PCOS). However, it should also be remembered that, currently, the downsides of these adipocytes are less clear and must be comprehensively evaluated. A balance between efficacy and safety must be reached. Hence, as illustrated above, brown and beige adipocyte-based therapies are full of promise, challenges, and pitfalls. A deeper understanding of their mechanisms of action, biological functions, and biogenesis is a prerequisite for eventually achieving effective and safe therapies for obesity, metabolic abnormalities, and a wider range of other diseases. Finally, in addition to the above scientific perspectives, safety and ethical issues should be considered and discussed carefully before brown and beige adipocytes can be applied clinically. Applicable laws and regulations governing related therapies will be warranted.

Author Contributions: Writing—original draft preparation, Y.C.; writing—review and editing, S.L. and S.Z.; supervision, X.H.; funding acquisition, X.H. All authors have read and agreed to the published version of the manuscript.

Funding: This research was funded by the Lo Kwee-Seong Biomedical Research Fund #7106480 and #7106481, and the Research Grants Council (RGC) General Research Fund #390659054.

Conflicts of Interest: The authors declare no conflicts of interest.

References

1. World Health Organization. Obesity and Overweight. Available online: https://www.who.int/news-room/fact-sheets/detail/obesity-and-overweight (accessed on 1 March 2024).
2. Sahoo, K.; Sahoo, B.; Choudhury, A.K.; Sofi, N.Y.; Kumar, R.; Bhadoria, A.S. Childhood obesity: Causes and consequences. *J. Fam. Med. Prim. Care* **2015**, *4*, 187–192. [CrossRef]
3. Blüher, M. Obesity: Global epidemiology and pathogenesis. *Nat. Rev. Endocrinol.* **2019**, *15*, 288–298. [CrossRef]
4. Bray, G.A.; Kim, K.K.; Wilding, J.P.H. Obesity: A chronic relapsing progressive disease process. A position statement of the World Obesity Federation. *Obes. Rev.* **2017**, *18*, 715–723. [CrossRef]
5. Ochner, C.N.; Tsai, A.G.; Kushner, R.F.; Wadden, T.A. Treating obesity seriously: When recommendations for lifestyle change confront biological adaptations. *Lancet Diabetes Endocrinol.* **2015**, *3*, 232–234. [CrossRef] [PubMed]
6. Zhao, S.; Nie, T.; Li, L.; Long, Q.; Gu, P.; Zhang, Y.; Sun, W.; Lin, Z.; Liu, Q.; Qi, Y.; et al. Androgen Receptor is a Negative Regulator of PRDM16 in Beige Adipocyte. *Adv. Sci.* **2023**, *10*, e2300070. [CrossRef]
7. Rosen, E.D.; Spiegelman, B.M. What we talk about when we talk about fat. *Cell* **2014**, *156*, 20–44. [CrossRef]
8. Janssen, I.; Katzmarzyk, P.T.; Ross, R. Waist circumference and not body mass index explains obesity-related health risk. *Am. J. Clin. Nutr.* **2004**, *79*, 379–384. [CrossRef]
9. Cannon, B.; Nedergaard, J. Brown adipose tissue: Function and physiological significance. *Physiol. Rev.* **2004**, *84*, 277–359. [CrossRef]
10. Wu, J.; Bostrom, P.; Sparks, L.M.; Ye, L.; Choi, J.H.; Giang, A.H.; Khandekar, M.; Virtanen, K.A.; Nuutila, P.; Schaart, G.; et al. Beige adipocytes are a distinct type of thermogenic fat cell in mouse and human. *Cell* **2012**, *150*, 366–376. [CrossRef]
11. Cypess, A.M.; Weiner, L.S.; Roberts-Toler, C.; Franquet Elia, E.; Kessler, S.H.; Kahn, P.A.; English, J.; Chatman, K.; Trauger, S.A.; Doria, A.; et al. Activation of human brown adipose tissue by a beta3-adrenergic receptor agonist. *Cell Metab.* **2015**, *21*, 33–38. [CrossRef]
12. Ishibashi, J.; Seale, P. Medicine. Beige can be slimming. *Science* **2010**, *328*, 1113–1114. [CrossRef]
13. Seale, P.; Bjork, B.; Yang, W.; Kajimura, S.; Chin, S.; Kuang, S.; Scime, A.; Devarakonda, S.; Conroe, H.M.; Erdjument-Bromage, H.; et al. PRDM16 controls a brown fat/skeletal muscle switch. *Nature* **2008**, *454*, 961–967. [CrossRef]
14. Timmons, J.A.; Wennmalm, K.; Larsson, O.; Walden, T.B.; Lassmann, T.; Petrovic, N.; Hamilton, D.L.; Gimeno, R.E.; Wahlestedt, C.; Baar, K. Myogenic gene expression signature establishes that brown and white adipocytes originate from distinct cell lineages. *Proc. Natl. Acad. Sci. USA* **2007**, *104*, 4401–4406. [CrossRef]

15. Liu, W.; Shan, T.; Yang, X.; Liang, S.; Zhang, P.; Liu, Y.; Liu, X.; Kuang, S. A heterogeneous lineage origin underlies the phenotypic and molecular differences of white and beige adipocytes. *J. Cell Sci.* **2013**, *126*, 3527–3532. [CrossRef]
16. Bartelt, A.; Heeren, J. Adipose tissue browning and metabolic health. *Nat. Rev. Endocrinol.* **2014**, *10*, 24. [CrossRef]
17. Ikeda, K.; Kang, Q.; Yoneshiro, T.; Camporez, P.J.; Maki, H.; Homma, M.; Shinoda, K.; Chen, Y.; Lu, X.; Maretich, P.; et al. UCP1-independent signaling involving SERCA2b-mediated calcium cycling regulates beige fat thermogenesis and systemic glucose homeostasis. *Nat. Med.* **2017**, *23*, 1454–1465. [CrossRef]
18. Kazak, L.; Chouchani, E.; Jedrychowski, M.; Erickson, B.; Shinoda, K.; Cohen, P.; Vetrivelan, R.; Lu, G.; Laznik-Bogoslavski, D.; Hasenfuss, S.J.C.; et al. A Creatine-Driven Substrate Cycle Enhances Energy Expenditure and Thermogenesis in Beige Fat. *Cell* **2015**, *163*, 643–655. [CrossRef]
19. Schweizer, S.; Oeckl, J.; Klingenspor, M. Substrate fluxes in brown adipocytes upon adrenergic stimulation and uncoupling protein 1 ablation. *Life Sci. Alliance* **2018**, *1*, e201800136. [CrossRef]
20. Cypess, A.M.; Lehman, S.; Williams, G.; Tal, I.; Rodman, D.; Goldfine, A.B.; Kuo, F.C.; Palmer, E.L.; Tseng, Y.H.; Doria, A.; et al. Identification and importance of brown adipose tissue in adult humans. *N. Engl. J. Med.* **2009**, *360*, 1509–1517. [CrossRef]
21. van Marken Lichtenbelt, W.D.; Vanhommerig, J.W.; Smulders, N.M.; Drossaerts, J.M.; Kemerink, G.J.; Bouvy, N.D.; Schrauwen, P.; Teule, G.J. Cold-activated brown adipose tissue in healthy men. *N. Engl. J. Med.* **2009**, *360*, 1500–1508. [CrossRef]
22. Becher, T.; Palanisamy, S.; Kramer, D.J.; Eljalby, M.; Marx, S.J.; Wibmer, A.G.; Butler, S.D.; Jiang, C.S.; Vaughan, R.; Schöder, H.; et al. Brown adipose tissue is associated with cardiometabolic health. *Nat. Med.* **2021**, *27*, 58–65. [CrossRef]
23. Seki, T.; Yang, Y.; Sun, X.; Lim, S.; Xie, S.; Guo, Z.; Xiong, W.; Kuroda, M.; Sakaue, H.; Hosaka, K.; et al. Brown-fat-mediated tumour suppression by cold-altered global metabolism. *Nature* **2022**, *608*, 421–428. [CrossRef]
24. Song, A.; Dai, W.; Jang, M.J.; Medrano, L.; Li, Z.; Zhao, H.; Shao, M.; Tan, J.; Li, A.; Ning, T.; et al. Low- and high-thermogenic brown adipocyte subpopulations coexist in *murine* adipose tissue. *J. Clin. Investig.* **2019**, *130*, 247–257. [CrossRef]
25. Schulz, T.J.; Huang, P.; Huang, T.L.; Xue, R.; McDougall, L.E.; Townsend, K.L.; Cypess, A.M.; Mishina, Y.; Gussoni, E.; Tseng, Y.H. Brown-fat paucity due to impaired BMP signalling induces compensatory browning of white fat. *Nature* **2013**, *495*, 379–383. [CrossRef]
26. Cohen, P.; Levy, J.D.; Zhang, Y.; Frontini, A.; Kolodin, D.P.; Svensson, K.J.; Lo, J.C.; Zeng, X.; Ye, L.; Khandekar, M.J.; et al. Ablation of PRDM16 and beige adipose causes metabolic dysfunction and a subcutaneous to visceral fat switch. *Cell* **2014**, *156*, 304–316. [CrossRef]
27. Sun, W.; Nie, T.; Li, K.; Wu, W.; Long, Q.; Feng, T.; Mao, L.; Gao, Y.; Liu, Q.; Gao, X.; et al. Hepatic CPT1A Facilitates Liver-Adipose Cross-Talk via Induction of FGF21 in Mice. *Diabetes* **2022**, *71*, 31–42. [CrossRef]
28. Chondronikola, M.; Volpi, E.; Børsheim, E.; Porter, C.; Annamalai, P.; Enerbäck, S.; Lidell, M.E.; Saraf, M.K.; Labbe, S.M.; Hurren, N.M.; et al. Brown adipose tissue improves whole-body glucose homeostasis and insulin sensitivity in humans. *Diabetes* **2014**, *63*, 4089–4099. [CrossRef]
29. Saito, M.; Okamatsu-Ogura, Y.; Matsushita, M.; Watanabe, K.; Yoneshiro, T.; Nio-Kobayashi, J.; Iwanaga, T.; Miyagawa, M.; Kameya, T.; Nakada, K.; et al. High incidence of metabolically active brown adipose tissue in healthy adult humans: Effects of cold exposure and adiposity. *Diabetes* **2009**, *58*, 1526–1531. [CrossRef]
30. Hanssen, M.J.W.; Hoeks, J.; Brans, B.; van der Lans, A.A.J.J.; Schaart, G.; van den Driessche, J.J.; Jörgensen, J.A.; Boekschoten, M.V.; Hesselink, M.K.C.; Havekes, B.; et al. Short-term cold acclimation improves insulin sensitivity in patients with type 2 diabetes mellitus. *Nat. Med.* **2015**, *21*, 863–865. [CrossRef]
31. El Hadi, H.; Di Vincenzo, A.; Vettor, R.; Rossato, M. Food Ingredients Involved in White-to-Brown Adipose Tissue Conversion and in Calorie Burning. *Front. Physiol.* **2019**, *9*, 1954. [CrossRef]
32. Choi, Y.; Yu, L. Natural Bioactive Compounds as Potential Browning Agents in White Adipose Tissue. *Pharm. Res.* **2021**, *38*, 549–567. [CrossRef] [PubMed]
33. Hachemi, I.; U-Din, M. Brown Adipose Tissue: Activation and Metabolism in Humans. *Endocrinol. Metab.* **2023**, *38*, 214–222. [CrossRef] [PubMed]
34. Zhao, J.; Unelius, L.; Bengtsson, T.; Cannon, B.; Nedergaard, J. Coexisting beta-adrenoceptor subtypes: Significance for thermogenic process in brown fat cells. *Am. J. Physiol.* **1994**, *267*, C969–C979. [CrossRef] [PubMed]
35. Cypess, A.M.; Chen, Y.C.; Sze, C.; Wang, K.; English, J.; Chan, O.; Holman, A.R.; Tal, I.; Palmer, M.R.; Kolodny, G.M.; et al. Cold but not sympathomimetics activates human brown adipose tissue in vivo. *Proc. Natl. Acad. Sci. USA* **2012**, *109*, 10001–10005. [CrossRef] [PubMed]
36. Blondin, D.P.; Nielsen, S.; Kuipers, E.N.; Severinsen, M.C.; Jensen, V.H.; Miard, S.; Jespersen, N.Z.; Kooijman, S.; Boon, M.R.; Fortin, M.; et al. Human Brown Adipocyte Thermogenesis Is Driven by beta2-AR Stimulation. *Cell Metab.* **2020**, *32*, 287–300.e7. [CrossRef]
37. Malik, M.; van Gelderen, E.M.; Lee, J.H.; Kowalski, D.L.; Yen, M.; Goldwater, R.; Mujais, S.K.; Schaddelee, M.P.; de Koning, P.; Kaibara, A.; et al. Proarrhythmic safety of repeat doses of mirabegron in healthy subjects: A randomized, double-blind, placebo-, and active-controlled thorough QT study. *Clin. Pharmacol. Ther.* **2012**, *92*, 696–706. [CrossRef] [PubMed]
38. Takasu, T.; Ukai, M.; Sato, S.; Matsui, T.; Nagase, I.; Maruyama, T.; Sasamata, M.; Miyata, K.; Uchida, H.; Yamaguchi, O. Effect of (R)-2-(2-aminothiazol-4-yl)-4′-{2-[(2-hydroxy-2-phenylethyl)amino]ethyl} acetanilide (YM178), a novel selective beta3-adrenoceptor agonist, on bladder function. *J. Pharmacol. Exp. Ther.* **2007**, *321*, 642–647. [CrossRef] [PubMed]

39. Loh, R.K.C.; Formosa, M.F.; La Gerche, A.; Reutens, A.T.; Kingwell, B.A.; Carey, A.L. Acute metabolic and cardiovascular effects of mirabegron in healthy individuals. *Diabetes Obes. Metab.* **2019**, *21*, 276–284. [CrossRef] [PubMed]
40. Cero, C.; Lea, H.J.; Zhu, K.Y.; Shamsi, F.; Tseng, Y.H.; Cypess, A.M. beta3-Adrenergic receptors regulate human brown/beige adipocyte lipolysis and thermogenesis. *JCI Insight* **2021**, *6*, e139160. [CrossRef]
41. Muller, T.D.; Finan, B.; Bloom, S.R.; D'Alessio, D.; Drucker, D.J.; Flatt, P.R.; Fritsche, A.; Gribble, F.; Grill, H.J.; Habener, J.F.; et al. Glucagon-like peptide 1 (GLP-1). *Mol. Metab.* **2019**, *30*, 72–130. [CrossRef]
42. Knudsen, L.B.; Nielsen, P.F.; Huusfeldt, P.O.; Johansen, N.L.; Madsen, K.; Pedersen, F.Z.; Thogersen, H.; Wilken, M.; Agerso, H. Potent derivatives of glucagon-like peptide-1 with pharmacokinetic properties suitable for once daily administration. *J. Med. Chem.* **2000**, *43*, 1664–1669. [CrossRef]
43. Torekov, S.S.; Madsbad, S.; Holst, J.J. Obesity—An indication for GLP-1 treatment? Obesity pathophysiology and GLP-1 treatment potential. *Obes. Rev.* **2011**, *12*, 593–601. [CrossRef]
44. Astrup, A.; Rössner, S.; Van Gaal, L.; Rissanen, A.; Niskanen, L.; Al Hakim, M.; Madsen, J.; Rasmussen, M.F.; Lean, M.E.J. Effects of liraglutide in the treatment of obesity: A randomised, double-blind, placebo-controlled study. *Lancet* **2009**, *374*, 1606–1616. [CrossRef]
45. Pi-Sunyer, X.; Astrup, A.; Fujioka, K.; Greenway, F.; Halpern, A.; Krempf, M.; Lau, D.C.W.; le Roux, C.W.; Violante Ortiz, R.; Jensen, C.B.; et al. A Randomized, Controlled Trial of 3.0 mg of Liraglutide in Weight Management. *N. Engl. J. Med.* **2015**, *373*, 11–22. [CrossRef]
46. Wilding, J.P.H.; Batterham, R.L.; Calanna, S.; Davies, M.; Van Gaal, L.F.; Lingvay, I.; McGowan, B.M.; Rosenstock, J.; Tran, M.T.D.; Wadden, T.A.; et al. Once-Weekly Semaglutide in Adults with Overweight or Obesity. *N. Engl. J. Med.* **2021**, *384*, 989–1002. [CrossRef]
47. Beiroa, D.; Imbernon, M.; Gallego, R.; Senra, A.; Herranz, D.; Villarroya, F.; Serrano, M.; Fernø, J.; Salvador, J.; Escalada, J.; et al. GLP-1 Agonism Stimulates Brown Adipose Tissue Thermogenesis and Browning Through Hypothalamic AMPK. *Diabetes* **2014**, *63*, 3346–3358. [CrossRef]
48. Vaittinen, M.; Ilha, M.; Herbers, E.; Wagner, A.; Virtanen, K.A.; Pietilainen, K.H.; Pirinen, E.; Pihlajamaki, J. Liraglutide demonstrates a therapeutic effect on mitochondrial dysfunction in human SGBS adipocytes in vitro. *Diabetes Res. Clin. Pract.* **2023**, *199*, 110635. [CrossRef]
49. Zhu, E.; Yang, Y.; Zhang, J.; Li, Y.; Li, C.; Chen, L.; Sun, B. Liraglutide suppresses obesity and induces brown fat-like phenotype via Soluble Guanylyl Cyclase mediated pathway in vivo and in vitro. *Oncotarget* **2016**, *7*, 81077–81089. [CrossRef]
50. Zhao, L.; Li, W.; Zhang, P.; Wang, D.; Yang, L.; Yuan, G. Liraglutide induced browning of visceral white adipose through regulation of miRNAs in high-fat-diet-induced obese mice. *Endocrine* **2024**. [CrossRef]
51. Dixit, G.; Prabhu, A. The pleiotropic peroxisome proliferator activated receptors: Regulation and therapeutics. *Exp. Mol. Pathol.* **2022**, *124*, 104723. [CrossRef]
52. Wilson-Fritch, L.; Nicoloro, S.; Chouinard, M.; Lazar, M.A.; Chui, P.C.; Leszyk, J.; Straubhaar, J.; Czech, M.P.; Corvera, S. Mitochondrial remodeling in adipose tissue associated with obesity and treatment with rosiglitazone. *J. Clin. Investig.* **2004**, *114*, 1281–1289. [CrossRef]
53. Hondares, E.; Mora, O.; Yubero, P.; de la Concepción, M.R.; Iglesias, R.; Giralt, M.; Villarroya, F. Thiazolidinediones and Rexinoids Induce Peroxisome Proliferator-Activated Receptor-Coactivator (PGC)-1α Gene Transcription: An Autoregulatory Loop Controls PGC-1α Expression in Adipocytes via Peroxisome Proliferator-Activated Receptor-γ Coactivation. *Endocrinology* **2006**, *147*, 2829–2838. [CrossRef]
54. Wilson-Fritch, L.; Burkart, A.; Bell, G.; Mendelson, K.; Leszyk, J.; Nicoloro, S.; Czech, M.; Corvera, S. Mitochondrial Biogenesis and Remodeling during Adipogenesis and in Response to the Insulin Sensitizer Rosiglitazone. *Mol. Cell. Biol.* **2003**, *23*, 1085–1094. [CrossRef]
55. Steppan, C.M.; Bailey, S.T.; Bhat, S.; Brown, E.J.; Banerjee, R.R.; Wright, C.M.; Patel, H.R.; Ahima, R.S.; Lazar, M.A. The hormone resistin links obesity to diabetes. *Nature* **2001**, *409*, 307–312. [CrossRef]
56. Oller do Nascimento, C.; Hunter, L.; Trayhurn, P. Regulation of haptoglobin gene expression in 3T3-L1 adipocytes by cytokines, catecholamines, and PPARγ. *Biochem. Biophys. Res. Commun.* **2004**, *313*, 702–708. [CrossRef]
57. Castriota, G.; Thompson, G.; Lin, Y.; Scherer, P.; Moller, D.; Berger, J. Peroxisome proliferator-activated receptor γ agonists inhibit adipocyte expression of α1-acid glycoprotein. *Cell Biol. Int.* **2007**, *31*, 586–591. [CrossRef]
58. Kung, J.; Henry, R.R. Thiazolidinedione safety. *Expert Opin. Drug Saf.* **2012**, *11*, 565–579. [CrossRef]
59. Kim, S.-H.; Choi, S.J.; Kim, Y.-C.; Kuh, H.-J. Anti-tumor activity of noble indirubin derivatives in human solid tumor models In Vitro. *Arch. Pharmacal Res.* **2009**, *32*, 915–922. [CrossRef]
60. Wei, G.; Sun, H.; Liu, J.-L.; Dong, K.; Liu, J.; Zhang, M. Indirubin, a small molecular deriving from connectivity map (CMAP) screening, ameliorates obesity-induced metabolic dysfunction by enhancing brown adipose thermogenesis and white adipose browning. *Nutr. Metab.* **2020**, *17*, 21. [CrossRef]
61. Jimenez-Aranda, A.; Fernandez-Vazquez, G.; Campos, D.; Tassi, M.; Velasco-Perez, L.; Tan, D.X.; Reiter, R.J.; Agil, A. Melatonin induces browning of inguinal white adipose tissue in Zucker diabetic fatty rats. *J. Pineal Res.* **2013**, *55*, 416–423. [CrossRef]
62. Qiu, Y.; Sun, Y.; Xu, D.; Yang, Y.; Liu, X.; Wei, Y.; Chen, Y.; Feng, Z.; Li, S.; Reyad-ul Ferdous, M.; et al. Screening of FDA-approved drugs identifies sutent as a modulator of UCP1 expression in brown adipose tissue. *EBioMedicine* **2018**, *37*, 344–355. [CrossRef]

63. Moisan, A.; Lee, Y.K.; Zhang, J.D.; Hudak, C.S.; Meyer, C.A.; Prummer, M.; Zoffmann, S.; Truong, H.H.; Ebeling, M.; Kiialainen, A.; et al. White-to-brown metabolic conversion of human adipocytes by JAK inhibition. *Nat. Cell Biol.* **2015**, *17*, 57–67. [CrossRef]
64. Rachid, T.L.; Penna-de-Carvalho, A.; Bringhenti, I.; Aguila, M.B.; Mandarim-de-Lacerda, C.A.; Souza-Mello, V. Fenofibrate (PPARalpha agonist) induces beige cell formation in subcutaneous white adipose tissue from diet-induced male obese mice. *Mol. Cell. Endocrinol.* **2015**, *402*, 86–94. [CrossRef]
65. Nie, B.; Nie, T.; Hui, X.; Gu, P.; Mao, L.; Li, K.; Yuan, R.; Zheng, J.; Wang, H.; Li, K.; et al. Brown Adipogenic Reprogramming Induced by a Small Molecule. *Cell Rep.* **2017**, *18*, 624–635. [CrossRef]
66. Nie, T.; Zhao, S.; Mao, L.; Yang, Y.; Sun, W.; Lin, X.; Liu, S.; Li, K.; Sun, Y.; Li, P.; et al. The natural compound, formononetin, extracted from Astragalus membranaceus increases adipocyte thermogenesis by modulating PPARγ activity. *Br. J. Pharmacol.* **2018**, *175*, 1439–1450. [CrossRef]
67. Mu, Q.; Fang, X.; Li, X.; Zhao, D.; Mo, F.; Jiang, G.; Yu, N.; Zhang, Y.; Guo, Y.; Fu, M.; et al. Ginsenoside Rb1 promotes browning through regulation of PPARgamma in 3T3-L1 adipocytes. *Biochem. Biophys. Res. Commun.* **2015**, *466*, 530–535. [CrossRef]
68. Aldiss, P.; Betts, J.; Sale, C.; Pope, M.; Budge, H.; Symonds, M.E. Exercise-induced 'browning' of adipose tissues. *Metabolism* **2018**, *81*, 63–70. [CrossRef]
69. Roberts, L.D.; Bostrom, P.; O'Sullivan, J.F.; Schinzel, R.T.; Lewis, G.D.; Dejam, A.; Lee, Y.K.; Palma, M.J.; Calhoun, S.; Georgiadi, A.; et al. beta-Aminoisobutyric acid induces browning of white fat and hepatic beta-oxidation and is inversely correlated with cardiometabolic risk factors. *Cell Metab.* **2014**, *19*, 96–108. [CrossRef]
70. Alvarez, R.; de Andrés, J.; Yubero, P.; Viñas, O.; Mampel, T.; Iglesias, R.; Giralt, M.; Villarroya, F. A Novel Regulatory Pathway of Brown Fat Thermogenesis: Retinoic acid is a transcriptional activator of the mitochondrial uncoupling protein gene. *J. Biol. Chem.* **1995**, *270*, 5666–5673. [CrossRef]
71. Puigserver, P.; Vázquez, F.; Bonet, M.L.; Picó, C.; Palou, A. In vitro and in vivo induction of brown adipocyte uncoupling protein (thermogenin) by retinoic acid. *Biochem. J.* **1996**, *317*, 827–833. [CrossRef]
72. McIlroy, G.D.; Tammireddy, S.R.; Maskrey, B.H.; Grant, L.; Doherty, M.K.; Watson, D.G.; Delibegović, M.; Whitfield, P.D.; Mody, N. Fenretinide mediated retinoic acid receptor signalling and inhibition of ceramide biosynthesis regulates adipogenesis, lipid accumulation, mitochondrial function and nutrient stress signalling in adipocytes and adipose tissue. *Biochem. Pharmacol.* **2016**, *100*, 86–97. [CrossRef] [PubMed]
73. Alvarez, R.; Checa, M.L.; Brun, S.; Viñas, O.; Mampel, T.; Iglesias, R.; Giralt, M.; Villarroya, F. Both retinoic-acid-receptor- and retinoid-X-receptor-dependent signalling pathways mediate the induction of the brown-adipose-tissue-uncoupling-protein-1 gene by retinoids. *Biochem. J.* **2000**, *345*, 91–97. [CrossRef] [PubMed]
74. Nishikata, I.; Sasaki, H.; Iga, M.; Tateno, Y.; Imayoshi, S.; Asou, N.; Nakamura, T.; Morishita, K. A novel *EVI1* gene family, *MEL1*, lacking a PR domain (MEL1S) is expressed mainly in t(1;3)(p36;q21)-positive AML and blocks G-CSF–induced myeloid differentiation. *Blood* **2003**, *102*, 3323–3332. [CrossRef]
75. Seale, P.; Kajimura, S.; Yang, W.; Chin, S.; Rohas, L.M.; Uldry, M.; Tavernier, G.; Langin, D.; Spiegelman, B.M. Transcriptional control of brown fat determination by PRDM16. *Cell Metab.* **2007**, *6*, 38–54. [CrossRef] [PubMed]
76. Kajimura, S.; Seale, P.; Tomaru, T.; Erdjument-Bromage, H.; Cooper, M.P.; Ruas, J.L.; Chin, S.; Tempst, P.; Lazar, M.A.; Spiegelman, B.M. Regulation of the brown and white fat gene programs through a PRDM16/CtBP transcriptional complex. *Genes Dev.* **2008**, *22*, 1397–1409. [CrossRef]
77. Liu, X.; Zhang, Y.; Chu, Y.; Zhao, X.; Mao, L.; Zhao, S.; Lin, S.; Hui, X.; Gu, P.; Xu, Y.; et al. The natural compound rutaecarpine promotes white adipocyte browning through activation of the AMPK-PRDM16 axis. *Biochem. Biophys. Res. Commun.* **2021**, *545*, 189–194. [CrossRef]
78. Liao, J.; Jiang, J.; Jun, H.; Qiao, X.; Emont, M.P.; Kim, D.-I.; Wu, J. HDAC3-Selective Inhibition Activates Brown and Beige Fat Through PRDM16. *Endocrinology* **2018**, *159*, 2520–2527. [CrossRef] [PubMed]
79. Song, N.J.; Choi, S.; Rajbhandari, P.; Chang, S.H.; Kim, S.; Vergnes, L.; Kwon, S.M.; Yoon, J.H.; Lee, S.; Ku, J.M.; et al. Prdm4 induction by the small molecule butein promotes white adipose tissue browning. *Nat. Chem. Biol.* **2016**, *12*, 479–481. [CrossRef]
80. Watanabe, M.; Houten, S.; Mataki, C.; Christoffolete, M.; Kim, B.; Sato, H.; Messaddeq, N.; Harney, J.; Ezaki, O.; Kodama, T.J.N. Bile acids induce energy expenditure by promoting intracellular thyroid hormone activation. *Nature* **2006**, *439*, 484–489. [CrossRef]
81. Liaset, B.; Hao, Q.; Jørgensen, H.; Hallenborg, P.; Du, Z.-Y.; Ma, T.; Marschall, H.-U.; Kruhøffer, M.; Li, R.; Li, Q.; et al. Nutritional regulation of bile acid metabolism is associated with improved pathological characteristics of the metabolic syndrome. *J. Biol. Chem.* **2011**, *286*, 28382–28395. [CrossRef]
82. Juvet, L.K.; Andresen, S.M.; Schuster, G.U.; Dalen, K.T.; Tobin, K.A.R.; Hollung, K.; Haugen, F.; Jacinto, S.; Ulven, S.M.; Bamberg, K.; et al. On the role of liver X receptors in lipid accumulation in adipocytes. *Mol. Endocrinol.* **2003**, *17*, 172–182. [CrossRef]
83. Seo, J.B.; Moon, H.M.; Kim, W.S.; Lee, Y.S.; Jeong, H.W.; Yoo, E.J.; Ham, J.; Kang, H.; Park, M.-G.; Steffensen, K.R.; et al. Activated Liver X Receptors Stimulate Adipocyte Differentiation through Induction of Peroxisome Proliferator-Activated Receptor γ Expression. *Mol. Cell. Biol.* **2004**, *24*, 3430–3444. [CrossRef]
84. Zheng, F.; Zhang, S.; Lu, W.; Wu, F.; Yin, X.; Yu, D.; Pan, Q.; Li, H. Regulation of Insulin Resistance and Adiponectin Signaling in Adipose Tissue by Liver X Receptor Activation Highlights a Cross-Talk with PPARγ. *PLoS ONE* **2014**, *9*, e101269. [CrossRef]

85. Shu, L.; Hoo, R.L.C.; Wu, X.; Pan, Y.; Lee, I.P.C.; Cheong, L.Y.; Bornstein, S.R.; Rong, X.; Guo, J.; Xu, A. A-FABP mediates adaptive thermogenesis by promoting intracellular activation of thyroid hormones in brown adipocytes. *Nat. Commun.* **2017**, *8*, 14147. [CrossRef]
86. Korach-André, M.; Archer, A.; Barros Rodrigo, P.; Parini, P.; Gustafsson, J.-Å. Both liver-X receptor (LXR) isoforms control energy expenditure by regulating Brown Adipose Tissue activity. *Proc. Natl. Acad. Sci. USA* **2011**, *108*, 403–408. [CrossRef]
87. Sheng, X.; Zhu, X.; Zhang, Y.; Cui, G.; Peng, L.; Lu, X.; Zang, Y.Q. Rhein Protects against Obesity and Related Metabolic Disorders through Liver X Receptor-Mediated Uncoupling Protein 1 Upregulation in Brown Adipose Tissue. *Int. J. Biol. Sci.* **2012**, *8*, 1375–1384. [CrossRef]
88. Sun, H.-J.; Chen, D.; Han, Y.; Zhou, Y.-B.; Wang, J.-J.; Chen, Q.; Li, Y.-H.; Gao, X.-Y.; Kang, Y.-M.; Zhu, G.-Q. Relaxin in paraventricular nucleus contributes to sympathetic overdrive and hypertension via PI3K-Akt pathway. *Neuropharmacology* **2016**, *103*, 247–256. [CrossRef]
89. Dominguez, E.; Galmozzi, A.; Chang, J.W.; Hsu, K.-L.; Pawlak, J.; Li, W.; Godio, C.; Thomas, J.; Partida, D.; Niessen, S.; et al. Integrated phenotypic and activity-based profiling links Ces3 to obesity and diabetes. *Nat. Chem. Biol.* **2014**, *10*, 113–121. [CrossRef]
90. Galmozzi, A.; Sonne, S.B.; Altshuler-Keylin, S.; Hasegawa, Y.; Shinoda, K.; Luijten, I.H.N.; Chang, J.W.; Sharp, L.Z.; Cravatt, B.F.; Saez, E.; et al. ThermoMouse: An In Vivo Model to Identify Modulators of UCP1 Expression in Brown Adipose Tissue. *Cell Rep.* **2014**, *9*, 1584–1593. [CrossRef]
91. Masuda, Y.; Haramizu, S.; Oki, K.; Ohnuki, K.; Watanabe, T.; Yazawa, S.; Kawada, T.; Hashizume, S.; Fushiki, T. Upregulation of uncoupling proteins by oral administration of capsiate, a nonpungent capsaicin analog. *J. Appl. Physiol.* **2003**, *95*, 2408–2415. [CrossRef]
92. Takeda, Y.; Dai, P. Capsaicin directly promotes adipocyte browning in the chemical compound-induced brown adipocytes converted from human dermal fibroblasts. *Sci. Rep.* **2022**, *12*, 6612. [CrossRef] [PubMed]
93. Haenisch, M.; Nguyen, T.; Fihn, C.A.; Goldstein, A.S.; Amory, J.K.; Treuting, P.; Brabb, T.; Paik, J. Investigation of an ALDH1A1-specific inhibitor for suppression of weight gain in a diet-induced mouse model of obesity. *Int. J. Obes.* **2021**, *45*, 1542–1552. [CrossRef] [PubMed]
94. Kiefer, F.W.; Vernochet, C.; O'Brien, P.; Spoerl, S.; Brown, J.D.; Nallamshetty, S.; Zeyda, M.; Stulnig, T.M.; Cohen, D.E.; Kahn, C.R.; et al. Retinaldehyde dehydrogenase 1 regulates a thermogenic program in white adipose tissue. *Nat. Med.* **2012**, *18*, 918–925. [CrossRef] [PubMed]
95. Sun, W.; Dong, H.; Balaz, M.; Slyper, M.; Drokhlyansky, E.; Colleluori, G.; Giordano, A.; Kovanicova, Z.; Stefanicka, P.; Balazova, L.; et al. snRNA-seq reveals a subpopulation of adipocytes that regulates thermogenesis. *Nature* **2020**, *587*, 98–102. [CrossRef] [PubMed]
96. Zhao, S.; Chu, Y.; Zhang, Y.; Zhou, Y.; Jiang, Z.; Wang, Z.; Mao, L.; Li, K.; Sun, W.; Li, P.; et al. Linifanib exerts dual anti-obesity effect by regulating adipocyte browning and formation. *Life Sci.* **2019**, *222*, 117–124. [CrossRef] [PubMed]
97. Seo, S.H.; Lee, D.; Lee, S.H.; Choi, K.Y. Blockade of CXXC5-dishevelled interaction inhibits adipogenic differentiation, obesity, and insulin resistance in mice. *Sci. Rep.* **2022**, *12*, 20669. [CrossRef] [PubMed]
98. Patel, A.; Dobbins, T.; Kong, X.; Patel, R.; Carter, G.; Harding, L.; Sparks, R.P.; Patel, N.A.; Cooper, D.R. Induction of beige-like adipocyte markers and functions in 3T3-L1 cells by Clk1 and PKCbetaII inhibitory molecules. *J. Cell. Mol. Med.* **2022**, *26*, 4183–4194. [CrossRef] [PubMed]
99. Reilly, S.M.; Chiang, S.-H.; Decker, S.J.; Chang, L.; Uhm, M.; Larsen, M.J.; Rubin, J.R.; Mowers, J.; White, N.M.; Hochberg, I.; et al. An inhibitor of the protein kinases TBK1 and IKK-ε improves obesity-related metabolic dysfunctions in mice. *Nat. Med.* **2013**, *19*, 313–321. [CrossRef] [PubMed]
100. Hoffmann, L.S.; Etzrodt, J.; Willkomm, L.; Sanyal, A.; Scheja, L.; Fischer, A.W.C.; Stasch, J.-P.; Bloch, W.; Friebe, A.; Heeren, J.; et al. Stimulation of soluble guanylyl cyclase protects against obesity by recruiting brown adipose tissue. *Nat. Commun.* **2015**, *6*, 7235. [CrossRef]
101. Yan, C.; Zeng, T.; Lee, K.; Nobis, M.; Loh, K.; Gou, L.; Xia, Z.; Gao, Z.; Bensellam, M.; Hughes, W.; et al. Peripheral-specific Y1 receptor antagonism increases thermogenesis and protects against diet-induced obesity. *Nat. Commun.* **2021**, *12*, 2622. [CrossRef]
102. Nie, T.; Hui, X.; Mao, L.; Nie, B.; Li, K.; Sun, W.; Gao, X.; Tang, X.; Xu, Y.; Jiang, B.; et al. Harmine Induces Adipocyte Thermogenesis through RAC1-MEK-ERK-CHD4 Axis. *Sci. Rep.* **2016**, *6*, 36382. [CrossRef]
103. Zhang, Y.; Li, X.; Zou, D.; Liu, W.; Yang, J.; Zhu, N.; Huo, L.; Wang, M.; Hong, J.; Wu, P.; et al. Treatment of type 2 diabetes and dyslipidemia with the natural plant alkaloid berberine. *J. Clin. Endocrinol. Metab.* **2008**, *93*, 2559–2565. [CrossRef]
104. Enerback, S.; Jacobsson, A.; Simpson, E.M.; Guerra, C.; Yamashita, H.; Harper, M.E.; Kozak, L.P. Mice lacking mitochondrial uncoupling protein are cold-sensitive but not obese. *Nature* **1997**, *387*, 90–94. [CrossRef]
105. Ukropec, J.; Anunciado, R.P.; Ravussin, Y.; Hulver, M.W.; Kozak, L.P. UCP1-independent thermogenesis in white adipose tissue of cold-acclimated Ucp1−/− mice. *J. Biol. Chem.* **2006**, *281*, 31894–31908. [CrossRef]
106. Long, J.Z.; Svensson, K.J.; Bateman, L.A.; Lin, H.; Kamenecka, T.; Lokurkar, I.A.; Lou, J.; Rao, R.R.; Chang, M.R.; Jedrychowski, M.P.; et al. The Secreted Enzyme PM20D1 Regulates Lipidated Amino Acid Uncouplers of Mitochondria. *Cell* **2016**, *166*, 424–435. [CrossRef]
107. Zoico, E.; Rubele, S.; De Caro, A.; Nori, N.; Mazzali, G.; Fantin, F.; Rossi, A.; Zamboni, M. Brown and Beige Adipose Tissue and Aging. *Front. Endocrinol.* **2019**, *10*, 368. [CrossRef]

108. Ouellet, V.; Routhier-Labadie, A.; Bellemare, W.; Lakhal-Chaieb, L.; Turcotte, E.; Carpentier, A.C.; Richard, D. Outdoor Temperature, Age, Sex, Body Mass Index, and Diabetic Status Determine the Prevalence, Mass, and Glucose-Uptake Activity of 18F-FDG-Detected BAT in Humans. *J. Clin. Endocrinol. Metab.* **2011**, *96*, 192–199. [CrossRef] [PubMed]
109. Ferren, L. Morphological Differentiation of Implanted Brown and White Fats. *Trans. Kans. Acad. Sci.* **1966**, *69*, 350–353. [CrossRef]
110. Liu, X.; Zheng, Z.; Zhu, X.; Meng, M.; Li, L.; Shen, Y.; Chi, Q.; Wang, D.; Zhang, Z.; Li, C.; et al. Brown adipose tissue transplantation improves whole-body energy metabolism. *Cell Res.* **2013**, *23*, 851–854. [CrossRef]
111. Stanford, K.I.; Middelbeek, R.J.W.; Townsend, K.L.; An, D.; Nygaard, E.B.; Hitchcox, K.M.; Markan, K.R.; Nakano, K.; Hirshman, M.F.; Tseng, Y.-H.; et al. Brown adipose tissue regulates glucose homeostasis and insulin sensitivity. *J. Clin. Investig.* **2012**, *123*, 215–223. [CrossRef]
112. Gunawardana, S.C.; Piston, D.W. Reversal of Type 1 Diabetes in Mice by Brown Adipose Tissue Transplant. *Diabetes* **2012**, *61*, 674–682. [CrossRef]
113. Dani, V.; Yao, X.; Dani, C. Transplantation of fat tissues and iPSC-derived energy expenditure adipocytes to counteract obesity-driven metabolic disorders: Current strategies and future perspectives. *Rev. Endocr. Metab. Disord.* **2022**, *23*, 103–110. [CrossRef]
114. Gunawardana, S.C.; Piston, D.W. Insulin-independent reversal of type 1 diabetes in nonobese diabetic mice with brown adipose tissue transplant. *Am. J. Physiol. Endocrinol. Metab.* **2015**, *308*, E1043–E1055. [CrossRef]
115. Gunawardana, S.C.; Piston, D.W. Insulin-Independent Reversal of Type-1 Diabetes Following Transplantation of Adult Brown Adipose Tissue Supplemented With IGF-1. *Transplant. Direct* **2019**, *5*, e500. [CrossRef]
116. Liu, X.; Wang, S.; You, Y.; Meng, M.; Zheng, Z.; Dong, M.; Lin, J.; Zhao, Q.; Zhang, C.; Yuan, X.; et al. Brown Adipose Tissue Transplantation Reverses Obesity in Ob/Ob Mice. *Endocrinology* **2015**, *156*, 2461–2469. [CrossRef]
117. Zhang, W.C.; Qin, F.; Wang, X.J.; Liu, Z.F.; Zhu, L.; Zeng, A.; Zhang, M.Z.; Yu, N.Z.; Long, X. Adipose-Derived Stromal Cells Attenuate Adipose Inflammation in Obesity through Adipocyte Browning and Polarization of M2 Macrophages. *Mediat. Inflamm.* **2019**, *2019*, 1731540. [CrossRef]
118. Shree, N.; Venkategowda, S.; Venkatranganna, M.V.; Datta, I.; Bhonde, R.R. Human adipose tissue mesenchymal stem cells as a novel treatment modality for correcting obesity induced metabolic dysregulation. *Int. J. Obes.* **2019**, *43*, 2107–2118. [CrossRef]
119. Singh, A.M.; Zhang, L.; Avery, J.; Yin, A.; Du, Y.; Wang, H.; Li, Z.; Fu, H.; Yin, H.; Dalton, S. Human beige adipocytes for drug discovery and cell therapy in metabolic diseases. *Nat. Commun.* **2020**, *11*, 2758. [CrossRef]
120. Takahashi, K.; Tanabe, K.; Ohnuki, M.; Narita, M.; Ichisaka, T.; Tomoda, K.; Yamanaka, S. Induction of Pluripotent Stem Cells from Adult Human Fibroblasts by Defined Factors. *Cell* **2007**, *131*, 861–872. [CrossRef]
121. Yu, J.; Vodyanik, M.A.; Smuga-Otto, K.; Antosiewicz-Bourget, J.; Frane, J.L.; Tian, S.; Nie, J.; Jonsdottir, G.A.; Ruotti, V.; Stewart, R.; et al. Induced pluripotent stem cell lines derived from human somatic cells. *Science* **2007**, *318*, 1917–1920. [CrossRef] [PubMed]
122. Taura, D.; Noguchi, M.; Sone, M.; Hosoda, K.; Mori, E.; Okada, Y.; Takahashi, K.; Homma, K.; Oyamada, N.; Inuzuka, M.; et al. Adipogenic differentiation of human induced pluripotent stem cells: Comparison with that of human embryonic stem cells. *FEBS Lett.* **2009**, *583*, 1029–1033. [CrossRef]
123. Ahfeldt, T.; Schinzel, R.T.; Lee, Y.K.; Hendrickson, D.; Kaplan, A.; Lum, D.H.; Camahort, R.; Xia, F.; Shay, J.; Rhee, E.P.; et al. Programming human pluripotent stem cells into white and brown adipocytes. *Nat. Cell Biol.* **2012**, *14*, 209–219. [CrossRef] [PubMed]
124. Nishio, M.; Yoneshiro, T.; Nakahara, M.; Suzuki, S.; Saeki, K.; Hasegawa, M.; Kawai, Y.; Akutsu, H.; Umezawa, A.; Yasuda, K.; et al. Production of functional classical brown adipocytes from human pluripotent stem cells using specific hemopoietin cocktail without gene transfer. *Cell Metab.* **2012**, *16*, 394–406. [CrossRef] [PubMed]
125. Mohsen-Kanson, T.; Hafner, A.L.; Wdziekonski, B.; Takashima, Y.; Villageois, P.; Carriere, A.; Svensson, M.; Bagnis, C.; Chignon-Sicard, B.; Svensson, P.A.; et al. Differentiation of human induced pluripotent stem cells into brown and white adipocytes: Role of Pax3. *Stem Cells* **2014**, *32*, 1459–1467. [CrossRef] [PubMed]
126. Hafner, A.L.; Contet, J.; Ravaud, C.; Yao, X.; Villageois, P.; Suknuntha, K.; Annab, K.; Peraldi, P.; Binetruy, B.; Slukvin, I.I.; et al. Brown-like adipose progenitors derived from human induced pluripotent stem cells: Identification of critical pathways governing their adipogenic capacity. *Sci. Rep.* **2016**, *6*, 32490. [CrossRef] [PubMed]
127. Zhang, L.; Avery, J.; Yin, A.; Singh, A.M.; Cliff, T.S.; Yin, H.; Dalton, S. Generation of Functional Brown Adipocytes from Human Pluripotent Stem Cells via Progression through a Paraxial Mesoderm State. *Cell Stem Cell* **2020**, *27*, 784–797.e11. [CrossRef] [PubMed]
128. Guenantin, A.C.; Briand, N.; Capel, E.; Dumont, F.; Morichon, R.; Provost, C.; Stillitano, F.; Jeziorowska, D.; Siffroi, J.P.; Hajjar, R.J.; et al. Functional Human Beige Adipocytes from Induced Pluripotent Stem Cells. *Diabetes* **2017**, *66*, 1470–1478. [CrossRef] [PubMed]
129. Su, S.; Guntur, A.R.; Nguyen, D.C.; Fakory, S.S.; Doucette, C.C.; Leech, C.; Lotana, H.; Kelley, M.; Kohli, J.; Martino, J.; et al. A Renewable Source of Human Beige Adipocytes for Development of Therapies to Treat Metabolic Syndrome. *Cell Rep.* **2018**, *25*, 3215–3228.e9. [CrossRef]
130. Zhang, J.; Liu, Y.; Chen, Y.; Yuan, L.; Liu, H.; Wang, J.; Liu, Q.; Zhang, Y. Adipose-Derived Stem Cells: Current Applications and Future Directions in the Regeneration of Multiple Tissues. *Stem Cells Int.* **2020**, *2020*, 8810813. [CrossRef]
131. Churko, J.M.; Burridge, P.W.; Wu, J.C. Generation of human iPSCs from human peripheral blood mononuclear cells using non-integrative Sendai virus in chemically defined conditions. *Methods Mol. Biol.* **2013**, *1036*, 81–88. [CrossRef]

132. Okumura, T.; Horie, Y.; Lai, C.-Y.; Lin, H.-T.; Shoda, H.; Natsumoto, B.; Fujio, K.; Kumaki, E.; Okano, T.; Ono, S.; et al. Robust and highly efficient hiPSC generation from patient non-mobilized peripheral blood-derived CD34+ cells using the auto-erasable Sendai virus vector. *Stem Cell Res. Ther.* **2019**, *10*, 185. [CrossRef] [PubMed]
133. Zhou, T.; Benda, C.; Duzinger, S.; Huang, Y.; Li, X.; Li, Y.; Guo, X.; Cao, G.; Chen, S.; Hao, L.; et al. Generation of induced pluripotent stem cells from urine. *J. Am. Soc. Nephrol.* **2011**, *22*, 1221–1228. [CrossRef]
134. Geuder, J.; Wange, L.E.; Janjic, A.; Radmer, J.; Janssen, P.; Bagnoli, J.W.; Müller, S.; Kaul, A.; Ohnuki, M.; Enard, W. A non-invasive method to generate induced pluripotent stem cells from primate urine. *Sci. Rep.* **2021**, *11*, 3516. [CrossRef] [PubMed]
135. Yuan, X.; Hu, T.; Zhao, H.; Huang, Y.; Ye, R.; Lin, J.; Zhang, C.; Zhang, H.; Wei, G.; Zhou, H.; et al. Brown adipose tissue transplantation ameliorates polycystic ovary syndrome. *Proc. Natl. Acad. Sci. USA* **2016**, *113*, 2708–2713. [CrossRef] [PubMed]
136. Ridler, C. Adipose tissue: BAT affects lipid metabolism. *Nat. Rev. Endocrinol.* **2016**, *12*, 435. [CrossRef] [PubMed]

biomedicines

MDPI

Review

The Interplay of Uterine Health and Obesity: A Comprehensive Review

Dina Šišljagić [1,2], Senka Blažetić [3,*], Marija Heffer [4], Mihaela Vranješ Delać [5] and Andrijana Muller [1,2]

1 Clinic of Gynecology and Obstetric, University Hospital Center Osijek, 31000 Osijek, Croatia; drdinasisljagic@yahoo.com (D.Š.); andrijana.muller@gmail.com (A.M.)
2 Department of Gynecology and Obstetrics, Faculty of Medicine, Josip Juraj Strossmayer University of Osijek, 31000 Osijek, Croatia
3 Department of Biology, Josip Juraj Strossmayer University of Osijek, 31000 Osijek, Croatia
4 Department of Medical Biology, School of Medicine, Josip Juraj Strossmayer University of Osijek, 31000 Osijek, Croatia; mheffer@mefos.hr
5 Faculty of Veterinary Medicine, University of Zagreb, 10000 Zagreb, Croatia; mvranjes@vef.hr
* Correspondence: senka@biologija.unios.hr

Abstract: Uterine physiology encompasses the intricate processes governing the structure, function, and regulation of the uterus, a pivotal organ within the female reproductive system. The escalating prevalence of obesity has emerged as a significant global health issue, profoundly impacting various facets of well-being, including female reproductive health. These effects extend to uterine structure and function, influencing reproductive health outcomes in women. They encompass alterations in uterine morphology, disruptions in hormonal signaling, and inflammatory processes. Insulin and leptin, pivotal hormones regulating metabolism, energy balance, and reproductive function, play crucial roles in this context. Insulin chiefly governs glucose metabolism and storage, while leptin regulates appetite and energy expenditure. However, in obesity, resistance to both insulin and leptin can develop, impacting uterine function. Inflammation and oxidative stress further exacerbate the development of uterine dysfunction in obesity. Chronic low-grade inflammation and heightened oxidative stress, characteristic of obesity, contribute to metabolic disruptions and tissue damage, including within the uterus. Obesity significantly disrupts menstrual cycles, fertility, and pregnancy outcomes in women. The accumulation of excess adipose tissue disrupts hormonal equilibrium, disturbs ovarian function, and fosters metabolic irregularities, all of which detrimentally impact reproductive health.

Keywords: uterus; obesity; insulin resistance; leptin resistance; female reproductive health; metabolic syndrome; adipokines; endometrial function; fertility; therapeutic interventions

Citation: Šišljagić, D.; Blažetić, S.; Heffer, M.; Vranješ Delać, M.; Muller, A. The Interplay of Uterine Health and Obesity: A Comprehensive Review. *Biomedicines* **2024**, *12*, 2801. https://doi.org/10.3390/biomedicines12122801

Academic Editors: Zaida Abad-Jiménez and Teresa Vezza

Received: 18 October 2024
Revised: 9 December 2024
Accepted: 9 December 2024
Published: 10 December 2024

1. Introduction

The uterus is a dynamic and complex organ central to female reproductive health. While its primary function is in reproduction, its roles extend beyond this, impacting overall health and disease states related to endocrine [1,2], immune [3–5], cardiovascular [6–8], neurological [9–11], and metabolic systems [12,13]. Understanding these roles is crucial for comprehensive women's health care, influencing treatment strategies and improving quality of life. Further research into these non-reproductive roles will continue to illuminate the uterus's full impact on overall health. Obesity, defined by an excessive accumulation of body fat, typically measured by a body mass index (BMI) of 30 or higher, is a chronic complex disease and one of the major factors impairing health. According to the WHO, in 2022, one in eight people in the world were living with obesity [14]. The rising prevalence of obesity poses significant challenges to female reproductive health. It impacts menstrual regularity, fertility, and pregnancy outcomes and increases the risk of reproductive cancers [15,16].

Pandey (2010) further emphasizes the association between obesity and gynecological issues, such as polycystic ovary syndrome and endometrial polyps [17]. Brown (2010) discusses the genetic contribution to obesity and its implications for reproduction. These studies collectively underscore the urgent need for interventions to address the growing burden of obesity on female reproductive health [18].

Insulin and leptin resistance are critical metabolic disturbances with profound implications for uterine function and female reproductive health. They contribute to a range of reproductive disorders, from menstrual irregularities and infertility to pregnancy complications [19,20] and increased cancer risk [21,22]. Addressing these resistances through lifestyle interventions, medical treatment, and ongoing research is essential for improving reproductive health outcomes. Understanding the mechanisms by which these resistances affect the uterus can help in developing targeted therapies and preventive strategies for women affected by these conditions.

The connection between obesity and uterine health is vital to understanding and managing various reproductive health issues. Overall, the research in this area can lead to better prevention, diagnosis, and treatment strategies, ultimately improving women's health and quality of life.

2. Impact of Obesity on Uterine Reproductive and Immune Functions

2.1. Effects of Obesity on Uterine Structure

The uterus, positioned within the pelvic cavity between the bladder and rectum, is a hollow, muscular organ. Comprising three primary layers—the endometrium (inner lining), myometrium (middle muscular layer), and perimetrium (outer layer)—it undergoes shape variations throughout a woman's life. These changes manifest during puberty, menstruation, pregnancy, and menopause [23,24]. Obesity affects uterine structure and function through a combination of hormonal imbalances [25], chronic inflammation [26], and metabolic disturbances [27]. These effects can lead to a range of reproductive health issues, including endometrial hyperplasia, irregular menstrual cycles [28], increased risk of uterine fibroids [29], endometrial cancer [30], and pregnancy complications [31].

The relationship between estrogen and obesity is bidirectional. Increased adipose tissue in obesity leads to higher levels of circulating estrogen synthesized and metabolized by the cytochrome P450 (CYP) superfamily of enzymes, specifically aromatase CYP19A1. This prolonged estrogen exposure, without the balancing effect of progesterone, can cause the endometrial lining to thicken excessively, resulting in endometrial hyperplasia [24,32]. This condition can lead to abnormal uterine bleeding and increase the risk of endometrial cancer [33]. Deficiency of estrogen leads to excessive fat accumulation and impairs adipocyte function; on the other hand, adipose tissue of obese individuals is characterized by altered expression of estrogen receptors and key enzymes involved in their synthesis [34]. Additionally, obesity is associated with increased androgen levels, likely mediated by insulin resistance, which may also have implications for uterine health [35]. This relationship will be further explored in Section 3.

Obesity-related hormonal imbalances, particularly elevated estrogen levels, are thought to contribute to the development and growth of uterine fibroids (uterine leiomyomas). These are benign tumor changes that originate from the smooth muscle layer of the uterus. Leiomyoma growth is a consequence of hormonal action that acts through their estrogen and progesterone receptors [36,37]. Fibroids can affect uterine structure and function by causing abnormal uterine bleeding, pelvic pain, and infertility, depending on their size and location [29].

2.2. Obesity and Uterine Reproductive Function

The uterus plays a crucial role in the menstrual cycle, as it is central to the preparation for a possible pregnancy each month. A complex interplay of hormones, mainly estrogen and progesterone, produced by the ovaries, governs the menstrual cycle in all phases—menstrual, follicular, ovulatory, and luteal. The uterus prepares for potential pregnancy by thickening the endometrial lining each month during the proliferative phase of the menstrual (uterine) cycle in response to hormonal changes, mainly estrogen and progesterone. If fertilization does not occur, the endometrial lining sheds, leading to menstrual bleeding. The uterus contracts to expel this lining, which can cause cramping. Chronic exposure to elevated levels of estrogen due to increased adipose tissue can lead to endometrial hyperplasia, characterized by abnormal thickening of the endometrial lining [38–40]. Endometrial hyperplasia is a risk factor for endometrial cancer, particularly in postmenopausal women [41,42]. Conditions such as polycystic ovary syndrome (PCOS), characterized by irregular menstrual cycles, ovarian dysfunction, and hyperandrogenism, are more prevalent in women with obesity and can further impair fertility [43–45]. Also, evidence shows that excessive weight can negatively impact reproductive health by altering endometrial gene expression and reducing endometrial receptivity [46]. The uterus must be free of abnormalities, like scarring or fibroids, to support a full-term pregnancy. Obese women more frequently experience prolonged pregnancies, and they have a higher chance of needing interventions during labor and delivery, such as cesarean section, due to factors such as macrosomia, labor dystocia, and inadequate uterine contractions [47–50]. Studies on uterine contractility in obese women have shown mixed results. Some studies indicated differences in contraction of myometrial strips from obese women, who contracted less frequently and with less force than those from nonobese women [51], while other studies reported no differences in contraction strength and frequency [52,53]. Also, the typically higher cholesterol levels in obese women alter myometrial myocyte cell membranes, particularly the caveolae, which inhibits oxytocin receptor function and increases K+ channel activity, resulting in preventing the uterus from contracting [54], while dysfunctional labor patterns and increased oxytocin use in obese women may not be attributed to differences in oxytocin expression [55]. A high-fat diet accumulates big fat droplets in the stromal layers of the uterus, ovary, and oviducts, with the consequence of altering their histological integrity, and in women, changes like hyperproliferative uterus, vacuolated ovarian tissue, and thickened oviduct walls occur [56]. Maternal obesity in humans and in high-fat diet (HFD) mice is associated with impaired uterine vascular remodeling at the maternal side of the placental–uterine interface, characterized by the presence of smooth muscle layers around arteries and narrower arterial conduits [57–60].

2.3. *Obesity and Uterine Immune Function*

Inflammation and oxidative stress play significant roles in uterine dysfunction associated with obesity by affecting endometrial health, reducing fertility, and increasing the risk of pregnancy complications [61]. Obesity is characterized by chronic low-grade inflammation, marked by elevated levels of inflammatory cytokines like TNF-α, IL-6, and CRP [62]. Systemic inflammation in obesity can also affect the uterus through circulation, as pro-inflammatory cytokines can cross the blood-uterine barrier and exert direct effects on uterine tissue [63]. Inflammation alters the gene expression and cellular signaling within the endometrium, impairing its ability to prepare for and support implantation [64]. Chronic inflammation may also interfere with endometrial receptivity, crucial for successful embryo implantation. Besides that, it is also linked to negatively impacting egg quality and embryo implantation in the uterus, raising the risk of miscarriage [65]. Pro-inflammatory cytokines can negatively alter the expression of adhesion molecules and cytokines involved in embryo implantation, potentially leading to implantation failure and infertility [66]. Chronic inflammation in the endometrium may also contribute to the development of endometrial disorders such as endometriosis and endometrial hyperplasia, which are associated with infertility and an increased risk of endometrial cancer [67]. Even excessive inflammation can be detrimental; insufficient immune activity may also impair uterine function and embryo invasion, underscoring the need for a delicate balance [68]. A controlled immune response is critical for processes such as endometrial remodeling during the menstrual cycle (especially during the proliferative phase of the uterine cycle) and supporting embryo implantation [69]. Uterine natural killer (uNK) cells in the endometrium have a crucial role in implantation and healthy pregnancy by producing cytokines and growth factors that promote uterine blood vessel growth and support embryo development [70]. Oxidative stress in the uterus can manipulate the immune system by both activating the inflammatory pathways and causing the immune cells, like macrophages, T cells, and natural killer (NK) cells, to change their behavior [61,71]. Obesity often leads to an imbalance between the production of ROS and the body's antioxidant defenses, resulting in oxidative stress. High levels of ROS can damage endometrial cells and impair uterine function [72]. Oxidative stress can disrupt the delicate processes of ovulation, implantation, and early embryo development by damaging the endometrial lining and cellular components, including lipids, proteins, and DNA, leading to cellular dysfunction and tissue damage in the uterus by inhibition of myometrial contraction, which contributes to dysfunctional labor [72,73]. Recent studies indicate a possible association between obesity and the endometrial microbiome [74], which is also part of the immune response. In obese mothers, myometrial arteries exhibit deficits in endothelial cell calcium signaling, and the eNOS system is modified to both contractile and relaxation responses [75].

3. Insulin Resistance and Uterine Health

3.1. Insulin Signaling in the Uterus

Insulin signaling within the uterus is one of the major factors controlling cellular metabolism and development, as well as the processes of cellular growth. Insulin and insulin-like growth factors 1 and 2 (IGF1 and IGF2) bind to their own receptors, resulting in the activation of two important pathways, PI3K-Akt and Ras-MAPK, important in regulating gene expression and the regulation of epithelial cell proliferation [76]. Insulin receptors are expressed in the endometrium, and aberrant insulin signaling may disrupt cellular proliferation, differentiation, and apoptosis processes [77]. Dysregulated insulin signaling may lead to abnormal endometrial growth and remodeling, contributing to conditions such as endometrial hyperplasia and dysfunctional uterine bleeding [78,79]. Endometrial proliferation and implantation in mice is regulated by insulin signaling, and insulin receptor and IGF1R are important for embryo implantation, while uterine receptivity is unaffected by insulin receptor ablation [76].

3.2. Mechanisms Linking Insulin Resistance and Uterine Dysfunction

Insulin resistance contributes to uterine dysfunction through several mechanisms: hormonal imbalance (increased androgens), direct impairment of endometrial cell function, chronic inflammation, oxidative stress, and altered progesterone signaling. These factors together disrupt the normal uterine environment, leading to menstrual irregularities, impaired fertility, and an increased risk of pregnancy complications. Insulin resistance can directly affect endometrial function by altering insulin signaling pathways in uterine tissue. The processes of angiogenesis and endometrial repair are not only strongly estrogen- and progesterone-dependent but also involve NO, platelets, immune cells, and numerous growth factors [80,81]. IR occurs in decrease of NO production, which leads to increase in the shear stress on endothelial cells because of diffuse vasoconstriction. Mutual action of IR, decreased NO and hyperglycemia generate the release of inflammatory cytokines and free radical formation, which leads to endothelial damage, surrounding tissue hypoxia, and possible cellular apoptosis [82]. Elevated insulin levels in vivo modulate apoptosis via the PI3K-Akt pathway, decrease receptors for bovine serum albumin 2 (BMP2), a key decidual marker, impair estrogen and progesterone signaling, further damage endometrial stromal cells, and enhance mitochondrial transmembrane potential, thereby disrupting endometrial decidualization [83]. Resistance-induced hyperinsulinemia can also stimulate the production of insulin-like growth factor 1 (IGF-1) [84], which may promote endometrial cell proliferation and increase the risk of endometrial cancer. Chronic inflammation alters the expression of genes involved in uterine receptivity and cellular function [85], further reducing the likelihood of successful implantation and increasing the risk of uterine dysfunction. Mechanisms of oxidative stress affect endometrial cell health, causing DNA damage, cell death, and an impaired response to hormonal signals [86], all of which are crucial for a functional and receptive endometrium. Inflammatory mediators, such as TNF-α and IL-6, can disrupt normal endometrial function, impair embryo implantation, and increase the risk of pregnancy complications [87]. The vascular changes allow different factors to leave the blood and form plaques in the blood vessel tunica intima layer. This thickening of the blood vessels leads to increased vascular resistance, together with chronic low-grade inflammation, and leads to atherosclerosis [82,88,89]. Insulin resistance can also affect how the uterus responds to progesterone, the hormone critical for maintaining a healthy endometrial lining during the second half of the menstrual cycle [90]. This can disrupt the menstrual cycle and lead to issues like endometrial hyperplasia or irregular shedding of the uterine lining. A common feature of obesity and conditions like PCOS is closely linked to uterine dysfunction [91]. Women with PCOS often experience insulin resistance, which is strongly linked to anovulation and thickened, dysfunctional endometrial lining, leading to fertility problems. Insulin-mediated glucose transporter factor-4 (GLUT4) facilitates glucose transport to endometrial glandular tissue. Fornes et al. demonstrated reduced levels of GLUT4, insulin pathway proteins, and their phosphorylation in the endometrial tissues of PCOS patients, indicating the presence of localized insulin resistance in the endometrium of these patients [92]. Reduced glucose transport function would cause glucose deficiency in the endometrial cells, which would affect the normal growth of the endometrial cells and lead to embryo implantation failure or increased risk of miscarriage.

Impaired endometrial receptivity in patients with PCOS in humans may be the reason endometrial stromal cells and excess TNF-α negatively affect insulin sensitivity and lead to abnormal energy metabolism by decreasing lipocalin signaling and blocking GLUT-4 membrane transport [93].

3.3. Role of Hyperinsulinemia in Uterine Pathology

In insulin resistance, the body compensates by producing more insulin, creating the state of hyperinsulinemia. Hyperinsulinemia plays a central role in the development of endometrial pathology and menstrual irregularities through its effects on estrogen production, androgen levels, and insulin-like growth factors. Elevated insulin levels stimulate the ovaries to produce excess androgens (male hormones like testosterone), which can disrupt the normal hormonal balance needed for healthy reproductive function. High androgen levels can alter the endometrial environment, leading to irregular menstrual cycles, poor endometrial receptivity, and difficulty with embryo implantation [94], which precedes the process of decidualization dependent on the steroid hormone progesterone acting through the nuclear progesterone receptor (PR) that works through insulin receptor substrate 2 (IRS2) expression [90]. Insulin acts as a growth factor in the endometrium, stimulating the proliferation of endometrial cells [95]. Hyperinsulinemia, a common feature of conditions such as PCOS and hyperinsulinemia type A, has been linked to endometrial dysfunction (hyperplasia [96] and endometrial cancer), menstrual irregularities [92,93], and infertility [97]. These findings suggest a potential role of hyperinsulinemia in the pathophysiology of endometrial pathology and menstrual irregularities. Insulin influences estrogen metabolism by increasing ovarian androgen production and decreasing the liver's production of sex hormone-binding globulin (SHBG) [98]. This results in higher circulating levels of free estrogen. Since estrogen stimulates the growth of the endometrial lining, chronic exposure to high estrogen levels without the counterbalance of progesterone (which normally happens in anovulatory cycles) can lead to endometrial thickening and hyperplasia. In addition to influencing estrogen, insulin itself acts as a growth-promoting hormone. In the endometrium, this promotes cellular growth and increases the risk of abnormal tissue development [99]. Combined with estrogen's effects, hyperinsulinemia accelerates endometrial growth, increasing the risk of abnormal pathology, including hyperplasia and possibly even endometrial cancer.

3.4. Pregnancy Complications and Fertility Treatments in Insulin-Resistant Individuals

Insulin resistance significantly impacts fertility and pregnancy outcomes. In individuals with IR, metabolic, hormonal, and inflammatory disturbances can affect ovulation, endometrial receptivity, and fetal development, leading to challenges in fertility treatments and increased pregnancy complications. Wells (1960) noted that insulin administration in pregnant diabetic rats improved fertility and maternal mortality, although fetal mortality remained higher [100]. Seely (2003) suggested that insulin resistance may play a role in pregnancy-induced hypertension, highlighting the need for interventions to reduce insulin resistance and potentially mitigate these risks [101]. Women with IR may require ovulation induction drugs like clomiphene citrate or letrozole to stimulate ovulation [102]. However, insulin resistance can make these treatments less effective, often requiring higher doses or additional medications to achieve successful ovulation. Medications like metformin (an insulin-sensitizer) are frequently used in fertility treatments for women with IR. Metformin helps reduce insulin levels, improve ovulation, and enhance the effectiveness of ovulation induction therapies. In women with PCOS, metformin is often prescribed to regulate cycles and improve pregnancy rates [103–105]. Despite ovulation induction, many insulin-resistant individuals experience implantation failure due to the disrupted uterine environment [106]. Insulin-sensitization treatments can improve outcomes, but this remains a significant challenge for fertility treatments. Insulin resistance and associated metabolic disturbances may impact the success rates of IVF treatments. Women with insulin resistance may have lower ovarian reserve, reduced oocyte quality, and decreased embryo implantation rates compared to non-insulin-resistant individuals. However, addressing insulin resistance prior to IVF may improve outcomes [107]. During pregnancy, insulin resistance increases the risk of complications such as miscarriage, gestational diabetes, pre-eclampsia, fetal growth abnormalities, and preterm birth. Malaza et al. reported more adverse pregnancy outcomes in women with pregestational diabetes compared to those with gestational diabetes. These complications include cesarean section, preterm birth, congenital anomalies, pre-eclampsia, neonatal hypoglycemia, macrosomia, neonatal intensive care unit admission, stillbirth, low Apgar score, large for gestational age infants, induction of labor, respiratory distress syndrome, and miscarriages [108]. This aligns with findings from the International Prospective Hyperglycemia and Adverse Pregnancy Outcomes (HAPO) study group [109]. Insulin, as an anabolic hormone, plays a key role in regulating fetal growth. Maternal hyperglycemia leads to fetal hyperglycemia and hyperinsulinemia, which stimulates anabolism and the development of muscle, adipose, and connective tissue. These processes contribute to increased storage of fetal fat and protein, resulting in macrosomia [110]. Careful management and monitoring throughout fertility treatments and pregnancy are essential to optimize outcomes for both the mother and baby. Standard therapy for gestational diabetes requiring insulin therapy [111,112] has shown better maternal and newborn outcomes compared to diet, oral anti-diabetic drugs, or insulin analogues, with a lower incidence of macrosomia [113]. In the management of diabetes during pregnancy, metformin monotherapy is effective in maintaining glycemic control [114] and is specifically recommended for patients with prediabetes or a BMI of 35 kg/m^2 [115].

4. Leptin Resistance and Uterine Health

Leptin resistance is a condition in which the body becomes less responsive to the hormone leptin, despite having elevated leptin levels. Leptin, primarily produced by fat cells, plays a key role in regulating energy balance, appetite, and reproductive functions. In the context of uterine function, leptin resistance has significant implications for reproductive health, particularly in individuals with obesity or metabolic disorders.

4.1. Leptin Signaling in the Uterus

Leptin, a hormone secreted by adipose (fat) tissue, serves as a key regulator of energy balance and reproduction, specifically by preparing the endometrium for implantation by influencing factors like angiogenesis (formation of new blood vessels) and cytokine regulation [116]. Leptin affects the uterus through the leptin receptors (Ob-R), belonging to the class I cytokine receptor family, that are expressed in the uterine epithelium, stroma, myometrium, and endometrial glands [117]. The presence of leptin receptors suggests that the uterus is responsive to leptin signaling, which can modulate various cellular processes involved in uterine function. When leptin attaches to Ob-R, it activates the JAK-STAT signaling pathway that promotes gene expression related to growth, differentiation, and inflammatory processes required for the maintenance of the uterine lining and for the embedding of the embryo [118]. Leptin signaling is essential for coordinating energy balance with reproductive function, particularly in the uterus. It regulates the menstrual cycle, enhances endometrial receptivity, supports implantation, and ensures proper pregnancy maintenance. Leptin signaling can also activate the MAPK pathway, influencing cell growth and survival in the uterine tissues. This pathway is important for controlling the structural integrity of the endometrium during the menstrual cycle and early pregnancy [119]. Another significant pathway affected by leptin is the phosphatidylinositol-3-kinase PI3K-Akt pathway, which is involved in cellular metabolism and survival [120] by regulating the metabolic demands of uterine cells and ensuring proper nutrient supply during pregnancy. Both mentioned pathways regulate the promotion of endometrial growth and regeneration [21]. Leptin resistance reduces the ability of the endometrium to respond to these signals, potentially leading to implantation failure and infertility [121]. Understanding and managing leptin signaling pathways is crucial for improving reproductive health and addressing fertility issues linked to metabolic imbalances.

4.2. *Disruption of Leptin Signaling in Obesity and Uterine Health*

In a healthy physiological state, leptin plays an essential role in regulating reproductive processes like regulation of the menstrual cycle and endometrial receptivity. Different studies indicated significant variations in serum leptin during the female menstrual cycle, but results are not uniform. Many previous studies have shown conflicting results related to the leptin levels during the menstrual cycle. Some researchers noted that the leptin concentration increases gradually from the first phase, reaching the highest concentration in the luteal phase [122,123], while others noted a highest concentration in the pre-ovulatory phase [124]. The disruption in leptin signaling, particularly leptin resistance seen in metabolic disorders like obesity, has profound consequences for uterine health and overall reproductive function, leading to significant reproductive dysfunctions, including infertility, poor pregnancy outcomes, and uterine pathologies [125]. Leptin is involved in the regulation of the hypothalamic–pituitary–gonadal (HPG) axis, influencing the secretion of hormones such as gonadotropin-releasing hormone (GnRH), luteinizing hormone (LH), and follicle-stimulating hormone (FSH). These hormones are critical for ovulation and the maintenance of a regular menstrual cycle. It also modulates the immune environment in the uterus to support early pregnancy [126]. Leptin resistance reduces the ability of leptin receptors (Ob-R) in the endometrium to respond to the hormone [116,127]. This impairs the regulatory function of leptin on cellular proliferation, immune defense, and the development of blood vessels within the uterine cavity structure, leading to a situation where the endometrium does not adequately react to hormonal stimuli towards its readiness for embryo attachment [64]. When this signaling is impaired, the uterine lining is less capable of supporting the attachment and growth of an embryo, leading to implantation failure and reduced fertility [128]. Leptin promotes the development of new blood vessels in the endometrium through its regulation of vascular endothelial growth factor (VEGF) [129]. In obesity, disrupted leptin signaling compromises this process, leading to poor uterine vascularization and a suboptimal environment for embryo implantation [130,131]. The impaired leptin signaling in obesity disrupts normal ovarian function and further contributes to infertility in PCOS patients [132]. Leptin resistance reduces the uterus's ability to support early pregnancy by compromising endometrial receptivity and placental development [133]. As a result, women with obesity and leptin resistance face a higher risk of early pregnancy loss. Addressing leptin resistance through weight management, lifestyle changes, and pharmacological interventions can help improve reproductive health and enhance fertility in individuals with obesity.

4.3. Potential Therapeutic Targets to Restore Leptin Sensitivity in the Context of Obesity

Restoring leptin sensitivity in obesity is a complex challenge, as leptin resistance often involves multiple pathways and factors. Correia (2007) and Santoro (2015) both highlight the potential of restoring leptin sensitivity as a treatment target, with Santoro emphasizing the need for new therapeutic tools [134,135]. GLP-1 receptors are widely distributed throughout the reproductive system, and both preclinical models and some clinical studies suggest that GLP-1 plays a key role in linking the reproductive and metabolic systems. GLP-1 also appears to have anti-inflammatory and anti-fibrotic effects in the gonads and endometrium, particularly in conditions like obesity, diabetes, and polycystic ovary syndrome (PCOS) [136]. Sáinz (2015) suggests that anti-inflammatory drugs or bioactive food components could help overcome leptin resistance [137], while Salazar (2019) underscores the importance of addressing leptin resistance in the context of type 2 diabetes [138]. These studies collectively underscore the need for further research to identify and develop effective therapeutic targets for restoring leptin sensitivity in obesity. Based on the in vitro studies, there are some potential therapeutic targets and strategies. Enhancing leptin signaling through the JAK2/STAT3 pathway can potentially restore sensitivity. This involves targeting molecules that modulate the Janus kinase 2 (JAK2) and signal transducer and activator of transcription 3 (STAT3) proteins [139]. Also, suppressing Suppressor of Cytokine Signaling (SOCS) proteins, especially SOCS3, which inhibit leptin signaling, could help restore leptin sensitivity [140]. Targeting inflammatory cytokines (e.g., TNF-α, IL-6) or signaling pathways (e.g., NF-κB) could reduce leptin resistance [141]. Improving insulin sensitivity can often help with leptin resistance. The gut microbiota plays a role in systemic inflammation and metabolic processes. Probiotics or prebiotics might help in restoring leptin sensitivity by influencing gut health [142]. Certain dietary components and bioactive compounds have been shown to modulate leptin signaling and improve leptin sensitivity. Omega-3 fatty acids, found in fish oil and certain plant sources, have anti-inflammatory properties and may enhance leptin signaling [143]. Polyphenols, such as resveratrol found in red grapes and berries, have been shown to enhance insulin sensitivity and modulate leptin signaling pathways. They achieve this by suppressing leptin expression in adipocytes, reducing hyperleptinemia, and alleviating leptin resistance [144]. Pharmacological leptin sensitizers include small molecules that produce modest weight loss when used alone but significantly amplify the anorectic effects of exogenous leptin. Examples include meta-chlorophenylpiperazine, a serotonin receptor agonist and reuptake inhibitor; metformin, a widely used anti-diabetic medication; and betulinic acid, which likely enhances leptin sensitivity through PTP1B inhibition [145–147]. Each of these strategies has its own set of challenges and potential benefits. Along with dietary manipulations to restore leptin sensitivity, exercise also decreases body weight and leptinemia and improves leptin sensitivity by activating leptin sensitive neurons in the VMH and increasing pSTAT3 in the VTA. This effect seems to be independent of fat mass loss [148,149]. Research is ongoing to better understand how these various factors interact and how best to target them therapeutically.

5. Future Directions and Conclusions

The relationship between uterine health and obesity is complex and deeply intertwined. Obesity influences uterine function through metabolic, hormonal, and inflammatory pathways. Obesity leads to insulin resistance and an imbalance in circulating adipokines, characterized by increased levels of leptin and decreased levels of adiponectin. Such changes can disrupt the normal physiological function of the uterus. In addition, such changes create an environment for chronic inflammation that has been associated with several conditions of the uterus, including endometrial hyperplasia, fibroids, and polycystic ovary syndrome (PCOS). Additionally, obesity-induced hormonal imbalances, such as elevated estrogen levels, can affect the uterine lining, increasing the risk of conditions like endometrial cancer.

The impact of obesity on uterine health is further compounded by alterations in the uterine microbiome and epigenetic modifications that may influence long-term uterine function and fertility. Obesity also affects uterine blood flow, contributing to pregnancy complications such as preeclampsia. These factors together contribute not only to poor fertility and pregnancy outcomes but also to the enhanced risk of reproductive disorders and cancers.

This would be beneficial in targeted therapies and prevention strategies to improve uterine health among obese individuals. As obesity is an important source of metabolic and hormonal disruption, addressing these problems could have the potential to improve reproductive health outcomes and reduce the burden of obesity-related uterine disorders.

Author Contributions: D.Š., S.B. and A.M., Conceptualization, D.Š. and S.B. writing—original draft preparation, D.Š., S.B., A.M., M.V.D. and M.H.; writing—review and editing, A.M. and M.H., supervision. All authors have read and agreed to the published version of the manuscript.

Funding: This research received no external funding.

Conflicts of Interest: The authors declare no conflicts of interest.

References

1. Riddick, D.H.; Daly, D.C.; Walters, C.A. The Uterus as an Endocrine Compartment. *Clin. Perinatol.* **1983**, *10*, 627–639. [CrossRef] [PubMed]
2. Kelleher, A.M.; DeMayo, F.J.; Spencer, T.E. Uterine Glands: Developmental Biology and Functional Roles in Pregnancy. *Endocr. Rev.* **2019**, *40*, 1424–1445. [CrossRef] [PubMed]
3. Feyaerts, D.; Joosten, I.; van der Molen, R.G. A Pregnancy to Remember: Trained Immunity of the Uterine Mucosae. *Mucosal Immunol.* **2021**, *14*, 539–541. [CrossRef] [PubMed]
4. Tur-Kaspa, I.; Gleicher, N. Immunology of the Uterus. In *The Uterus: Pathology, Diagnosis, and Management*; Altchek, A., Deligdisch, L., Eds.; Springer: New York, NY, USA, 1991; pp. 27–38. ISBN 978-1-4613-9086-2.
5. Agostinis, C.; Mangogna, A.; Bossi, F.; Ricci, G.; Kishore, U.; Bulla, R. Uterine Immunity and Microbiota: A Shifting Paradigm. *Front. Immunol.* **2019**, *10*, 2387. [CrossRef]
6. Smith, R.; Imtiaz, M.; Banney, D.; Paul, J.W.; Young, R.C. Why the Heart Is like an Orchestra and the Uterus Is like a Soccer Crowd. *Am. J. Obstet. Gynecol.* **2015**, *213*, 181–185. [CrossRef]
7. Prefumo, F.; Sharma, R.; Brecker, S.J.D.; Gaze, D.C.; Collinson, P.O.; Thilaganathan, B. Maternal Cardiac Function in Early Pregnancies with High Uterine Artery Resistance. *Ultrasound Obs. Gynecol.* **2007**, *29*, 58–64. [CrossRef]
8. Uteroplacental Blood Flow, Cardiac Function, and Pregnancy Outcome in Women With Congenital Heart Disease | Circulation. Available online: https://www.ahajournals.org/doi/10.1161/CIRCULATIONAHA.113.002810 (accessed on 9 October 2024).
9. Uterus Plays a Role in Brain Function, Animal Study Shows. Available online: https://www.nia.nih.gov/news/uterus-plays-role-brain-function-animal-study-shows (accessed on 9 October 2024).
10. Koebele, S.V.; Bernaud, V.E.; Northup-Smith, S.N.; Willeman, M.N.; Strouse, I.M.; Bulen, H.L.; Schrier, A.R.; Newbern, J.M.; DeNardo, D.F.; Mayer, L.P.; et al. Gynecological Surgery in Adulthood Imparts Cognitive and Brain Changes in Rats: A Focus on Hysterectomy at Short-, Moderate-, and Long-Term Intervals after Surgery. *Horm. Behav.* **2023**, *155*, 105411. [CrossRef]
11. Koebele, S.V.; Palmer, J.M.; Hadder, B.; Melikian, R.; Fox, C.; Strouse, I.M.; DeNardo, D.F.; George, C.; Daunis, E.; Nimer, A.; et al. Hysterectomy Uniquely Impacts Spatial Memory in a Rat Model: A Role for the Nonpregnant Uterus in Cognitive Processes. *Endocrinology* **2019**, *160*, 1–19. [CrossRef]
12. Olusi, A.M.; Rabiu, K.A.; Oduola-Owoo, B.B.; Rasheed, M.W.; Oduola-Owoo, L.T.; Windapo, O.B. Relationship between Metabolic Syndrome and Uterine Leiomyoma: A Case Control Study. *Niger. Health J.* **2024**, *24*, 1058–1069. [CrossRef]
13. Salcedo, A.C.; Shehata, H.; Berry, A.; Riba, C. Insulin Resistance and Other Risk Factors of Cardiovascular Disease amongst Women with Abnormal Uterine Bleeding. *J. Metab. Health* **2022**, *5*, 7. [CrossRef]
14. Obesity and Overweight. Available online: https://www.who.int/news-room/fact-sheets/detail/obesity-and-overweight (accessed on 10 October 2024).
15. Zheng, L.; Yang, L.; Guo, Z.; Yao, N.; Zhang, S.; Pu, P. Obesity and Its Impact on Female Reproductive Health: Unraveling the Connections. *Front. Endocrinol.* **2024**, *14*, 1326546. [CrossRef] [PubMed]
16. Schon, S.B.; Cabre, H.E.; Redman, L.M. The Impact of Obesity on Reproductive Health and Metabolism in Reproductive-Age Females. *Fertil. Steril.* **2024**, *122*, 194–203. [CrossRef] [PubMed]
17. Pandey, S.; Bhattacharya, S. Impact of Obesity on Gynecology. *Womens Health* **2010**, *6*, 107–117. [CrossRef] [PubMed]
18. Brown, K.; Apuzzio, J.; Weiss, G. Maternal Obesity and Associated Reproductive Consequences. *Womens Health* **2010**, *6*, 197–203. [CrossRef] [PubMed]

19. Thong, E.P.; Codner, E.; Laven, J.S.E.; Teede, H. Diabetes: A Metabolic and Reproductive Disorder in Women. *Lancet Diabetes Endocrinol.* **2020**, *8*, 134–149. [CrossRef]
20. Metwally, M.; Li, T.C.; Ledger, W.L. The Impact of Obesity on Female Reproductive Function. *Obes. Rev.* **2007**, *8*, 515–523. [CrossRef]
21. Słabuszewska-Jóźwiak, A.; Lukaszuk, A.; Janicka-Kośnik, M.; Wdowiak, A.; Jakiel, G. Role of Leptin and Adiponectin in Endometrial Cancer. *Int. J. Mol. Sci.* **2022**, *23*, 5307. [CrossRef]
22. Sidorkiewicz, I.; Jóźwik, M.; Niemira, M.; Krętowski, A. Insulin Resistance and Endometrial Cancer: Emerging Role for microRNA. *Cancers* **2020**, *12*, 2559. [CrossRef]
23. Ameer, M.A.; Fagan, S.E.; Sosa-Stanley, J.N.; Peterson, D.C. Anatomy, Abdomen and Pelvis: Uterus. In *StatPearls*; StatPearls Publishing: Treasure Island, FL, USA, 2024.
24. Gasner, A.; Aatsha, P.A. Physiology, Uterus. In *StatPearls [Internet]*; StatPearls Publishing: Treasure Island, FL, USA, 2023.
25. Ylli, D.; Sidhu, S.; Parikh, T.; Burman, K.D. Endocrine Changes in Obesity. In *Endotext [Internet]*; MDText.com, Inc.: South Dartmouth, MA, USA, 2022.
26. Parisi, F.; Milazzo, R.; Savasi, V.M.; Cetin, I. Maternal Low-Grade Chronic Inflammation and Intrauterine Programming of Health and Disease. *Int. J. Mol. Sci.* **2021**, *22*, 1732. [CrossRef]
27. Yong, W.; Wang, J.; Leng, Y.; Li, L.; Wang, H. Role of Obesity in Female Reproduction. *Int. J. Med. Sci.* **2023**, *20*, 366–375. [CrossRef]
28. Itriyeva, K. The Effects of Obesity on the Menstrual Cycle. *Curr. Probl. Pediatr. Adolesc. Health Care* **2022**, *52*, 101241. [CrossRef] [PubMed]
29. Qin, H.; Lin, Z.; Vásquez, E.; Luan, X.; Guo, F.; Xu, L. Association between Obesity and the Risk of Uterine Fibroids: A Systematic Review and Meta-Analysis. *J. Epidemiol. Community Health* **2021**, *75*, 197–204. [CrossRef] [PubMed]
30. Harvey, S.V.; Wentzensen, N.; Bertrand, K.; Black, A.; Brinton, L.A.; Chen, C.; Costas, L.; Dal Maso, L.; De Vivo, I.; Du, M.; et al. Associations of Life Course Obesity with Endometrial Cancer in the Epidemiology of Endometrial Cancer Consortium (E2C2). *Int. J. Epidemiol.* **2023**, *52*, 1086–1099. [CrossRef] [PubMed]
31. Obesity in Pregnancy: Risks and Management | AAFP. Available online: https://www.aafp.org/pubs/afp/issues/2018/0501/p559.html (accessed on 10 October 2024).
32. Mooney, S.S.; Sumithran, P. Does Weight Loss in Women with Obesity Induce Regression of Endometrial Hyperplasia? A Systematic Review. *Eur. J. Obstet. Gynecol. Reprod. Biol.* **2023**, *288*, 49–55. [CrossRef]
33. Overweight and Obesity | Cancer Australia. Available online: https://www.canceraustralia.gov.au/cancer-types/endometrial-cancer/awareness/lifestyle/overweight-and-obesity (accessed on 10 October 2024).
34. Kuryłowicz, A. Estrogens in Adipose Tissue Physiology and Obesity-Related Dysfunction. *Biomedicines* **2023**, *11*, 690. [CrossRef]
35. Ding, H.; Zhang, J.; Zhang, F.; Zhang, S.; Chen, X.; Liang, W.; Xie, Q. Resistance to the Insulin and Elevated Level of Androgen: A Major Cause of Polycystic Ovary Syndrome. *Front. Endocrinol.* **2021**, *12*, 741764. [CrossRef]
36. Borahay, M.A.; Asoglu, M.R.; Mas, A.; Adam, S.; Kilic, G.S.; Al-Hendy, A. Estrogen Receptors and Signaling in Fibroids: Role in Pathobiology and Therapeutic Implications. *Reprod. Sci.* **2017**, *24*, 1235–1244. [CrossRef]
37. Ali, M.; Ciebiera, M.; Vafaei, S.; Alkhrait, S.; Chen, H.-Y.; Chiang, Y.-F.; Huang, K.-C.; Feduniw, S.; Hsia, S.-M.; Al-Hendy, A. Progesterone Signaling and Uterine Fibroid Pathogenesis; Molecular Mechanisms and Potential Therapeutics. *Cells* **2023**, *12*, 1117. [CrossRef]
38. Barboza, I.C.; Depes, D.D.B.; Vianna, I.; Patriarca, M.T.; Arruda, R.M.; Martins, J.A.; Lopes, R.G.C. Analysis of Endometrial Thickness Measured by Transvaginal Ultrasonography in Obese Patients. *Einstein* **2014**, *12*, 164–167. [CrossRef]
39. Lu, L.; Risch, H.; Irwin, M.L.; Mayne, S.T.; Cartmel, B.; Schwartz, P.; Rutherford, T.; Yu, H. Long-Term Overweight and Weight Gain in Early Adulthood in Association with Risk of Endometrial Cancer. *Int. J. Cancer* **2011**, *129*, 1237–1243. [CrossRef]
40. Viola, A.S.; Gouveia, D.; Andrade, L.; Aldrighi, J.M.; Viola, C.F.M.; Bahamondes, L. Prevalence of Endometrial Cancer and Hyperplasia in Non-Symptomatic Overweight and Obese Women. *Aust. N. Z. J. Obs. Gynaecol.* **2008**, *48*, 207–213. [CrossRef] [PubMed]
41. Wang, L.; Wei, W.; Cai, M. A Review of the Risk Factors Associated with Endometrial Hyperplasia During Perimenopause. *IJWH* **2024**, *16*, 1475–1482. [CrossRef] [PubMed]
42. Endometrial Hyperplasia—An Overview | ScienceDirect Topics. Available online: https://www.sciencedirect.com/topics/medicine-and-dentistry/endometrial-hyperplasia (accessed on 10 October 2024).
43. Singh, S.; Pal, N.; Shubham, S.; Sarma, D.K.; Verma, V.; Marotta, F.; Kumar, M. Polycystic Ovary Syndrome: Etiology, Current Management, and Future Therapeutics. *J. Clin. Med.* **2023**, *12*, 1454. [CrossRef] [PubMed]
44. Hajam, Y.A.; Rather, H.A.; Neelam; Kumar, R.; Basheer, M.; Reshi, M.S. A Review on Critical Appraisal and Pathogenesis of Polycystic Ovarian Syndrome. *Endocr. Metab. Sci.* **2024**, *14*, 100162. [CrossRef]
45. Barber, T.M.; Franks, S. Obesity and Polycystic Ovary Syndrome. *Clin. Endocrinol.* **2021**, *95*, 531–541. [CrossRef]
46. Gonnella, F.; Konstantinidou, F.; Donato, M.; Gatta, D.M.P.; Peserico, A.; Barboni, B.; Stuppia, L.; Nothnick, W.B.; Gatta, V. The Molecular Link between Obesity and the Endometrial Environment: A Starting Point for Female Infertility. *Int. J. Mol. Sci.* **2024**, *25*, 6855. [CrossRef]
47. Bjorklund, J.; Wiberg-Itzel, E.; Wallstrom, T. Is There an Increased Risk of Cesarean Section in Obese Women after Induction of Labor? A Retrospective Cohort Study. *PLoS ONE* **2022**, *17*, e0263685. [CrossRef]
48. Kissler, K.; Hurt, K.J. The Pathophysiology of Labor Dystocia: Theme with Variations. *Reprod. Sci.* **2023**, *30*, 729–742. [CrossRef]

49. Bogaerts, A.; Witters, I.; Van den Bergh, B.R.H.; Jans, G.; Devlieger, R. Obesity in Pregnancy: Altered Onset and Progression of Labour. *Midwifery* **2013**, *29*, 1303–1313. [CrossRef]
50. Azaïs, H.; Leroy, A.; Ghesquiere, L.; Deruelle, P.; Hanssens, S. Effects of Adipokines and Obesity on Uterine Contractility. *Cytokine Growth Factor. Rev.* **2017**, *34*, 59–66. [CrossRef]
51. Zhang, J.; Bricker, L.; Wray, S.; Quenby, S. Poor Uterine Contractility in Obese Women. *BJOG* **2007**, *114*, 343–348. [CrossRef] [PubMed]
52. Higgins, C.A.; Martin, W.; Anderson, L.; Blanks, A.M.; Norman, J.E.; McConnachie, A.; Nelson, S.M. Maternal Obesity and Its Relationship with Spontaneous and Oxytocin-Induced Contractility of Human Myometrium in Vitro. *Reprod. Sci.* **2010**, *17*, 177–185. [CrossRef] [PubMed]
53. Chin, J.R.; Henry, E.; Holmgren, C.M.; Varner, M.W.; Branch, D.W. Maternal Obesity and Contraction Strength in the First Stage of Labor. *Am. J. Obs. Gynecol.* **2012**, *207*, 129.e1–129.e6. [CrossRef] [PubMed]
54. Carvajal, J.A.; Oporto, J.I. The Myometrium in Pregnant Women with Obesity. *Curr. Vasc. Pharmacol.* **2021**, *19*, 193–200. [CrossRef] [PubMed]
55. Grotegut, C.A.; Gunatilake, R.P.; Feng, L.; Heine, R.P.; Murtha, A.P. The Influence of Maternal Body Mass Index on Myometrial Oxytocin Receptor Expression in Pregnancy. *Reprod. Sci.* **2013**, *20*, 1471–1477. [CrossRef]
56. Gao, X.; Li, Y.; Ma, Z.; Jing, J.; Zhang, Z.; Liu, Y.; Ding, Z. Obesity Induces Morphological and Functional Changes in Female Reproductive System through Increases in NF-κB and MAPK Signaling in Mice. *Reprod. Biol. Endocrinol.* **2021**, *19*, 148. [CrossRef]
57. Baltayeva, J.; Konwar, C.; Castellana, B.; Mara, D.L.; Christians, J.K.; Beristain, A.G. Obesogenic Diet Exposure Alters Uterine Natural Killer Cell Biology and Impairs Vasculature Remodeling in Mice†. *Biol. Reprod.* **2020**, *102*, 63–75. [CrossRef]
58. Hayes, E.K.; Tessier, D.R.; Percival, M.E.; Holloway, A.C.; Petrik, J.J.; Gruslin, A.; Raha, S. Trophoblast Invasion and Blood Vessel Remodeling Are Altered in a Rat Model of Lifelong Maternal Obesity. *Reprod. Sci.* **2014**, *21*, 648–657. [CrossRef]
59. Perdu, S.; Castellana, B.; Kim, Y.; Chan, K.; DeLuca, L.; Beristain, A.G. Maternal Obesity Drives Functional Alterations in Uterine NK Cells. *JCI Insight* **2016**, *1*, e85560. [CrossRef]
60. Roberts, K.A.; Riley, S.C.; Reynolds, R.M.; Barr, S.; Evans, M.; Statham, A.; Hor, K.; Jabbour, H.N.; Norman, J.E.; Denison, F.C. Placental Structure and Inflammation in Pregnancies Associated with Obesity. *Placenta* **2011**, *32*, 247–254. [CrossRef]
61. Jiménez-Osorio, A.S.; Carreón-Torres, E.; Correa-Solís, E.; Ángel-García, J.; Arias-Rico, J.; Jiménez-Garza, O.; Morales-Castillejos, L.; Díaz-Zuleta, H.A.; Baltazar-Tellez, R.M.; Sánchez-Padilla, M.L.; et al. Inflammation and Oxidative Stress Induced by Obesity, Gestational Diabetes, and Preeclampsia in Pregnancy: Role of High-Density Lipoproteins as Vectors for Bioactive Compounds. *Antioxidants* **2023**, *12*, 1894. [CrossRef] [PubMed]
62. Khanna, D.; Khanna, S.; Khanna, P.; Kahar, P.; Patel, B.M. Obesity: A Chronic Low-Grade Inflammation and Its Markers. *Cureus* **2022**, *14*, e22711. [CrossRef] [PubMed]
63. St-Germain, L.E.; Castellana, B.; Baltayeva, J.; Beristain, A.G. Maternal Obesity and the Uterine Immune Cell Landscape: The Shaping Role of Inflammation. *Int. J. Mol. Sci.* **2020**, *21*, 3776. [CrossRef] [PubMed]
64. Blanco-Breindel, M.F.; Singh, M.; Kahn, J. Endometrial Receptivity. In *StatPearls*; StatPearls Publishing: Treasure Island, FL, USA, 2024.
65. Zavatta, A.; Parisi, F.; Mandò, C.; Scaccabarozzi, C.; Savasi, V.M.; Cetin, I. Role of Inflammaging on the Reproductive Function and Pregnancy. *Clin. Rev. Allergy Immunol.* **2023**, *64*, 145–160. [CrossRef]
66. Salamun, V.; Bokal, E.V.; Maver, A.; Papler, T.B. Transcriptome Study of Receptive Endometrium in Overweight and Obese Women Shows Important Expression Differences in Immune Response and Inflammatory Pathways in Women Who Do Not Conceive. *PLoS ONE* **2021**, *16*, e0261873. [CrossRef]
67. Ye, J.; Peng, H.; Huang, X.; Qi, X. The Association between Endometriosis and Risk of Endometrial Cancer and Breast Cancer: A Meta-Analysis. *BMC Womens Health* **2022**, *22*, 455. [CrossRef]
68. Wu, H.-M.; Chen, L.-H.; Hsu, L.-T.; Lai, C.-H. Immune Tolerance of Embryo Implantation and Pregnancy: The Role of Human Decidual Stromal Cell- and Embryonic-Derived Extracellular Vesicles. *Int. J. Mol. Sci.* **2022**, *23*, 13382. [CrossRef]
69. Ma, H.; Cai, S.; Yang, L.; Wang, L.; Ding, J.; Li, L.; Li, H.; Huang, C.; Diao, L. How Do Pre-Pregnancy Endometrial Macrophages Contribute to Pregnancy? *J. Reprod. Immunol.* **2022**, *154*, 103736. [CrossRef]
70. Lédée, N.; Petitbarat, M.; Prat-Ellenberg, L.; Dray, G.; Vaucoret, V.; Kazhalawi, A.; Rodriguez-Pozo, A.; Habeichi, N.; Ruoso, L.; Cassuto, N.G.; et al. The Next Frontier in ART: Harnessing the Uterine Immune Profile for Improved Performance. *Int. J. Mol. Sci.* **2023**, *24*, 11322. [CrossRef]
71. Sharma, S. Natural Killer Cells and Regulatory T Cells in Early Pregnancy Loss. *Int. J. Dev. Biol.* **2014**, *58*, 219–229. [CrossRef]
72. Lu, J.; Wang, Z.; Cao, J.; Chen, Y.; Dong, Y. A Novel and Compact Review on the Role of Oxidative Stress in Female Reproduction. *Reprod. Biol. Endocrinol.* **2018**, *16*, 80. [CrossRef] [PubMed]
73. Kaltsas, A.; Zikopoulos, A.; Moustakli, E.; Zachariou, A.; Tsirka, G.; Tsiampali, C.; Palapela, N.; Sofikitis, N.; Dimitriadis, F. The Silent Threat to Women's Fertility: Uncovering the Devastating Effects of Oxidative Stress. *Antioxidants* **2023**, *12*, 1490. [CrossRef] [PubMed]
74. King, S.; Osei, F.; Marsh, C. Prevalence of Pathogenic Microbes within the Endometrium in Normal Weight vs. Obese Women with Infertility. *Reprod. Med.* **2024**, *5*, 90–96. [CrossRef]
75. Prendergast, C.; Wray, S. Human Myometrial Artery Function and Endothelial Cell Calcium Signalling Are Reduced by Obesity: Can This Contribute to Poor Labour Outcomes? *Acta Physiol.* **2019**, *227*, e13341. [CrossRef] [PubMed]

76. Sekulovski, N.; Whorton, A.E.; Shi, M.; Hayashi, K.; MacLean, J.A. Insulin Signaling Is an Essential Regulator of Endometrial Proliferation and Implantation in Mice. *FASEB J.* **2021**, *35*, e21440. [CrossRef]
77. Chen, M.; Li, J.; Zhang, B.; Zeng, X.; Zeng, X.; Cai, S.; Ye, Q.; Yang, G.; Ye, C.; Shang, L.; et al. Uterine Insulin Sensitivity Defects Induced Embryo Implantation Loss Associated with Mitochondrial Dysfunction-Triggered Oxidative Stress. *Oxid. Med. Cell Longev.* **2021**, *2021*, 6655685. [CrossRef]
78. Jain, V.; Chodankar, R.R.; Maybin, J.A.; Critchley, H.O.D. Uterine Bleeding: How Understanding Endometrial Physiology Underpins Menstrual Health. *Nat. Rev. Endocrinol.* **2022**, *18*, 290–308. [CrossRef]
79. Singh, G.; Cue, L.; Puckett, Y. Endometrial Hyperplasia. In *StatPearls*; StatPearls Publishing: Treasure Island, FL, USA, 2024.
80. Chodankar, R.; Critchley, H.O.D. Biomarkers in Abnormal Uterine Bleeding†. *Biol. Reprod.* **2019**, *101*, 1155–1166. [CrossRef]
81. Ciarmela, P.; Islam, M.S.; Reis, F.M.; Gray, P.C.; Bloise, E.; Petraglia, F.; Vale, W.; Castellucci, M. Growth Factors and Myometrium: Biological Effects in Uterine Fibroid and Possible Clinical Implications. *Hum. Reprod. Update* **2011**, *17*, 772–790. [CrossRef]
82. Ormazabal, V.; Nair, S.; Elfeky, O.; Aguayo, C.; Salomon, C.; Zuñiga, F.A. Association between Insulin Resistance and the Development of Cardiovascular Disease. *Cardiovasc. Diabetol.* **2018**, *17*, 122. [CrossRef]
83. Zhang, C.; Yang, C.; Li, N.; Liu, X.; He, J.; Chen, X.; Ding, Y.; Tong, C.; Peng, C.; Yin, H.; et al. Elevated Insulin Levels Compromise Endometrial Decidualization in Mice with Decrease in Uterine Apoptosis in Early-Stage Pregnancy. *Arch. Toxicol.* **2019**, *93*, 3601–3615. [CrossRef] [PubMed]
84. Friedrich, N.; Thuesen, B.; Jørgensen, T.; Juul, A.; Spielhagen, C.; Wallaschofksi, H.; Linneberg, A. The Association Between IGF-I and Insulin Resistance. *Diabetes Care* **2012**, *35*, 768–773. [CrossRef] [PubMed]
85. Cicinelli, E.; Vitagliano, A.; Loizzi, V.; De Ziegler, D.; Fanelli, M.; Bettocchi, S.; Nardelli, C.; Trojano, G.; Cicinelli, R.; Minervini, C.F.; et al. Altered Gene Expression Encoding Cytochines, Grow Factors and Cell Cycle Regulators in the Endometrium of Women with Chronic Endometritis. *Diagnostics* **2021**, *11*, 471. [CrossRef] [PubMed]
86. Didziokaite, G.; Biliute, G.; Gudaite, J.; Kvedariene, V. Oxidative Stress as a Potential Underlying Cause of Minimal and Mild Endometriosis-Related Infertility. *Int. J. Mol. Sci.* **2023**, *24*, 3809. [CrossRef]
87. Kwon, M.J.; Kim, J.H.; Kim, K.J.; Ko, E.J.; Lee, J.Y.; Ryu, C.S.; Ha, Y.H.; Kim, Y.R.; Kim, N.K. Genetic Association between Inflammatory-Related Polymorphism in STAT3, IL-1β, IL-6, TNF-α and Idiopathic Recurrent Implantation Failure. *Genes* **2023**, *14*, 1588. [CrossRef]
88. Do, H.D.; Lohsoonthorn, V.; Jiamjarasrangsi, W.; Lertmaharit, S.; Williams, M.A. Prevalence of Insulin Resistance and Its Relationship with Cardiovascular Disease Risk Factors among Thai Adults over 35 Years Old. *Diabetes Res. Clin. Pract.* **2010**, *89*, 303–308. [CrossRef]
89. De Boer, I.H.; Katz, R.; Chonchol, M.B.; Fried, L.F.; Ix, J.H.; Kestenbaum, B.; Mukamal, K.J.; Peralta, C.A.; Siscovick, D.S. Insulin Resistance, Cystatin C, and Mortality Among Older Adults. *Diabetes Care* **2012**, *35*, 1355–1360. [CrossRef]
90. Neff, A.M.; Yu, J.; Taylor, R.N.; Bagchi, I.C.; Bagchi, M.K. Insulin Signaling Via Progesterone-Regulated Insulin Receptor Substrate 2 Is Critical for Human Uterine Decidualization. *Endocrinology* **2019**, *161*, bqz021. [CrossRef]
91. Barber, T.M.; Hanson, P.; Weickert, M.O.; Franks, S. Obesity and Polycystic Ovary Syndrome: Implications for Pathogenesis and Novel Management Strategies. *Clin. Med. Insights Reprod. Health* **2019**, *13*, 1179558119874042. [CrossRef]
92. Fornes, R.; Ormazabal, P.; Rosas, C.; Gabler, F.; Vantman, D.; Romero, C.; Vega, M. Changes in the Expression of Insulin Signaling Pathway Molecules in Endometria from Polycystic Ovary Syndrome Women with or without Hyperinsulinemia. *Mol. Med.* **2010**, *16*, 129–136. [CrossRef]
93. Oróstica, L.; García, P.; Vera, C.; García, V.; Romero, C.; Vega, M. Effect of TNF-α on Molecules Related to the Insulin Action in Endometrial Cells Exposed to Hyperandrogenic and Hyperinsulinic Conditions Characteristics of Polycystic Ovary Syndrome. *Reprod. Sci.* **2018**, *25*, 1000–1009. [CrossRef] [PubMed]
94. Yusuf, A.N.M.; Amri, M.F.; Ugusman, A.; Hamid, A.A.; Wahab, N.A.; Mokhtar, M.H. Hyperandrogenism and Its Possible Effects on Endometrial Receptivity: A Review. *Int. J. Mol. Sci.* **2023**, *24*, 12026. [CrossRef] [PubMed]
95. Dai, C.; Li, N.; Song, G.; Yang, Y.; Ning, X. Insulin-like Growth Factor 1 Regulates Growth of Endometrial Carcinoma through PI3k Signaling Pathway in Insulin-Resistant Type 2 Diabetes. *Am. J. Transl. Res.* **2016**, *8*, 3329–3336. [PubMed]
96. Shan, W.; Ning, C.; Luo, X.; Zhou, Q.; Gu, C.; Zhang, Z.; Chen, X. Hyperinsulinemia Is Associated with Endometrial Hyperplasia and Disordered Proliferative Endometrium: A Prospective Cross-Sectional Study. *Gynecol. Oncol.* **2014**, *132*, 606–610. [CrossRef]
97. Giudice, L.C. Endometrium in PCOS: Implantation and Predisposition to Endocrine CA. *Best. Pract. Res. Clin. Endocrinol. Metab.* **2006**, *20*, 235–244. [CrossRef]
98. Xing, C.; Zhang, J.; Zhao, H.; He, B. Effect of Sex Hormone-Binding Globulin on Polycystic Ovary Syndrome: Mechanisms, Manifestations, Genetics, and Treatment. *Int. J. Womens Health* **2022**, *14*, 91–105. [CrossRef]
99. Yu, K.; Huang, Z.-Y.; Xu, X.-L.; Li, J.; Fu, X.-W.; Deng, S.-L. Estrogen Receptor Function: Impact on the Human Endometrium. *Front. Endocrinol.* **2022**, *13*, 827724. [CrossRef]
100. Eriksson, U.J.; Swenne, I. Diabetes in Pregnancy: Fetal Macrosomia, Hyperinsulinism, and 1slet Hyperplasia in the Offspring of Rats Subjected to Temporary Protein-Energy Malnutrition Early in Life. *Pediatr. Res.* **1993**, *34*, 791–795. [CrossRef]
101. Seely, E.W.; Solomon, C.G. Insulin Resistance and Its Potential Role in Pregnancy-Induced Hypertension. *J. Clin. Endocrinol. Metab.* **2003**, *88*, 2393–2398. [CrossRef]
102. Franik, S.; Le, Q.-K.; Kremer, J.A.; Kiesel, L.; Farquhar, C. Aromatase Inhibitors (Letrozole) for Ovulation Induction in Infertile Women with Polycystic Ovary Syndrome. *Cochrane Database Syst. Rev.* **2022**, *2022*, CD010287. [CrossRef]

103. Johnson, N.P. Metformin Use in Women with Polycystic Ovary Syndrome. *Ann. Transl. Med.* **2014**, *2*, 56. [CrossRef] [PubMed]
104. Elnashar, A.M. The Role of Metformin in Ovulation Induction: Current Status. *Middle East. Fertil. Soc. J.* **2011**, *16*, 175–181. [CrossRef]
105. Attia, G.M.; Almouteri, M.M.; Alnakhli, F.T. Role of Metformin in Polycystic Ovary Syndrome (PCOS)-Related Infertility. *Cureus* **2023**, *15*, e44493. [CrossRef] [PubMed]
106. Lei, R.; Chen, S.; Li, W. Advances in the Study of the Correlation between Insulin Resistance and Infertility. *Front. Endocrinol.* **2024**, *15*, 1288326. [CrossRef] [PubMed]
107. Wang, H.; Zhang, Y.; Fang, X.; Kwak-Kim, J.; Wu, L. Insulin Resistance Adversely Affect IVF Outcomes in Lean Women Without PCOS. *Front. Endocrinol.* **2021**, *12*, 734638. [CrossRef]
108. Malaza, N.; Masete, M.; Adam, S.; Dias, S.; Nyawo, T.; Pheiffer, C. A Systematic Review to Compare Adverse Pregnancy Outcomes in Women with Pregestational Diabetes and Gestational Diabetes. *Int. J. Environ. Res. Public Health* **2022**, *19*, 10846. [CrossRef]
109. HAPO Study Cooperative Research Group. The Hyperglycemia and Adverse Pregnancy Outcome (HAPO) Study. *Int. J. Gynaecol. Obstet. Off. Organ. Int. Fed. Gynaecol. Obstet.* **2002**, *78*, 69–77. [CrossRef]
110. Ornoy, A.; Becker, M.; Weinstein-Fudim, L.; Ergaz, Z. Diabetes during Pregnancy: A Maternal Disease Complicating the Course of Pregnancy with Long-Term Deleterious Effects on the Offspring. A Clinical Review. *Int. J. Mol. Sci.* **2021**, *22*, 2965. [CrossRef]
111. Affres, H.; Senat, M.-V.; Letourneau, A.; Deruelle, P.; Coustols-Valat, M.; Bouchghoul, H.; Bouyer, J. Glyburide Therapy for Gestational Diabetes: Glycaemic Control, Maternal Hypoglycaemia, and Treatment Failure. *Diabetes Metab.* **2021**, *47*, 101210. [CrossRef]
112. Balsells, M.; García-Patterson, A.; Solà, I.; Roqué, M.; Gich, I.; Corcoy, R. Glibenclamide, Metformin, and Insulin for the Treatment of Gestational Diabetes: A Systematic Review and Meta-Analysis. *BMJ* **2015**, *350*, h102. [CrossRef]
113. Subiabre, M.; Silva, L.; Toledo, F.; Paublo, M.; López, M.A.; Boric, M.P.; Sobrevia, L. Insulin Therapy and Its Consequences for the Mother, Foetus, and Newborn in Gestational Diabetes Mellitus. *Biochim. Biophys. Acta (BBA)-Mol. Basis Dis.* **2018**, *1864*, 2949–2956. [CrossRef] [PubMed]
114. Priya, G.; Kalra, S. Metformin in the Management of Diabetes during Pregnancy and Lactation. *Drugs Context* **2018**, *7*, 212523. [CrossRef] [PubMed]
115. Aroda, V.R.; Ratner, R.E. Metformin and Type 2 Diabetes Prevention. *Diabetes Spectr.* **2018**, *31*, 336–342. [CrossRef] [PubMed]
116. Jafari-Gharabaghlou, D.; Vaghari-Tabari, M.; Oghbaei, H.; Lotz, L.; Zarezadeh, R.; Rastgar Rezaei, Y.; Ranjkesh, M.; Nouri, M.; Fattahi, A.; Nikanfar, S.; et al. Role of Adipokines in Embryo Implantation. *Endocr. Connect.* **2021**, *10*, R267–R278. [CrossRef]
117. González, R.R.; Caballero-Campo, P.; Jasper, M.; Mercader, A.; Devoto, L.; Pellicer, A.; Simon, C. Leptin and Leptin Receptor Are Expressed in the Human Endometrium and Endometrial Leptin Secretion Is Regulated by the Human Blastocyst. *J. Clin. Endocrinol. Metab.* **2000**, *85*, 4883–4888. [CrossRef]
118. Malik, M.; Britten, J.; DeAngelis, A.; Catherino, W.H. Cross-Talk between Janus Kinase-Signal Transducer and Activator of Transcription Pathway and Transforming Growth Factor Beta Pathways and Increased collagen1A1 Production in Uterine Leiomyoma Cells. *F&S Sci.* **2020**, *1*, 206–220. [CrossRef]
119. Makieva, S.; Giacomini, E.; Ottolina, J.; Sanchez, A.M.; Papaleo, E.; Viganò, P. Inside the Endometrial Cell Signaling Subway: Mind the Gap(s). *Int. J. Mol. Sci.* **2018**, *19*, 2477. [CrossRef]
120. Garcia-Galiano, D.; Borges, B.C.; Allen, S.J.; Elias, C.F. PI3K Signaling in Leptin Receptor Cells: Role in Growth and Reproduction. *J. Neuroendocr.* **2019**, *31*, e12685. [CrossRef]
121. Massimiani, M.; Lacconi, V.; La Civita, F.; Ticconi, C.; Rago, R.; Campagnolo, L. Molecular Signaling Regulating Endometrium–Blastocyst Crosstalk. *Int. J. Mol. Sci.* **2019**, *21*, 23. [CrossRef]
122. Asimakopoulos, B.; Milousis, A.; Gioka, T.; Kabouromiti, G.; Gianisslis, G.; Troussa, A.; Simopoulou, M.; Katergari, S.; Tripsianis, G.; Nikolettos, N. Serum Pattern of Circulating Adipokines throughout the Physiological Menstrual Cycle. *Endocr. J.* **2009**, *56*, 425–433. [CrossRef]
123. Ajala, O.M.; Ogunro, P.S.; Elusanmi, G.F.; Ogunyemi, O.E.; Bolarinde, A.A. Changes in Serum Leptin during Phases of Menstrual Cycle of Fertile Women: Relationship to Age Groups and Fertility. *Int. J. Endocrinol. Metab.* **2013**, *11*, 27–33. [CrossRef] [PubMed]
124. Ahrens, K.; Mumford, S.L.; Schliep, K.C.; Kissell, K.A.; Perkins, N.J.; Wactawski-Wende, J.; Schisterman, E.F. Serum Leptin Levels and Reproductive Function during the Menstrual Cycle. *Am. J. Obs. Gynecol.* **2014**, *210*, 248.e1–248.e9. [CrossRef] [PubMed]
125. Elias, C.F.; Purohit, D. Leptin Signaling and Circuits in Puberty and Fertility. *Cell Mol. Life Sci.* **2012**, *70*, 841–862. [CrossRef] [PubMed]
126. Briffa, J.F.; McAinch, A.J.; Romano, T.; Wlodek, M.E.; Hryciw, D.H. Leptin in Pregnancy and Development: A Contributor to Adulthood Disease? *Am. J. Physiol.-Endocrinol. Metabolism.* 2015, 308, pp. E335–E350. Available online: https://journals.physiology.org/doi/full/10.1152/ajpendo.00312.2014 (accessed on 10 October 2024).
127. Kim, T.H.; Bae, N.; Kim, T.; Hsu, A.L.; Hunter, M.I.; Shin, J.-H.; Jeong, J.-W. Leptin Stimulates Endometriosis Development in Mouse Models. *Biomedicines* **2022**, *10*, 2160. [CrossRef]
128. Teh, W.-T.; McBain, J.; Rogers, P. What Is the Contribution of Embryo-Endometrial Asynchrony to Implantation Failure? *J. Assist. Reprod. Genet.* **2016**, *33*, 1419–1430. [CrossRef]
129. Curtis, G.H.; Reeve, R.E.; Crespi, E.J. Leptin Signaling Promotes Blood Vessel Formation in the *Xenopus* Tail during the Embryo-Larval Transition. *Dev. Biol.* **2024**, *512*, 26–34. [CrossRef]

130. Zenclussen, A.C.; Hämmerling, G.J. Cellular Regulation of the Uterine Microenvironment That Enables Embryo Implantation. *Front. Immunol.* **2015**, *6*, 321. [CrossRef]
131. Obradovic, M.; Sudar-Milovanovic, E.; Soskic, S.; Essack, M.; Arya, S.; Stewart, A.J.; Gojobori, T.; Isenovic, E.R. Leptin and Obesity: Role and Clinical Implication. *Front. Endocrinol.* **2021**, *12*, 585887. [CrossRef]
132. Sharma, Y.; Galvão, A.M. Maternal Obesity and Ovarian Failure: Is Leptin the Culprit? *Anim. Reprod.* **2023**, *19*, e20230007. [CrossRef]
133. Muhammad, T.; Wan, Y.; Lv, Y.; Li, H.; Naushad, W.; Chan, W.-Y.; Lu, G.; Chen, Z.-J.; Liu, H. Maternal Obesity: A Potential Disruptor of Female Fertility and Current Interventions to Reduce Associated Risks. *Obes. Rev.* **2023**, *24*, e13603. [CrossRef]
134. Santoro, A.; Mattace Raso, G.; Meli, R. Drug Targeting of Leptin Resistance. *Life Sci.* **2015**, *140*, 64–74. [CrossRef] [PubMed]
135. Correia, M.L.G.; Haynes, W.G. Lessons from Leptin's Molecular Biology: Potential Therapeutic Actions of Recombinant Leptin and Leptin-Related Compounds. *Mini Rev. Med. Chem.* **2007**, *7*, 31–38. [CrossRef]
136. Jensterle, M.; Janez, A.; Fliers, E.; DeVries, J.H.; Vrtacnik-Bokal, E.; Siegelaar, S.E. The Role of Glucagon-like Peptide-1 in Reproduction: From Physiology to Therapeutic Perspective. *Hum. Reprod. Update* **2019**, *25*, 504–517. [CrossRef] [PubMed]
137. Sáinz, N.; González-Navarro, C.J.; Martínez, J.A.; Moreno-Aliaga, M.J. Leptin Signaling as a Therapeutic Target of Obesity. *Expert. Opin. Ther. Targets* **2015**, *19*, 893–909. [CrossRef] [PubMed]
138. Salazar, J.; Chávez-Castillo, M.; Rojas, J.; Ortega, A.; Nava, M.; Pérez, J.; Rojas, M.; Espinoza, C.; Chacin, M.; Herazo, Y.; et al. Is "Leptin Resistance" Another Key Resistance to Manage Type 2 Diabetes? *Curr. Diabetes Rev.* **2020**, *16*, 733–749. [CrossRef]
139. Su, H.; Jiang, L.; Carter-Su, C.; Rui, L. Glucose Enhances Leptin Signaling through Modulation of AMPK Activity. *PLoS ONE* **2012**, *7*, e31636. [CrossRef]
140. Pedroso, J.A.B.; Silveira, M.A.; Lima, L.B.; Furigo, I.C.; Zampieri, T.T.; Ramos-Lobo, A.M.; Buonfiglio, D.C.; Teixeira, P.D.S.; Frazão, R.; Donato, J., Jr. Changes in Leptin Signaling by SOCS3 Modulate Fasting-Induced Hyperphagia and Weight Regain in Mice. *Endocrinology* **2016**, *157*, 3901–3914. [CrossRef]
141. Pérez-Pérez, A.; Sánchez-Jiménez, F.; Vilariño-García, T.; Sánchez-Margalet, V. Role of Leptin in Inflammation and Vice Versa. *Int. J. Mol. Sci.* **2020**, *21*, 5887. [CrossRef]
142. Cerdó, T.; García-Santos, J.A.; Bermúdez, M.G.; Campoy, C. The Role of Probiotics and Prebiotics in the Prevention and Treatment of Obesity. *Nutrients* **2019**, *11*, 635. [CrossRef]
143. Calder, P.C. Omega-3 Fatty Acids and Inflammatory Processes. *Nutrients* **2010**, *2*, 355. [CrossRef]
144. Franco, J.G.; Dias-Rocha, C.P.; Fernandes, T.P.; Albuquerque Maia, L.; Lisboa, P.C.; Moura, E.G.; Pazos-Moura, C.C.; Trevenzoli, I.H. Resveratrol Treatment Rescues Hyperleptinemia and Improves Hypothalamic Leptin Signaling Programmed by Maternal High-Fat Diet in Rats. *Eur. J. Nutr.* **2016**, *55*, 601–610. [CrossRef] [PubMed]
145. Yan, C.; Yang, Y.; Saito, K.; Xu, P.; Wang, C.; Hinton, A.O., Jr.; Yan, X.; Wu, Q.; Tong, Q.; Elmquist, J.K.; et al. Meta-Chlorophenylpiperazine Enhances Leptin Sensitivity in Diet-Induced Obese Mice. *Br. J. Pharmacol.* **2015**, *172*, 3510–3521. [CrossRef] [PubMed]
146. Kim, Y.-W.; Kim, J.-Y.; Park, Y.-H.; Park, S.-Y.; Won, K.-C.; Choi, K.-H.; Huh, J.-Y.; Moon, K.-H. Metformin Restores Leptin Sensitivity in High-Fat–Fed Obese Rats With Leptin Resistance. *Diabetes* **2006**, *55*, 716–724. [CrossRef] [PubMed]
147. Choi, Y.-J.; Park, S.-Y.; Kim, J.-Y.; Won, K.-C.; Kim, B.-R.; Son, J.-K.; Lee, S.-H.; Kim, Y.-W. Combined Treatment of Betulinic Acid, a PTP1B Inhibitor, with Orthosiphon Stamineus Extract Decreases Body Weight in High-Fat–Fed Mice. *J. Med. Food* **2013**, *16*, 2–8. [CrossRef]
148. Shapiro, A.; Cheng, K.-Y.; Gao, Y.; Seo, D.; Anton, S.; Carter, C.S.; Zhang, Y.; Tumer, N.; Scarpace, P.J. The Act of Voluntary Wheel Running Reverses Dietary Hyperphagia and Increases Leptin Signaling in Ventral Tegmental Area of Aged Obese Rats. *Gerontology* **2010**, *57*, 335–342. [CrossRef]
149. Kang, S.; Kim, K.B.; Shin, K.O. Exercise Training Improve Leptin Sensitivity in Peripheral Tissue of Obese Rats. *Biochem. Biophys. Res. Commun.* **2013**, *435*, 454–459. [CrossRef]

biomedicines

MDPI

Systematic Review

Impact of Endocrine Disrupting Pesticide Use on Obesity: A Systematic Review

Marcelino Pérez-Bermejo [1,*], Cristian Barrezueta-Aguilar [2], Javier Pérez-Murillo [1], Ignacio Ventura [3], María Ester Legidos-García [1], Francisco Tomás-Aguirre [1], Manuel Tejeda-Adell [1], Miriam Martínez-Peris [1], Belén Marí-Beltrán [1] and María Teresa Murillo-Llorente [1]

1 SONEV Research Group, School of Medicine and Health Sciences, Catholic University of Valencia San Vicente Mártir, C/Quevedo nº 2, 46001 Valencia, Spain; javierperezmurillo@mail.ucv.es (J.P.-M.); ester.legidos@ucv.es (M.E.L.-G.); paco.tomas@ucv.es (F.T.-A.); manuel.tejeda@ucv.es (M.T.-A.); miriam.martinez@ucv.es (M.M.-P.); belen.mari@mail.ucv.es (B.M.-B.); mt.murillo@ucv.es (M.T.M.-L.)

2 Department of Nutrition. School of Medicine and Health Sciences, Catholic University of Valencia San Vicente Mártir, C/Quevedo nº 2, 46001 Valencia, Spain; crisbaa@mail.ucv.es

3 Molecular and Mitochondrial Medicine Research Group, School of Medicine and Health Sciences, Catholic University of Valencia San Vicente Mártir, C/Quevedo nº 2, 46001 Valencia, Spain; ignacio.ventura@ucv.es

* Correspondence: marcelino.perez@ucv.es; Tel.: +34-620984639

Abstract: Background/Objectives: Endocrine disruptors are substances capable of altering the functions of the endocrine system. There is evidence that some pesticides can be endocrine disruptors and, among some of their effects, we find alterations in pubertal development and in the function of the thyroid gland, which could be related to a greater tendency of obesity. The aim was to evaluate the evidence from clinical and preclinical studies on the association between pesticides used in agriculture and found in plant-based foods with overweight/obesity. **Methods:** This is a systematic review of articles on the impact of the use of endocrine disrupting pesticides on obesity, conducted according to the PRISMA-2020 guidelines. **Results:** There was evidence that some pesticides, such as chlorpyrifos, pyrethroids, and neonicotinoids, may promote obesity and other anthropometric changes by altering lipid and glucose metabolism, modifying genes, or altering hormone levels such as leptin. Other studies suggest that perinatal exposure to chlorpyrifos or pesticides such as vinclozolin may alter lipid metabolism and promote weight gain in adulthood, whereas other pesticides such as boscalib, captan, thiacloprid, and ziram were not associated with changes in weight. Exposure to pesticides such as vinclozolin may be associated with a higher prevalence of overweight/obesity in later generations. **Conclusions:** The few studies that do not show these associations have methodological limitations in data collection with confounding variables. Further studies are needed to provide more and higher quality evidence to determine the true effect of these substances on obesity.

Keywords: pesticide; endocrine disruptor; overweight; obesity

Citation: Pérez-Bermejo, M.; Barrezueta-Aguilar, C.; Pérez-Murillo, J.; Ventura, I.; Legidos-García, M.E.; Tomás-Aguirre, F.; Tejeda-Adell, M.; Martínez-Peris, M.; Marí-Beltrán, B.; Murillo-Llorente, M.T. Impact of Endocrine Disrupting Pesticide Use on Obesity: A Systematic Review. *Biomedicines* **2024**, *12*, 2677. https://doi.org/10.3390/biomedicines12122677

Academic Editors: Teresa Vezza and Zaida Abad-Jiménez

Received: 26 October 2024
Revised: 19 November 2024
Accepted: 22 November 2024
Published: 24 November 2024

1. Introduction

Obesity is considered to be a worldwide pandemic that leads to an increase in medical costs and thus becomes a public health problem [1]. Initially, obesity was considered as an imbalance between energy intake and energy expenditure [2–4], sedentary lifestyle, and genetics [4], although the heritability remains a mystery [5] and the origins are multifactorial [6]. Currently, the obesity epidemic is also associated with the increased production of environmental chemicals [7], also called environmental obesogens [8], used mainly in agriculture, as disease vector control, helping to prevent harmful effects caused by fungi, bacteria, or even pests, using pesticides, insecticides, and herbicides [9], or endocrine disruptors (ED), which interfere in different processes.

An ED is an exogenous substance or mixture of substances that alters the functions of the endocrine system and consequently causes adverse health effects in an intact organism or its progeny [10]. EDs include natural and synthetic substances that are widely present in the environment (water, air, food, pesticides) [11]. These substances increase fat mass and participate in adipogenesis, even leading to metabolic disorders. For this reason, they are also called metabolic disruptors [12].

These chemicals that can act as EDs include polychlorinated biphenyls (PCBs), bisphenol A (BPA), and hexachlorobenzene (HCB) found in some pesticides such as pentatrichloronitrobenzene and pentachlorophenol, triclosan, polybrominated diphenyl ethers (PBDEs), dichlorodiphenyltrichloroethane (DDT), or chlordane. Dichlorodiphenyltrichloroethane, DDT degradation (DDE), and hexachlorobenzene (B-HCH) increase the likelihood of metabolic syndrome [13]. They also increase body mass index [14–16], total HDL and LDL cholesterol levels [17], and even abdominal waist index [15,18,19]; the latter adverse effect may also occur with excess malathion [20].

Not only agrochemicals can be obesogenic, but products such as fructose [21,22] and some antidiabetic drugs can behave as such [21]; alcohol, chemotherapeutic drugs, dietary fatty acids, and industrial chemicals are also metabolic chemical disruptors [22].

EDs have been implicated in a variety of adverse effects, including neurotoxicity [23], autism spectrum disorder and developmental delays in children after prenatal exposure [24], impaired behavior, learning, memory, attention, sensation and neurodevelopment [25], depression and anxiety in children [26], reduced fertility in women due to increased polycystic ovarian syndrome or premature ovarian failure, and even breast [27] or vaginal cancer [11,28]. EDs can also affect fertility in men [29], especially in the case of some pesticides [30]. They are also associated with endocrine disruption, causing thyroid dysfunction [31], affecting both iodine uptake and thyroid hormone metabolism [32], and interfering with insulin physiology [31], becoming a risk factor for the development of diabetes [27]. They are also associated with the alteration of corticoid function; EDs can act on glucocorticoid receptors, mineralocorticoids, and steroidogenic pathways. Finally, they are associated with induction of oxidative stress, the alteration of arachidonic acid metabolism, or porphyria [33].

Phytosanitary products are pesticides used to maintain the health of crops. This group includes herbicides, fungicides, insecticides, acaricides, plant growth regulators, and repellents [34]. According to the bibliography of the TEDX list (The Endocrine Disruption Exchange), 512 pesticides may have the capacity to alter the endocrine system [35] and finally, according to the European Commission, it is estimated that there are 162 phytochemicals with endocrine disrupting activity [36]. Thus, the enormous presence of EDs in agriculture and their potential health risk is evident.

Many regions of the planet are currently experiencing water scarcity. This is due to both biophysical factors and factors that increase the demand for water, resulting in an increase in water stress [37]. Wastewater has become an option as a non-conventional water source to alleviate the water demands of the agricultural sector. However, the discharge of untreated or inadequately treated wastewater could have harmful effects on human health, the environment, and economic activity [38]. In rich countries, 70% of wastewater is treated, while in upper-middle-income, lower-middle-income, and poor countries, this figure drops to 38%, 28%, and 8%, respectively [38]. It is estimated that about 10% of the world's irrigated area uses untreated wastewater [37]. Improper treatment prior to use as crop irrigation can introduce a range of contaminants, including EDs.

Pollution of irrigation water is not only caused by the reuse of wastewater, but also by the agricultural activity itself. Chemigation is a technique for the application of phytosanitary products that consists of incorporating these products dissolved in the irrigation water itself [38]. The problem arises when pesticides persist in crops and food and contaminate the environment (rivers, lakes, groundwater, soil. . .). Therefore, the presence of these EDs would indicate both the persistence of these substances and their current use despite their prohibition [39].

Although there are countries where these substances have been banned, levels can still be maintained both in the environment and in the body itself due to the excessively long half-life [40], and numerous efforts are being made to reduce the unhealthy effects of the different types of agrochemicals by implementing product designs that do not present high toxicity [9].

There is evidence that some pesticides have endocrine disrupting effects. These effects include alterations in pubertal development and thyroid function [41], which may be associated with increased susceptibility to obesity. If we add that there are reports of food contamination with these substances due to their use in agriculture [42–45], it is pertinent to know what effects their consumption could have on our health. Therefore, this systematic review aims to identify those pesticides present in foods of plant origin that could have obesogenic effects on the organism.

2. Materials and Methods

This systematic review was conducted according to the Preferred Reporting Items for Systematic Reviews and Meta-Analysis (PRISMA) criteria, which was registered in the records of the Open Science Framework (https://osf.io/hcpsg, accessed 26 October 2024). A literature search was conducted in the PubMed and Web of Science databases in April 2024 using the following terms: pesticide, 2,4-dichlorophenoxyacetic acid, cypermethrin, dithiocarbamate, malathion, bifenthrin, dimethoate, fipronil, lambda-cyhalothrin, chlorpyrifos, imidacloprid, pyraclostrobin, DDT, overweight, and obesity. All pesticides included in the search equation are considered endocrine disruptors as set out in the European Commission's working document defining criteria for the identification of endocrine disruptors in the context of the application of the Plant Protection Products Regulation and the Biocidal Products Regulation [36]. An initial screening was performed by reading titles and abstracts, followed by a review of the full text of eligible studies.

We selected experimental animal studies, in vitro studies, and human observational studies published since 2000 that investigated environmental or prenatal, perinatal, or postnatal dietary exposure to pesticides, in Spanish or English, and that examined the association of such exposure with overweight or obesity and other anthropometric changes, excluding non-original reports, review articles, conference abstracts, editorials, commentaries, and experimental or pilot studies.

The quality of evidence was assessed using the Agency for Healthcare Research and Quality (ARRQ) scale for human studies, the Systematic Review Center for Laboratory animal Experimentation (SYRCLE) scale for animal studies, and the ToxRTool for in vitro studies.

3. Results

As shown in the search diagram (Figure 1), 634 articles were retrieved: 196 from PubMed and 438 from Web of Science. After removing duplicates, 527 articles remained. There were 324 articles eliminated after title reading, leaving 203 for full text review. All articles that did not meet the inclusion criteria were excluded. Finally, 36 articles were included in this systematic review: 5 human cross-sectional studies, 24 animal studies, and 7 in vitro studies.

3.1. Human Studies

Table 1 shows the five studies selected for this review, all of which were cross-sectional. According to the AHRQ quality rating scale, the quality of evidence of the selected studies was moderate. The studies were conducted in the USA and Thailand. The pesticides evaluated in these studies were 2,4,5-dichlorophenol (2,4-D and 2,5-D), neonicotinoids (imidachlorid, acetamiprid, and clothianidin), glyphosate, diuron, chlorpyrifos, permethrin, mancozeb, maneb, carbendazim, thiophanate, and benomyl.

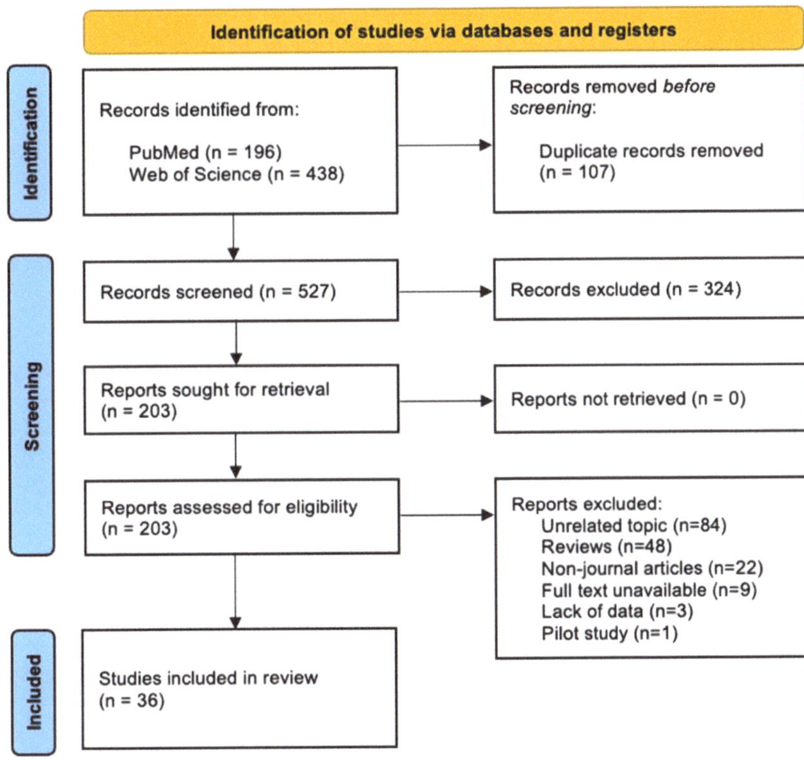

Figure 1. PRISMA flowchart of study selection process.

Table 1. Characteristics of the human studies included in the systematic review.

Author, Year, Country	Type of Study	Age/Sex/Sample	Agent/Source of Exposure	Assessment Parameters	Main Results	Quality Score
Buser MC et al. [46], 2014, USA	Cross-sectional study	6–19 years/ both/n = 1298	2,5-dichlorophenol, 2,4-dichlorophenol/ environmental	zBMI, waist circumference and obesity	Positive association between 2,5-D, 2,4-D with BMI, Hip Circumference, and obesity in children and adolescents.	Moderate (III)
Wei Y et al. [47], 2014, USA	Cross-sectional study	20–85 years/ Both/n = 2963	2,5-dichlorophenol, 2,4-dichlorophenol/ environmental	Urinary pesticide concentrations, BMI	Urinary concentrations of 2,5-D were associated with obesity, but there was no significant association with 2,4-D.	Moderate (III)
Twum C et al. [48], 2011, USA	Cross-sectional study	6–19 years/ Both/n = 6770	2,5-dichlorophenol, 2,4-dichlorophenol/ environmental	BMI	The prevalence of obesity was significantly associated with 2,5-D, but no such association was demonstrated with 2,4-D.	Moderate (III)
Godbole AM et al. [49], 2022, USA	Cross-sectional study	>19 years/ Both/n = 1675	Imidacloprid, acetamiprid, clothianidin/ environmental	BMI, % body fat	Acetamiprid was associated with decreased IMG, % fat, Waist Hip Index, and BMI. Imidacloprid was associated with increased rates of overweight/obesity.	Moderate (III)
Noppakun K et al. [50] 2021 Thailand	Cross-sectional study	≥20 years/ Both/n = 20,295	Insecticides, herbicides, fungicides, rodenticides, mollusci-cides/environmental	BMI, waist circumference	Of the 35 pesticides studied, 22 were associated with increased prevalence of obesity.	Moderate (III)

3.1.1. 2,4-Dichlorophenol and 2,5-Dichlorophenol

The relationship between urinary concentrations of 2,4-D and 2,5-D and weight gain and obesity was evaluated. These studies used data from participants in the National Health and Nutrition Examination Survey (NHANES) [46–48]. Two studies evaluated children and adolescents aged 6–19 years [46,48], while the other study evaluated the possible association in adults aged 20–85 years [47]. Participants with obesity were found to have higher urinary concentrations of 2,4-D and 2,5-D. Higher concentrations of these pesticides were associated with increased BMI and waist circumference [46–48]. As in children, the adult study showed a higher prevalence of obesity with higher urinary levels of 2,4-D and 2,5-D [47]. However, the study by Noppakun et al. found no significant association between this pesticide and obesity [50].

3.1.2. Neonicotinoids: Imidacloprid, Acetamiprid, and Clothianidin

An association was found between detectable levels of imidacloprid and an 11% higher prevalence of overweight/obesity and a positive association with lean mass index. However, the association was not statistically significant [49]. Similarly, another study did not find a statistically significant association either [50]. However, an inverse association was observed between acetamiprid and fat mass, fat percentage, waist circumference, and BMI. In addition, there was no association between clothianidin and an increase in any measure of adiposity [49].

3.1.3. Other Pesticides

Glyphosate, diuron, permethrin, mancozeb, and maneb were not significantly associated with obesity prevalence. However, carbendazim, thiophanate, benomyl, metalaxyl, propineb, and chlorpyrifos showed a statistically significant association with obesity prevalence [50].

3.2. Animal Studies

A total of 24 experimental animal studies were selected for this review (Table 2). All articles were experimental studies evaluating different pesticides and their effects. Of these 24 studies, the following 17 used mice: 13 studies used C57BL/6 mice (7 with the C57BL/6J substrain, 1 with the C5BL/6N substrain and 5 without specifying the substrain), 4 used ICR mice, 1 study used TR apoE3 mice, and 1 used BALB/c mice. The following 7 studies used rats: 3 with Wistar rats, 2 with Long-Evans rats, 1 with Sprague Dawley rats, and 1 did not specify the species.

Of the included studies, 13 used male animals, 4 used female animals, and 7 used mixed populations. The sample sizes ranged from 4 to 133 animals. The studied pesticides were deltamethrin, chlorothalonil, chlorpyrifos, vinclozolin, imidacloprid, bifenthrin, permethrin, lambda-cyhalothrin, malathion, difenoconazole, cyrochlorothalonil, cyrochlorothalonil, vinclozolin, difenoconazole, cyromazine, pirimicarb, chemoclamine, thiram, ziram, glyphosate, metolachlor, thiamethoxam, propamocarb, cypermethrin, boscalid, captan, and thiacloprid. The routes of exposure to these pesticides were gastroesophageal, perinatal, and subcutaneous. The most frequently investigated pesticide was the organophosphate pesticide chlorpyrifos (9 out of 24, 37.5%). For the remaining pesticides, only one study was found, except for imidacloprid, which was found in two studies. In this way, it was possible to observe whether there were changes in appetite, as this could contribute to weight gain and thus to overweight and obesity.

Finally, to evaluate the effect of pesticides on obesity, BMI, waist circumference, weight gain, and fat mass were the main measures used. In addition, several studies assessed biochemical parameters such as insulin levels, blood glucose, serum leptin, thyroid hormones, and lipid profile, as well as biomarkers and indicators of lipid metabolism. All studies were of moderate- to low-quality.

Table 2. Characteristics of the animal studies included in the systematic review.

Author, Year, Country	Species/ Age/Sex/ Sample	Compound/Dose/ Route of Administration/ Duration	Treatment	Control	Parameters of Evaluation	Main Results
Tsakiridis E et al. [51], 2023, Canada	Mice C57BL/6J/ 8 weeks/males/ N = 40	Deltamethrin/ 0; 0.10; 1.0 and 10 mg/kg/ orally/7 days	High- or normal-fat diet with Deltamethrin dose	Same diet without exposure to deltamethrin	Weight, fat mass, lean mass, glucose tolerance, and insulin tolerance	Deltamethrin inhibited UCP1 expression, but did not alter markers of thermogenesis or increase development of obesity and insulin resistance.
Meng Z et al. [52], 2023, China	Mice C57BL/6 and ICR/ 5 weeks/males/ N = 40	Chlorothalonil/ 0.2 mg/L/ oral route/12 weeks.	Ad lib feeding and hydration with chlorothalonil dissolved in water	The water was only deionized	TG, LDL, HDL, AST, ALT, glucose, metagenomic DNA from gut microbiota	Chlorothalonil may alter bile acid metabolism and lead to glycolipid metabolic disorders in the liver.
Wang B et al. [53], 2021, Canada	Mice C57BL/6J/ 7 weeks/males/ N = 10	Chlorpyrifos/ 0.5 mg/kg/ oral route/11 weeks.	High- and normal-fat diet and ad lib hydration with chlorpyrifos	Same diet with no exposure to chlorpyrifos	Weight, body composition, glucose and insulin tolerance, food thermogenesis, mitochondrion	Chlorpyrifos affects mitochondrial function and food thermogenesis promoting increased obesity.
Ben Maamar M et al. [54], 2020, USA	Rats Sprague Dawley/ 70–90 days/both	Vinclozolin/ F0 exposed at 100 mg/kg/day/ intraperitoneal route/6 days	Intraperitoneal injection during pregnancy	DMSO injection during pregnancy	BMI, abdominal adiposity, adipocyte size	There was a higher incidence of pathologies in F4, including obesity.
Sun Q et al. [55], 2017, USA	Mice C57BL/6J/ 5 weeks/females/ N = 4–7	Imidacloprid/ 0, 0.06, 0.6, and 6 mg/kg/day/ oral route/12 weeks.	High- or low-fat diet and hydration ad lib with imidacloprid	Same diet with no exposure to imidacloprid	Weight, insulin, glucose tolerance	Mice treated with imidacloprid significantly increased weight, adiposity, and insulin levels.
Wei C et al. [56], 2019, China	Mice C57BL/6/ 2 months/females/ N = 20	Bifenthrin/ 0.6 mg/kg/ oral route/6 weeks.	Fed corn oil with dissolved bifenthrin and standard ad lib diet	Fed on pure corn oil and standard diet ad libitum	Weight, fat mass, adipocyte size, protein expression	Bifenthrin treatment significantly increased body weight and fat mass.
Yang, D. et al. [57], 2023, China	Mice ICR/ 4–6 weeks/males/ N = 30	Lambda-cyhalothrin/ 0.4 and 2 mg/kg/ oral route/22 weeks.	Diet with low and high doses of lambda-cyhalothrin	Same diet without exposure to lambda-cyhalothrin	Indicators of lipid metabolism, lipid profile, weight	Results suggest that TBI may induce obesity, dyslipidemia, and hepatic steatosis.
Lassiter TL et al. [58], 2008, USA	Rats Long-Evans/ 20–100 days/both/ N = 20	Chlorpyrifos/ 1, 2.5, 4 mL/kg/ perinatal/gestation route	F0 fed with rat feed and water ad lib and chlorpyrifos dissolved in oil	Same diet with pure corn oil	Weight, height, volume, BMI, weight/volume ratio, leptin	Exposure to CPF caused weight gain in males.
Simoni-Berra MA et al. [59], 2023, Mexico	Mice BALB/c/ 4 weeks/males/ N = 20	Malathion/ 10 ppm/ oral route/180 days	Ad lib fed with malathion or malathion + probiotics	Same diet without exposure to malathion	Weight, glucose	Low doses of malathion induced an increase in weight and glucose levels.
Meggs WJ et al. [60], 2007, USA	Rats Long-Evans/ 6 months/females/ N = 20	Chlorpyrifos/ 5 mg/kg/day/ subcutaneous route/ 4 months	Daily injection of chlorpyrifos dissolved in DSMO	Pure DSMO injection	Body and organ weight	CPF-treated mice showed significant weight and fat gain.

Table 2. *Cont.*

Author, Year, Country	Species/ Age/Sex/ Sample	Compound/Dose/ Route of Administration/ Duration	Treatment	Control	Parameters of Evaluation	Main Results
Zhang H et al. [61], 2022, China	Mice C57BL/6/ 6 weeks/males/ N = 88	Difenoconazole/ 30 and 100 mg/kg/day/ oral route/28 day	Ad lib feeding with difenoconazole dissolved in corn oil	Same diet with pure corn oil	Lipid profile, intestinal permeability, microbiota, hepatic TG levels	Exposure to difenoconazole was associated with increased lipid accumulation in the liver, affected intestinal permeability, and microbiota.
Svingen T et al. [62], 2018, Denmark	Rats Wistar/ 5–6 months/both/ N= 70	Cyromazine, MCPB, pirimicarb, chemoclamine, thiram, ziram/ doses 5, 16 and 37.5%/ intrauterine route/ Pregnancy	F0 with ad lib feeding and pesticides dissolved in corn oil	Same diet with pure corn oil	Glucose tolerance, insulin, weight	Exposure to high doses was associated with gene alteration in adipose tissue. In males, some degree of weight regain was shown.
Xiao X et al. [63], 2018, USA	Mice C57BL/6J/ 3 weeks/males/ N = 4–8	Permethrin/ 50 µg/kg/d/day/ oral route/12 wk.	High-fat and low-fat diet with dissolved permethrin	Same diet without exposure to permethrin	Glucose, insulin, leptin, TG, cholesterol, weight, fat mass	Permethrin treatment significantly increased body weight and fat mass.
Fang B et al. [64], 2018, China	Rats/ 8 weeks/both/ N = 36	Chlorpyrifos/ 0.3 and 3 mg/kg/day/ oral route/9 weeks	High-fat and low-fat diet with different doses of chlorpyrifos	Same diet with pure DMSO	Weight, glucose, lipids, cytokines, and intestinal microbiome	Chronic exposure to chlorpyrifos was associated with the abundance of opportunistic pathogens and bacteria associated with obese and diabetic phenotypes.
Peris-Sampedro F et al. [65], 2015, Spain	Mice C57BL/6N and TR/ 7 months/males/ N = 40	Chlorpyrifos/ 2 mg/kg/day/ oral route/8 weeks.	Mouse feed ad lib supplemented with chlorpyrifos	Same diet without exposure to chlorpyrifos	Weight, diet, lipids, glucose, total cholesterol	There was a relationship between CPF and obesity in apoE3 mice, although this group is already vulnerable to developing obesity when treated with CPF.
Djekkoun N et al. [66], 2022, France	Rats Wistar/ 2 months/females/ N = 16	Chlorpyrifos/ 10 mg/kg/ oral route/5 days	Standard ad libitum diet in F0 with chlorpyrifos	Same diet without exposure to chlorpyrifos	Glycemia, lipid profile, microbiota	Exposure to chlorpyrifos with a high-fat diet induced dysmetabolism and an imbalance of gut microbiota.
GuibourdenchEM et al. [67], 2021, France	Rats Wistar/ 7 months/both/ N = 67	Chlorpyrifos/ 1 mg/kg/day/ oral route/16 weeks.	Administration of chlorpyrifos dissolved in oil and feeding ad lib	Administration of pure oil and same feeding	Weight, lipid profile, glucose, mRNA of proteins related to lipid and glucose metabolism	Maternal exposure to CPF + high-fat diet was associated with metabolic changes in the offspring and altered lipid and glucose metabolism.
Liang Y et al. [68], 2019, China	Mice C57BL/6 and CD1/ 3 weeks/males/ N = 40	Chlorpyrifos/ 5 mg/kg/day/ oral route/12 weeks.	High-fat and normal-fat diet with chlorpyrifos dissolved in corn oil	Same diet with pure corn oil	Glucose, insulin, lipopolysaccharides, weight, microbiota	Chlorpyrifos caused a breakdown of the intestinal barrier, increased lipid entry, a mild inflammatory state, and increased tendency to gain fat.
Guardia-Escote L et al. [69], 2020, Spain	Mice C57BL/6/ both/ N = 133	Chlorpyrifos/ 1 mg/kg/ oral route/5 days	Oral administration of chlorpyrifos with micropipette + high-fat diet	Same diet with placebo	DNA methylation, leptin, growth factor, weight	Postnatal exposure to CPF caused metabolic alterations in adulthood.

Table 2. *Cont.*

Author, Year, Country	Species/ Age/Sex/ Sample	Compound/Dose/ Route of Administration/ Duration	Treatment	Control	Parameters of Evaluation	Main Results
Yang D et al. [70], 2021, USA	Mice ICR/ 6–8 weeks/males/ N = 30	Thiamethoxam/ 4 and 20 mg/kg/ oral route/12 weeks	Ad lib oral feeding with oral administration of TMX	Same diet with phosphate-buffered saline in DSMO 1%	Lipid profile, glucose, tissue index	TMX exposure caused dyslipidemia and fatty liver disease.
Sun Q et al. [71], 2016, USA	Mice C57BL/6J/ 5 semester/males/ N = 30	Imidacloprid/ 0.07, 0.7, and 7 mg/kg/day/ oral route/12 weeks	High-fat and low-fat diet plus imidacloprid administration	Same diet without exposure to imidacloprid	Weight, glucose, insulin, leptin, lipid profile	Imidacloprid was associated with weight gain and adiposity.
Wu S et al. [72], 2018, China	Mice ICR/ 5 weeks/males/ N = 32	Propamocarb/ 0.5, 5, and 50 mg/kg/day/ oral route/4 weeks	Ad lib feeding with propamocarb dissolved in water	Same diet, but with deionized water.	Genes related to lipid metabolism, weight, hepatic TGs	Propamocarb exposure altered transcription of hepatic genes responsible for lipid regulation.
Jin Y et al. [73], 2014 China	Mice C57BL/6J/ 3 weeks/males/ N = 45	Cypermethrin/ 50 µg/kg/ oral route/20 weeks	High-energy diet with cypermethrin administration	Basal diet or high-energy diet without cypermethrin	Weight, lipid profile, hepatic TGs	There were no significant changes in weight. Hepatic lipid accumulation and TG content were significantly increased in the CYP-HFD group.
Smith, L. et al. [74], 2020, France	Mice C57BL/6J 8 weeks/both/ N = 30	Boscalid, captan, chlorpyrifos, thiacloprid, and ziram/ 0.25 mg/kg/day intrauterine/ gestational route	F0 ad lib feeding + administration of pesticide mixture	Same diet without pesticide exposure	Weight, blood glucose, adipose tissue, cholesterol, insulin, urinary, and fecal metabolomes	Perinatal pesticide exposure did not affect body weight or energy homeostasis in 6- and 14-week-old mice.

3.2.1. Organophosphates: Chlorpyrifos and Malathion

Chlorpyrifos exposure has been associated with increased body mass, adiposity, and impaired glucose tolerance and insulin sensitivity in male C57BL/6 mice [53,65,67–69], CD-1 mice [68], and rats [60] when combined with a high-fat diet. However, this effect was not observed in Wistar rats [64,67]. In addition, a difference was observed between mice with apoE2-TR, apoE3-TR, and apoE4-TR genotypes, with apoE3-TR [65] and apoE4-TR [69] mice showing a greater effect after pesticide exposure. However, this CPF-associated weight gain was not affected by diet type or was not significant [64,68].

Chlorpyrifos has been observed to alter the gut microbiota in C57BL/6 mice, CD-1 mice [68], and Wistar rats [64]. This altered microbiota could have a significant effect on weight gain, serum free lipopolysaccharide concentrations, and fasting glucose. Furthermore, this alteration could promote body weight gain, insulin resistance, and glucose intolerance. In addition, altered intestinal permeability and increased lipopolysaccharide levels were observed, which could induce chronic inflammation and promote the development of insulin resistance and obesity [68]. In addition, chlorpyrifos exposure may alter leptin levels, which could be associated with increased fat deposition, and a strong correlation was found between increased leptin levels and weight gain in CPF-exposed mice [65]. However, another study found that exposure reduced plasma leptin concentrations and had no effect on ghrelin levels [64]. CPF reduced diet-induced thermogenesis and was also observed to alter the synthesis of proteins related to lipid metabolism, with an increase in fat deposition in brown adipose tissue [53].

Similarly, perinatal exposure (during pregnancy) was associated with lower birth weight [66,67] and more rapid weight gain during puberty in males, and males had an altered relationship between body weight and leptin levels in adulthood [58]. Perinatal exposure increased gene expression of lipid metabolism regulatory proteins such as the Srebf1/Srebp1c gene (encoding a protein involved in the regulation of lipogenesis) and the Pparg gene (encoding a protein related to lipid storage) [67]. Malathion exposure has been shown to increase body weight and serum glucose levels in mice [59].

3.2.2. Pyrethroids: Bifenthrin, Permethrin, Lambda-Cyhalothrin, Deltamethrin, and Cypermethrin

The effect of pyrethroids on body weight change has been studied in C57BL/6 mice [51,56,63,72] and in male ICR mice [70]. Deltamethrin was shown to have no significant effect on body mass, adipose tissue (although it was associated with a reduction in fat mass), or insulin tolerance [51]. In contrast, mice exposed to bifenthrin [56] and permethrin [63] had significantly increased body weight, fat mass, and serum cholesterol levels. In addition, permethrin reduced GLUT4 gene expression in muscle [52], whereas the effects of bifenthrin were sex-specific [56].

Exposure to lambda-cyhalothrin [70] and cypermethrin [73] was associated with increased plasma concentrations of free fatty acids, and increased cholesterol was also associated with lambda-cyhalothrin, so it could induce overweight and hyperlipidemia. At the hepatic level, an increase in hepatic lipid accumulation and an increase in PPARγ protein levels (which promote hepatic lipid synthesis) were observed after exposure to lambda-cyhalothrin [70] and a decrease in PPARα (regulators of fatty acid oxidation) after treatment with permethrin [63]. Similarly, cypermethrin treatment was associated with increased hepatic triglyceride levels [73].

3.2.3. Neonicotinoids: Imidacloprid and Thiamethoxam

Exposure to imidacloprid combined with a high-fat diet was observed to cause significant increases in body weight and adipose tissue in male [71] and female [55] mice. Increased epididymal and retroperitoneal adipose tissue in males [71], increased omental adipose tissue in females [55], and increased adipocyte size were observed in both sexes and in males exposed to thiamethoxam. In contrast, thiamethoxam had no significant effect on body weight.

With respect to serum markers, exposure to imidacloprid [71] and thiamethoxam [57] increased levels of TG, free fatty acids, and cholesterol in male mice (in the case of thiamethoxam, there was only a significant change in LDL cholesterol). In females, there were higher TG levels but no significant effects on free fatty acid levels [55]. Imidacloprid exposure was associated with increased serum glucose levels in males and increased insulin and leptin levels in both sexes.

In adipocytes, imidacloprid exposure increased fat accumulation and altered lipid and glucose metabolism, suggesting that it may promote adipogenesis in both sexes. At the hepatic level, thiamethoxam induced tissue damage, fat accumulation, and steatosis. In turn, the effects of imidacloprid and thiamethoxam suggest that it may contribute to a reduction in fatty acid oxidation in the liver [55].

3.2.4. Chlorothalonil

In C57BL/6 and ICR mice treated with low doses of chlorothalonil, there was a significant increase in body weight and in the weight of liver tissue and white adipose tissue. In terms of serum parameters, ICR mice showed an increase in glucose and insulin levels and consequently in the insulin resistance index. In addition, an increase in hepatic lipid accumulation was observed, which was associated with liver damage, and there was an increase in adipocyte size in white adipose tissue [52].

Regarding the effects on the microbiota, an alteration in the intestinal microbiota, both in structure and composition, was observed in mice exposed to the pesticide. This change was associated with an alteration in bile acid metabolism. The study also shows how the change in microbiota may be causally related to obesity in chlorothalonil-treated mice [52].

3.2.5. Vinclozolin

Fetal exposure to vinclozolin in mice was not associated with significant changes. In the F4 litter of exposed mice (transgenerational exposure), a decrease in body weight and a higher prevalence of obesity were observed. Thus, it is suggested that exposure to this pesticide could promote transgenerational epigenetic management of various diseases (including obesity) and sperm epimutations [55].

3.2.6. Difenoconazole

Body weight was significantly decreased in mice exposed to difenoconazole. However, liver weight increased, and increased lipid deposition and hepatic TG levels were observed. At the serum level and glucose and cholesterol levels, both HDL and LDL, were reduced. Thus, it is suggested that it alters glucose and lipid metabolism. However, it is suggested that although it alters metabolism, it does not have an effect that contributes to obesity [61].

3.2.7. Propamocarb

Male mice exposed to propamocarb showed no significant changes in body weight, body weight gain, fat content, or changes in hepatic bile levels. In addition, it was observed that hepatic TGs were reduced, which was associated with an alteration in hepatic lipid metabolism due to propamocarb exposure. Exposure to high doses of propamocarb was associated with metabolic changes related to altered gut microbiota and microbial metabolites [72].

3.2.8. Effect of Pesticide Mixtures

Perinatal exposure to a combination of six pesticides (ziram, chlorpyrifos, thiacloprid, boscalid, thiophanate, and captan) showed no significant differences in body weight and glucose tolerance. An alteration in the activity of the gut microbiota was demonstrated, but this did not show an alteration in body weight gain, subcutaneous adipose tissue, cholesterol levels, or glucose tolerance [74]. The effect of perinatal exposure to a mixture of six other pesticides (cyromazine, MCPB, pirimicarb, quinoclamine, thiram, and ziram) was studied and the offspring showed lower birth weight and reduced insulin and glucagon production. However, in adulthood, these parameters were equal to those of the control group, suggesting a possible capacity for recovery. Finally, an alteration in leptin levels was observed in female rats [62].

3.3. In Vitro Studies

We found seven in vitro studies that evaluated the effects of pesticides on the anatomy and physiology of hepatocytes and adipocytes and how such effects may contribute to overweight and obesity (Table 3). The pesticides reviewed in these articles were β-cypermethrin, chlorpyrifos, endosulfan, cis-bifenthrin, fipronil, imidacloprid, quizalofop-p-ethyl, glyphosate, 2,4-D, isoxaflutole, dicamba, quizalofop, and propaquizafop. Only one article was found for each of the pesticides reviewed in the case of endosulfan, fipronil, quizalofop-p-ethyl, dicamba, quizalofop, and propaquizafop. Therefore, the level of evidence in these cases is very low. All articles were rated as reliable without restrictions according to the ToxRTool scale.

Table 3. Characteristics of the in vitro studies included in the systematic review.

Author, Year, Country	Compound/ Concentration/Time	Type of Tissue	Control	Main Results	Conclusions	Quality Index
He B et al. [75], 2020 China	β-cypermethrin/ 25, 50, and 100 μM/ 2, 4 and 8 days	3T3-L1 adipocytes	Treated with DMEM/10% FBS	Treatment with β-cypermethrin increased ROS levels, autophagy, and adipogenesis.	β-cypermethrin promotes adipogenesis.	16-CSR
Sun Q et al. [76], 2016, USA	Fipronil/ 0.1, 1, and 10 μM/ 8 days	3T3-L1 adipocytes	Treated with DMSO	Fipronil was associated with fat accumulation in 3T3-L1 adipocytes and lipogenesis.	Fipronil alters adipogenesis and increases lipid accumulation.	15-CSR
Park Y et al. [77], 2013, USA	Imidacloprid/ 10 and 20 μM/ 8 days	3T3-L1 adipocytes	Treated with DMSO	Imidacloprid treatment enhanced adipocyte lipid accumulation and lipogenesis regulators.	Imidacloprid could alter adipogenesis and increase fat accumulation.	15-CSR
Biserni M et al. [78], 2019, United Kingdom	Quizalofop-p-ethyl, glyphosate, 2,4-D, isoxaflutole, dicamba, quizalofop, propaquizafop/ 0.1, 1, 10, and 100 μM/ 8 days	3T3-L1 adipocytes	Treated with Dexamethasone	Only QpE showed a statistically significant TG accumulation. In the case of Isoxaflutole and dicamba, the effect was smaller.	The lipid-accumulating capacity of QpE suggests a possible obesogenic capacity.	18-CSR
LIM S et al. [79], 2016, Korea	Fenoxycarb, pyriproxyfen/ 5, 10, 25, 50, and 100 μM/	3T3-L1 adipocytes	Treated with DMEM/10% FBS	Fenoxycarb stimulated PPARγ and FATP1 activity and expression in 3T3-L1 adipocytes. Pyriproxyfen increased lipid deposition to a lesser extent.	Fenoxycarb may promote lipid accumulation in adipocytes.	15-CSR
Xiang D et al. [80], 2018, China	Cis-bifenthrin/ 0.001, 0.01, 0.1, and 1 μM/ 24 h.	Hepatocytes	Treated with DMSO	HepG2 cells incubated with bifenthrin showed a significant accumulation of TG.	Pyrethroid-induced toxicity could alter lipid metabolism.	16-CSR
Blanco J et al. [81], 2020, Spain	Chlorpyrifos/ 25, 50, 100, and 200 μM/ 24 h.	3T3-L1 adipocytes	Treated with DMSO	Chlorpyrifos promotes adipogenesis by increasing the number of 3T3-L1 preadipocytes and improving their lipid storage capacity.	Chlorpyrifos may contribute to increased incidence of obesity.	17-CSR

3.3.1. Effect on Adipocytes

- β-Cypermethrin

It was observed that exposure of 3T3-L1 adipocytes to β-CYP for 2–4 days had no significant effect on these cells. However, cells exposed to β-CYP exhibited 20% higher levels of reactive oxygen species (ROS) and a reduction in mitochondrial membrane potential

(MMP) levels in differentiated adipocytes. These data suggest that the pesticide induces autophagy and adipogenesis by increasing oxidative stress [75].

- Chlorpyrifos

It was observed that exposure of 3T3-L1 preadipocytes to CPF induced a significant increase in the number of differentiated adipocytes and their internal lipid storage capacity. In addition, there was an increase in the expression of PPARγ and C/EBPα transcription factors involved in the function of adipogenesis [81].

- Fipronil

Treatment with the pesticide fipronil significantly increased triglyceride content in adipocytes. It also affected differentiation and lipid metabolism in adipocytes. A significant increase in proteins involved in lipid metabolism and storage was observed in the cytoplasm of adipocytes, resulting in increased lipogenesis and lipid accumulation [76].

- Imidacloprid

Adipocytes treated with imidacloprid showed an increased number of fat droplets compared to the control group. Triglyceride accumulation was increased by 91% and 116% at imidacloprid concentrations of 10 and 20 μM, respectively. Increases in molecular markers related to adipocyte differentiation (CCAAT) and enzymes responsible for lipogenesis (acetyl-CoA carboxylase and fatty acid synthase) were also observed. However, imidacloprid exposure did not affect lipid mobilization in 3T3-L1 adipocytes [77].

- Quizalofop-p-ethyl, glyphosate, 2,4-D, isoxaflutole, dicamba, quizalofop, and propaquizafop

Exposure to QpE for eight days promoted lipid accumulation in adipocytes. However, exposure to QpE metabolites was not able to induce lipid accumulation. Considering that QpE is rapidly metabolized after ingestion, this could indicate that they would not be able to induce lipid accumulation after ingestion. However, in the same study, the commercial formulation of QpE was found to be more potent than the active ingredient. Other pesticides were evaluated in the same study. Treatment with glyphosate, propaquizafop, and 2,4-D showed no effect on adipocytes, whereas dicamba and isoxaflutole increased lipid accumulation [78].

- Fenoxycarb and Pyriproxyfen

It was observed that exposure to fenoxycarb increased lipid accumulation during 3T3-L1 adipocyte differentiation. It also reduced adipocyte viability at high concentrations. An increase in the expression of PPARγ and FATP1, which are involved in fat transport in adipocytes, was induced. In the same study, the effect of pyriproxyfen was evaluated and it was observed that it also induced an increase in lipid deposition, although to a lesser extent than fenoxycarb [79].

3.3.2. Effect on Hepatocytes

- Cis-Hifenthrin

HepG2 cells were exposed to cis-bifenthrin for 24 h and an increase in intracellular triglyceride levels was observed in these cells. It has also been suggested that there may be an alteration in lipid metabolism by cis-bifenthrin due to the increased activation of PXR and PPARγ [80].

4. Discussion

In this systematic review on the obesogenic effects of agricultural pesticides, 36 articles were collected, including observational studies in humans and experimental studies in animals and in vitro. The collected articles evaluated changes and alterations in body weight and other anthropometric parameters, as well as metabolic changes that promote fat accumulation and adipogenesis.

The results obtained show that some pesticides could promote the development of obesity, the most studied being chlorpyrifos. However, contradictory results have been obtained in some cases [50,72,74,75], suggesting the possibility of a potential obesogenic effect of some pesticides, but without obtaining significant results.

In some animal studies, because of the methodology developed, the quality of the evidence on the obesogenic potential of some pesticides is still too low to extrapolate the risk to humans. Data have been obtained on the effect of these pesticides in promoting obesity through exposure during pregnancy or following exposure of parents or other progenitors, in addition to their own consumption [54,62,66,74].

Despite the great diversity of results, it is evident that some pesticides might promote obesity by altering lipid and glucose metabolism, modifying genes, or altering hormone levels such as leptin. Notably, some pesticides showed no effect or even a reduction in weight gain and lipid levels, such as propamocarb [72] and difenoconazole [61].

The endocrine-disrupting effects of many pesticides, such as altering estrogenic, androgenic, and thyroid functions, have been observed [4,12,28]. These alterations could have significant metabolic effects, potentially leading to weight gain and early pubertal development, which is associated with increased obesity risk in adulthood [82].

The review highlights several limitations, including the moderate to low level of evidence, the predominance of animal and in vitro studies, and the reliance on cross-sectional observational studies in humans. There are several important limitations to extrapolating results from animal and laboratory studies to humans. These include biological differences, the discrepancy between the controlled conditions of experiments, and the complexity of human reality, genetic variability, and differences in dose and exposure. In addition, ethical issues play a critical role. Although it is essential to complement animal and laboratory studies with human clinical trials to validate the results and ensure their applicability and safety, this is not feasible in this specific case, so human studies are mainly observational. The diversity of samples studied also limits the ability to draw clear conclusions. Additionally, most studies focus on individual pesticides, whereas the combined effects of multiple pesticides, which are commonly found in food, remain under-researched [43,44].

In this review, only two studies have been found that study the effect on health of a mixture of pesticides [62,74] and, in both, the data do not show an effect on obesity. However, more research is needed in this aspect, since there is the possibility of an additive or synergistic action of pesticides. This is also the point of the PAN report, which mentions that the individual effect of pesticides or mixed with other pesticides is not considered [43].

The need for more research on endocrine disrupting pesticides is well documented. The Endocrine Society and the International Pollutant Elimination Network (IPEN) [83] have highlighted the profound threats to human health posed by endocrine disrupting chemicals (EDCs), including pesticides, and have called for urgent action and further research to understand their full impact. Various agencies such as the Environmental Protection Agency (EPA) [84], which has re-established its Endocrine Disruptor Screening Program, and the National Institutes of Health (NIH) [85] have actively discussed new research focused on interventions to reduce exposure to endocrine disruptors, highlighting the continued need for research and public input.

In addition, there is a growing call to address misleading claims about the safety of endocrine disrupting pesticides, such as the European Food Safety Authority's (EFSA) guidance on the identification of endocrine disrupting substances in pesticides [86] or Pesticide Facts' claims that current scientific evidence shows that pesticides are not associated with endocrine disruption at relevant levels of exposure [87]. The aforementioned comprehensive report by The Endocrine Society and IPEN [83] calls for urgent action to address these threats. The World Health Organization (WHO) has also recognized EDCs as a public health priority, emphasizing the need for improved regulatory strategies and public education to protect vulnerable populations. The Endocrine Society's position statement further emphasizes the importance of public disclosure to ensure that people can make informed decisions and be protected from the adverse effects of EDCs [88].

5. Conclusions

This systematic review shows the association between pesticide use and a greater tendency to be overweight and obese. While some studies suggest this association, clear and conclusive evidence is lacking. Pesticides such as chlorpyrifos, pyrethroids, and neonicotinoids were associated with weight gain and other anthropometric changes, while others such as boscalib and captan were not.

The findings underscore the urgent need for policy action, further research, and public awareness. From a policy perspective, the findings underscore the need to review regulations governing pesticide use and health effects. Government agencies and regulators need to consider these findings to develop stronger regulations and promote good agricultural practices that protect both the environment and human health.

From a research perspective, there is an urgent need for more studies to obtain reliable results on the relationship between pesticide use and obesity. Long-term exposure effects and the impact of lifestyle factors, such as high-fat diets, on the obesogenic potential of pesticides need to be investigated.

Public awareness is essential to drive change. Educating the public about the potential health risks associated with pesticide exposure can lead to more informed choices and increased demand for safer agricultural practices. This awareness can also put pressure on policymakers to implement stricter regulations and support further research.

Author Contributions: Conceptualization, M.P.-B., M.T.M.-L. and C.B.-A.; methodology, M.P.-B., M.T.M.-L., M.E.L.-G. and M.T.-A.; investigation, M.P.-B., C.B.-A., M.E.L.-G., M.M.-P., J.P.-M., B.M.-B., I.V., F.T.-A. and M.T.M.-L.; data curation, M.P.-B., M.T.M.-L. and C.B.-A.; writing—original draft preparation, M.P.-B., C.B.-A., M.E.L.-G., M.M.-P., J.P.-M., M.T.-A., I.V., F.T.-A., B.M.-B. and M.T.M.-L.; writing—review and editing, M.P.-B., C.B.-A., M.E.L.-G., M.M.-P., J.P.-M., M.T.-A., I.V., F.T.-A., B.M.-B. and M.T.M.-L.; supervision, M.P.-B. and M.T.M.-L. All authors have read and agreed to the published version of the manuscript.

Funding: This research and the APC were funded by the Consellería de Innovación, Universidades, Ciencia y Sociedad Digital de la Generalitat Valenciana, grant number CIGE/2022/168.

Data Availability Statement: No new data were created or analyzed in this study.

Conflicts of Interest: The authors declare no conflicts of interest.

References

1. Hales, C.M.; Carroll, M.D.; Fryar, C.D.; Ogden, C.L. *Prevalence of Obesity and Severe Obesity Among Adults: United States, 2017–2018*; NCHS Data Brief. No. 360; Centers for Disease Control and Prevention, National Center for Health Statistics: Atlanta, GA, USA, 2020; pp. 1–8. [PubMed]
2. Hall, K.D.; Heymsfield, S.B.; Kemnitz, J.W.; Klein, S.; Schoeller, D.A.; Speakman, J.R. Energy balance and its components: Implications for body weight regulation. *Am. J. Clin. Nutr.* **2012**, *95*, 989–994. [CrossRef] [PubMed]
3. Bosy-Westphal, A. Regulation of energy balance-classical concepts and novel insights. *Eur. J. Clin. Nutr.* **2017**, *71*, 293. [CrossRef] [PubMed]
4. Veiga-Lopez, A.; Pu, Y.; Gingrich, J.; Padmanabhan, V. Obesogenic Endocrine Disrupting Chemicals: Identifying Knowledge Gaps. *Trends Endocrinol. Metab.* **2018**, *29*, 607–625. [CrossRef] [PubMed]
5. Waalen, J. The genetics of human obesity. *Transl. Res.* **2014**, *164*, 293–301. [CrossRef]
6. Heindel, J.J.; Blumberg, B. Environmental Obesogens: Mechanisms and Controversies. *Annu. Rev. Pharmacol. Toxicol.* **2019**, *59*, 89–106. [CrossRef]
7. Snedeker, S.M.; Hay, A.G. Do interactions between gut ecology and environmental chemicals contribute to obesity and diabetes? *Environ. Health Perspect.* **2012**, *120*, 332–339. [CrossRef]
8. Martínez-Esquivel, A.; Trujillo-Silva, D.J.; Cilia-López, V.G. Impact of environmental pollution on the obesogenic environment. *Nutr. Rev.* **2022**, *80*, 1787–1799. [CrossRef]
9. Sparks, T.C.; Lorsbach, B.A. Perspectives on the agrochemical industry and agrochemical discovery. *Pest. Manag. Sci.* **2017**, *73*, 672–677. [CrossRef]
10. International Programme on Chemical Safety. Global Assessment on the State of the Science of Endocrine Disruptors [Internet]. Report No.: WHO/PCS/EDC/02.2. World Health Organization. Available online: https://apps.who.int/iris/handle/10665/67357 (accessed on 15 May 2024).

11. Anne, B.; Raphael, R. Endocrine Disruptor Chemicals. In *Endotext [Internet]*; Feingold, K.R., Anawalt, B., Blackman, M.R., Eds.; MDText.com, Inc.: South Dartmouth, MA, USA, 2000. Available online: http://www.ncbi.nlm.nih.gov/books/NBK569327/ (accessed on 15 May 2024).
12. Heindel, J.J.; Blumberg, B.; Cave, M.; Machtinger, R.; Mantovani, A.; Mendez, M.A.; Nadal, A.; Palanza, P.; Panzica, G.; Sargis, R.; et al. Metabolism disrupting chemicals and metabolic disorders. *Reprod. Toxicol.* **2017**, *68*, 3–33. [CrossRef]
13. Dusanov, S.; Ruzzin, J.; Kiviranta, H.; Klemsdal, T.; Retterstøl, L.; Rantakokko, P.; Airaksinen, R.; Djurovic, S.; Tonstad, S. Associations between persistent organic pollutants and metabolic syndrome in morbidly obese individuals. *Nutr. Metab. Cardiovasc. Dis.* **2018**, *28*, 735–742. [CrossRef]
14. La Merrill, M.A.; Lind, P.M.; Salihovic, S.; van Bavel, B.; Lind, L. The association between p,p′-DDE levels and left ventricular mass is mainly mediated by obesity. *Environ. Res.* **2018**, *160*, 541–546. [CrossRef] [PubMed]
15. Dirinck, E.; Jorens, P.G.; Covaci, A.; Geens, T.; Roosens, L.; Neels, H.; Mertens, I.; van Gaal, L. Obesity and persistent organic pollutants: Possible obesogenic effect of organochlorine pesticides and polychlorinated biphenyls. *Obesity* **2011**, *19*, 709–714. [CrossRef] [PubMed]
16. Bachelet, D.; Truong, T.; Verner, M.-A.; Arveux, P.; Kerbrat, P.; Charlier, C.; Guihenneuc-Jouyaux, C.; Guénel, P. Determinants of serum concentrations of 1,1-dichloro-2,2-bis(p-chlorophenyl)ethylene and polychlorinated biphenyls among French women in the CECILE study. *Environ. Res.* **2011**, *111*, 861–870. [CrossRef] [PubMed]
17. Arrebola-Moreno, A.; Marfil-Alvarez, R.; Catena, A.; García-Retamero, R.; Arrebola, J.; Melgares-Moreno, R.; Ramirez-Hernández, J.; Kaski, J. Body mass index and myocardium at risk in patients with acute coronary syndrome. *Rev. Clin. Esp.* **2014**, *214*, 113–120. [CrossRef]
18. Lee, D.H.; Lind, L.; Jacobs, D.R., Jr.; Salihovic, S.; van Bavel, B.; Lind, P.M. Associations of persistent organic pollutants with abdominal obesity in the elderly: The Prospective Investigation of the Vasculature in Uppsala Seniors (PIVUS) study. *Environ. Int.* **2012**, *40*, 170–178. [CrossRef]
19. Lee, D.H.; Lind, P.M.; Jacobs, D.R., Jr.; Salihovic, S.; van Bavel, B.; Lind, L. Background exposure to persistent organic pollutants predicts stroke in the elderly. *Environ. Int.* **2012**, *47*, 115–120. [CrossRef]
20. Raafat, N.; Abass, M.A.; Salem, H.M. Malathion exposure and insulin resistance among a group of farmers in Al-Sharkia governorate. *Clin. Biochem.* **2012**, *45*, 1591–1595. [CrossRef]
21. Janesick, A.S.; Blumberg, B. Obesogens: An emerging threat to public health. *Am. J. Obstet. Gynecol.* **2016**, *214*, 559–565. [CrossRef]
22. Joshi-Barve, S.; Kirpich, I.; Cave, M.C.; Marsano, L.S.; McClain, C.J. Alcoholic, Nonalcoholic, and Toxicant-Associated Steatohepatitis: Mechanistic Similarities and Differences. *Cell Mol. Gastroenterol. Hepatol.* **2015**, *1*, 356–367. [CrossRef]
23. Mostafalou, S.; Abdollahi, M. Pesticides: An update of human exposure and toxicity. *Arch. Toxicol.* **2017**, *91*, 549–599. [CrossRef]
24. Shelton, J.F.; Geraghty, E.M.; Tancredi, D.J.; Delwiche, L.D.; Schmidt, R.; Ritz, B.; Hansen, R.L.; Hertz-Picciotto, I. Neurodevelopmental disorders and prenatal residential proximity to agricultural pesticides: The CHARGE study. *Environ. Health Perspect.* **2014**, *122*, 1103–1109. [CrossRef] [PubMed]
25. Mellanen, P.; Petänen, T.; Lehtimäki, J.; Mäkelä, S.; Bylund, G.; Holmbom, B.; Mannila, E.; Oikari, A.; Santti, R. Wood-derived estrogens: Studies in vitro with breast cancer cell lines and in vivo in trout. *Toxicol. Appl. Pharmacol.* **1996**, *136*, 381–388. [CrossRef] [PubMed]
26. Harley, K.G.; Gunier, R.B.; Kogut, K.; Johnson, C.; Bradman, A.; Calafat, A.M.; Eskenazi, B. Prenatal and early childhood bisphenol A concentrations and behavior in school-aged children. *Environ. Res.* **2013**, *126*, 43–50. [CrossRef] [PubMed]
27. Costa, E.M.; Spritzer, P.M.; Hohl, A.; Bachega, T.A. Effects of endocrine disruptors in the development of the female reproductive tract. *Arq. Bras. Endocrinol. Metabol.* **2014**, *58*, 153–161. [CrossRef]
28. Pombo, M.; Castro-Feijóo, L. Endocrine disruptors. *J. Pediatr. Endocrinol. Metab.* **2005**, *18* (Suppl. 1), 1145–1155. [CrossRef]
29. Whitten, P.L.; Lewis, C.; Russell, E.; Naftolin, F. Phytoestrogen influences on the development of behavior and gonadotropin function. *Proc. Soc. Exp. Biol. Med.* **1995**, *208*, 82–86. [CrossRef]
30. Nicolopoulou-Stamati, P.; Pitsos, M.A. The impact of endocrine disrupters on the female reproductive system. *Hum. Reprod. Update* **2001**, *7*, 323–330. [CrossRef]
31. Kabir, E.R.; Rahman, M.S.; Rahman, I. A review on endocrine disruptors and their possible impacts on human health. *Environ. Toxicol. Pharmacol.* **2015**, *40*, 241–258. [CrossRef]
32. Patrick, L. Thyroid disruption: Mechanism and clinical implications in human health. *Altern. Med. Rev.* **2009**, *14*, 326–346, Erratum in: *Altern Med Rev.* **2010**, *15*, 58. [PubMed]
33. Lelli, S.M.; Ceballos, N.R.; Mazzetti, M.B.; Aldonatti, C.A.; San Martín de Viale, L.C. Hexachlorobenzene as hormonal disruptor—Studies about glucocorticoids: Their hepatic receptors, adrenal synthesis and plasma levels in relation to impaired gluconeogenesis. *Biochem. Pharmacol.* **2007**, *73*, 873–879. [CrossRef]
34. Plaguicidas | EFSA [Internet]. Available online: https://www.efsa.europa.eu/es/topics/topic/pesticides (accessed on 16 May 2024).
35. Search the TEDX List [Internet]. TEDX—The Endocrine Disruption Exchange. Available online: https://endocrinedisruption.org/interactive-tools/tedx-list-of-potential-endocrine-disruptors/search-the-tedx-list (accessed on 15 May 2024).

36. European Commission. Commission Staff Working Document. Impact Assessment. Defining Criteria for Identifying Endocrine Disruptors in the Context of the Implementation of the Plant Protection Products Regulation and Biocidal Products Regulation. Main Report. Accompanying the Document Communication from the Commission to the European Parliament and the Council on Endocrine Disruptors and the Draft Commission Acts Setting out Scientific Criteria for Their Determination in the Context of the EU Legislation on Plant Protection Products and Biocidal Products. Available online: https://eur-lex.europa.eu/legal-content/en/TXT/?uri=CELEX:52016SC0211 (accessed on 16 May 2024).
37. El Estado Mundial de la Agricultura y la Alimentación 2020 [Internet]. Available online: https://www.fao.org/3/cb1447es/online/cb1447es.html#chapter-2_2 (accessed on 15 May 2024).
38. Técnicas de Quimigación. Aplicación de Productos Fitosanitarios Mediante Riego por Goteo en Hortícolas Bajo Invernadero | SERVIFAPA—Plataforma de Asesoramiento y Transferencia del Conocimiento Agrario y Pesquero en Andalucía. Available online: https://www.juntadeandalucia.es/agriculturaypesca/ifapa/servifapa/registro-servifapa/e1b52be2-15f2-489a-bdcc-6b244fd94154 (accessed on 17 May 2024).
39. Ministerio para la Transición Ecológica y el Reto Demográfico. *Ríos Hormonados [Internet]*; Centro Nacional de Educación Ambiental: Madrid, Spain, 2018. Available online: https://www.miteco.gob.es/es/ceneam/recursos/pag-web/rios-hormonados.aspx (accessed on 16 May 2024).
40. Bornman, M.S.; Aneck-Hahn, N.H.; de Jager, C.; Wagenaar, G.M.; Bouwman, H.; Barnhoorn, I.E.; Patrick, S.M.; Vandenberg, L.N.; Kortenkamp, A.; Blumberg, B.; et al. Endocrine Disruptors and Health Effects in Africa: A Call for Action. *Environ. Health Perspect.* **2017**, *125*, 085005. [CrossRef] [PubMed]
41. Meléndez, I.F.S.; Almazán, R.C.; Alba, J.A.D.; García, H.M.D.; Larragoitia, J.C. Calidad del agua de riego en suelos agrícolas y cultivos del valle de san luis Potosí, México. *Rev. Int. Contam. Ambiental.* **2011**, *27*, 103–113.
42. OMS Residuos de Plaguicidas en los Alimentos [Internet]. Available online: https://www.who.int/es/news-room/fact-sheets/detail/pesticide-residues-in-food (accessed on 15 May 2024).
43. Endocrine Disrupting Pesticides in Food [Internet]. PAN Europe. Available online: https://www.pan-europe.info/resources/reports/2017/10/endocrine-disrupting-pesticides-food (accessed on 15 May 2024).
44. Ecologistas en Acción. Informe "Directo a tus Hormonas: Guía de Alimentos Disruptores". Available online: https://www.ecologistasenaccion.org/169891/informe-directo-a-tus-hormonas-guia-de-alimentos-disruptores/ (accessed on 16 May 2024).
45. Ecologistas en Acción. Informe "Residuos de Plaguicidas en Alimentos". Available online: https://www.ecologistasenaccion.org/207149/informe-residuos-de-plaguicidas-en-alimentos/ (accessed on 17 May 2024).
46. Buser, M.C.; Murray, H.E.; Scinicariello, F. Association of urinary phenols with increased body weight measures and obesity in children and adolescents. *J. Pediatr.* **2014**, *165*, 744–749. [CrossRef] [PubMed]
47. Wei, Y.; Zhu, J.; Nguyen, A. Urinary concentrations of dichlorophenol pesticides and obesity among adult participants in the U.S. National Health and Nutrition Examination Survey (NHANES) 2005–2008. *Int. J. Hyg. Environ. Health* **2014**, *217*, 294–299. [CrossRef]
48. Twum, C.; Wei, Y. The association between urinary concentrations of dichlorophenol pesticides and obesity in children. *Rev. Environ. Health* **2011**, *26*, 215–219. [CrossRef]
49. Godbole, A.M.; Moonie, S.; Coughenour, C.; Zhang, C.; Chen, A.; Vuong, A.M. Exploratory analysis of the associations between neonicotinoids and measures of adiposity among US adults: NHANES 2015–2016. *Chemosphere* **2022**, *300*, 134450. [CrossRef]
50. Noppakun, K.; Juntarawijit, C. Association between pesticide exposure and obesity: A cross-sectional study of 20,295 farmers in Thailand. *F1000Research* **2021**, *10*, 445. [CrossRef]
51. Tsakiridis, E.E.; Morrow, M.R.; Desjardins, E.M.; Wang, D.; Llanos, A.; Wang, B.; Wade, M.G.; Morrison, K.M.; Holloway, A.C.; Steinberg, G.R. Effects of the pesticide deltamethrin on high fat diet-induced obesity and insulin resistance in male mice. *Food Chem. Toxicol.* **2023**, *176*, 113763. [CrossRef]
52. Meng, Z.; Yan, S.; Sun, W.; Yan, J.; Teng, M.; Jia, M.; Tian, S.; Zhou, Z.; Zhu, W. Chlorothalonil induces obesity in mice by regulating host gut microbiota and bile acids metabolism via FXR pathways. *J. Hazard. Mater.* **2023**, *452*, 131310. [CrossRef]
53. Wang, B.; Tsakiridis, E.E.; Zhang, S.; Llanos, A.; Desjardins, E.M.; Yabut, J.M.; Green, A.E.; Day, E.A.; Smith, B.K.; Lally, J.S.V.; et al. The pesticide chlorpyrifos promotes obesity by inhibiting diet-induced thermogenesis in brown adipose tissue. *Nat. Commun.* **2021**, *12*, 5163. [CrossRef]
54. Ben Maamar, M.; King, S.E.; Nilsson, E.; Beck, D.; Skinner, M.K. Epigenetic transgenerational inheritance of parent-of-origin allelic transmission of outcross pathology and sperm epimutations. *Dev. Biol.* **2020**, *458*, 106–119. [CrossRef] [PubMed]
55. Sun, Q.; Qi, W.; Xiao, X.; Yang, S.-H.; Kim, D.; Yoon, K.S.; Clark, J.M.; Park, Y. Imidacloprid Promotes High Fat Diet-Induced Adiposity in Female C57BL/6J Mice and Enhances Adipogenesis in 3T3-L1 Adipocytes via the AMPKα-Mediated Pathway. *J. Agric. Food Chem.* **2017**, *65*, 6572–6581. [CrossRef] [PubMed]
56. Wei, C.; Wang, X.; Yao, X.; Xi, F.; He, Y.; Xu, Y.; Ma, L.; Chen, X.; Zhao, C.; Du, R.; et al. Bifenthrin Induces Fat Deposition by Improving Fatty Acid Uptake and Inhibiting Lipolysis in Mice. *J. Agric. Food Chem.* **2019**, *67*, 14048–14055. [CrossRef]
57. Yang, D.; Sun, X.; Wei, X.; Zhang, B.; Fan, X.; Du, H.; Zhu, R.; Oh, Y.; Gu, N. Lambda-cyhalothrin induces lipid accumulation in mouse liver is associated with AMPK inactivation. *Food Chem. Toxicol.* **2023**, *172*, 113563. [CrossRef]
58. Lassiter, T.L.; Brimijoin, S. Rats gain excess weight after developmental exposure to the organophosphorothionate pesticide, chlorpyrifos. *Neurotoxicol Teratol.* **2008**, *30*, 125–130. [CrossRef]

59. Simoni-Berra, M.A.; Yáñez-Santos, J.A.; Girón-Ortiz, J.A.; Huerta-Lara, M.; Cedillo-Ramírez, M.L. Effect of probiotics on glucose levels and weight gain in mice exposed to low doses of malathion. *Gac. Med. Mex.* **2023**, *159*, 44–48. [CrossRef]
60. Meggs, W.J.; Brewer, K.L. Weight gain associated with chronic exposure to chlorpyrifos in rats. *J. Med. Toxicol.* **2007**, *3*, 89–93. [CrossRef]
61. Zhang, H.; Yang, G.; Bao, Z.; Jin, Y.; Wang, J.; Chen, J.; Qian, M. Stereoselective effects of fungicide difenoconazole and its four stereoisomers on gut barrier, microbiota, and glucolipid metabolism in male mice. *Sci. Total Environ.* **2022**, *805*, 150454. [CrossRef]
62. Svingen, T.; Ramhøj, L.; Mandrup, K.; Christiansen, S.; Axelstad, M.; Vinggaard, A.M.; Hass, U. Effects on metabolic parameters in young rats born with low birth weight after exposure to a mixture of pesticides. *Sci. Rep.* **2018**, *8*, 305. [CrossRef]
63. Xiao, X.; Sun, Q.; Kim, Y.; Yang, S.-H.; Qi, W.; Kim, D.; Yoon, K.S.; Clark, J.M.; Park, Y. Exposure to permethrin promotes high fat diet-induced weight gain and insulin resistance in male C57BL/6J mice. *Food Chem. Toxicol.* **2018**, *111*, 405–416. [CrossRef]
64. Fang, B.; Li, J.W.; Zhang, M.; Ren, F.Z.; Pang, G.F. Chronic chlorpyrifos exposure elicits diet-specific effects on metabolism and the gut microbiome in rats. *Food Chem. Toxicol.* **2018**, *111*, 144–152. [CrossRef] [PubMed]
65. Peris-Sampedro, F.; Cabré, M.; Basaure, P.; Reverte, I.; Domingo, J.L.; Teresa Colomina, M. Adulthood dietary exposure to a common pesticide leads to an obese-like phenotype and a diabetic profile in apoE3 mice. *Environ. Res.* **2015**, *142*, 169–176. [CrossRef]
66. Djekkoun, N.; Depeint, F.; Guibourdenche, M.; Sabbouri, H.E.K.E.; Corona, A.; Rhazi, L.; Gay-Queheillard, J.; Rouabah, L.; Hamdad, F.; Bach, V.; et al. Chronic Perigestational Exposure to Chlorpyrifos Induces Perturbations in Gut Bacteria and Glucose and Lipid Markers in Female Rats and Their Offspring. *Toxics* **2022**, *10*, 138. [CrossRef]
67. Guibourdenche, M.; Sabbouri, H.E.K.E.; Bonnet, F.; Djekkoun, N.; Khorsi-Cauet, H.; Corona, A.; Guibourdenche, J.; Bach, V.; Anton, P.M.; Gay-Quéheillard, J. Perinatal exposure to chlorpyrifos and/or a high-fat diet is associated with liver damage in male rat offspring. *Cells Dev.* **2021**, *166*, 203678. [CrossRef]
68. Liang, Y.; Zhan, J.; Liu, D.; Luo, M.; Han, J.; Liu, X.; Liu, C.; Cheng, Z.; Zhou, Z.; Wang, P. Organophosphorus pesticide chlorpyrifos intake promotes obesity and insulin resistance through impacting gut and gut microbiota. *Microbiome* **2019**, *7*, 19. [CrossRef]
69. Guardia-Escote, L.; Blanco, J.; Basaure, P.; Biosca-Brull, J.; Verkaik-Schakel, R.N.; Cabré, M.; Peris-Sampedro, F.; Pérez-Fernández, C.; Sánchez-Santed, F.; Plösch, T.; et al. Sex and Exposure to Postnatal Chlorpyrifos Influence the Epigenetics of Feeding-Related Genes in a Transgenic APOE Mouse Model: Long-Term Implications on Body Weight after a High-Fat Diet. *Int. J. Environ. Res. Public Health* **2020**, *18*, 184. [CrossRef]
70. Yang, D.; Zhang, X.; Yue, L.; Hu, H.; Wei, X.; Guo, Q.; Zhang, B.; Fan, X.; Xin, Y.; Oh, Y.; et al. Thiamethoxam induces nonalcoholic fatty liver disease in mice via methionine metabolism disturb via nicotinamide N-methyltransferase overexpression. *Chemosphere* **2021**, *273*, 129727. [CrossRef]
71. Sun, Q.; Xiao, X.; Kim, Y.; Kim, D.; Yoon, K.S.; Clark, J.M.; Park, Y. Imidacloprid Promotes High Fat Diet-Induced Adiposity and Insulin Resistance in Male C57BL/6J Mice. *J. Agric. Food Chem.* **2016**, *64*, 9293–9306. [CrossRef]
72. Wu, S.; Jin, C.; Wang, Y.; Fu, Z.; Jin, Y. Exposure to the fungicide propamocarb causes gut microbiota dysbiosis and metabolic disorder in mice. *Environ. Pollut.* **2018**, *237*, 775–783. [CrossRef]
73. Jin, Y.; Lin, X.; Miao, W.; Wu, T.; Shen, H.; Chen, S.; Li, Y.; Pan, Q.; Fu, Z. Chronic exposure of mice to environmental endocrine-disrupting chemicals disturbs their energy metabolism. *Toxicol. Lett.* **2014**, *225*, 392–400. [CrossRef]
74. Smith, L.; Klément, W.; Dopavogui, L.; de Bock, F.; Lasserre, F.; Barretto, S.; Lukowicz, C.; Fougerat, A.; Polizzi, A.; Schaal, B.; et al. Perinatal exposure to a dietary pesticide cocktail does not increase susceptibility to high-fat diet-induced metabolic perturbations at adulthood but modifies urinary and fecal metabolic fingerprints in C57Bl6/J mice. *Environ. Int.* **2020**, *144*, 106010. [CrossRef] [PubMed]
75. He, B.; Wang, X.; Jin, X.; Xue, Z.; Zhu, J.; Wang, C.; Jin, Y.; Fu, Z. β-Cypermethrin promotes the adipogenesis of 3T3-L1 cells via inducing autophagy and shaping an adipogenesis-friendly microenvironment. *Acta Biochim. Biophys. Sin.* **2020**, *52*, 821–831. [CrossRef] [PubMed]
76. Sun, Q.; Qi, W.; Yang, J.J.; Yoon, K.S.; Clark, J.M.; Park, Y. Fipronil promotes adipogenesis via AMPKα-mediated pathway in 3T3-L1 adipocytes. *Food Chem. Toxicol.* **2016**, *92*, 217–223. [CrossRef] [PubMed]
77. Park, Y.; Kim, Y.; Kim, J.; Yoon, K.S.; Clark, J.; Lee, J.; Park, Y. Imidacloprid, a neonicotinoid insecticide, potentiates adipogenesis in 3T3-L1 adipocytes. *J. Agric. Food Chem.* **2013**, *61*, 255–259. [CrossRef]
78. Biserni, M.; Mesnage, R.; Ferro, R.; Wozniak, E.; Xenakis, T.; Mein, C.A.; Antoniou, M.N. Quizalofop-p-Ethyl Induces Adipogenesis in 3T3-L1 Adipocytes. *Toxicol. Sci.* **2019**, *170*, 452–461. [CrossRef]
79. Lim, S.; Choi, H.; Park, S.S.; Kim, M.; Kim, E. Fenoxycarb promotes adipogenesis in 3T3-L1 preadipocytes. *Entomol. Res.* **2016**, *46*, 80–84. [CrossRef]
80. Xiang, D.; Chu, T.; Li, M.; Wang, Q.; Zhu, G. Effects of pyrethroid pesticide cis-bifenthrin on lipogenesis in hepatic cell line. *Chemosphere* **2018**, *201*, 840–849. [CrossRef]
81. Blanco, J.; Guardia-Escote, L.; Mulero, M.; Basaure, P.; Biosca-Brull, J.; Cabré, M.; Colomina, M.T.; Domingo, J.L.; Sánchez, D.J. Obesogenic effects of chlorpyrifos and its metabolites during the differentiation of 3T3-L1 preadipocytes. *Food Chem. Toxicol.* **2020**, *137*, 111171. [CrossRef]
82. Prentice, P.; Viner, R.M. Pubertal timing and adult obesity and cardiometabolic risk in women and men: A systematic review and meta-analysis. *Int. J. Obes.* **2013**, *37*, 1036–1043. [CrossRef]

83. Gore, A.C.; La Merrill, M.A.; Patisaul, H.B.; Sargis, R. Endocrine Disrupting Chemicals: Threats to Human Health. The Endocrine Society and IPEN. February 2024. Available online: https://ipen.org/sites/default/files/documents/edc_report-2024-final-compressed.pdf (accessed on 17 May 2024).
84. Environmental Protection Agency (EPA). Endocrine Disruptor Screening Program. Available online: https://www.epa.gov/endocrine-disruption (accessed on 6 September 2024).
85. The National Institutes of Health (NIH). The Health Effects of Endocrine-Disrupting Chemicals. Available online: https://irp.nih.gov/catalyst/32/4/the-health-effects-of-endocrine-disrupting-chemicals (accessed on 6 September 2024).
86. European Chemical Agency (ECHA) and European Food Safety Authority (EFSA) with the technical support of the Joint Research Centre (JRC); Andersson, N.; Arena, M.; Auteri, D.; Barmaz, S.; Grignard, E.; Kienzler, A.; Lepper, P.; Lostia, A.M.; Munn, S.; et al. Guidance for the identification of endocrine disruptors in the context of Regulations (EU) No 528/2012 and (EC) No 1107/2009. *EFSA J.* **2018**, *16*, e05311. [CrossRef]
87. World Health Organization. Reducing Health Risks from Endocrine-Disrupting Chemicals. Available online: https://www.who.int/europe/health-topics/chemical-safety/reducing-health-risks-from-endocrine-disrupting-chemicals#tab=tab_1 (accessed on 7 September 2024).
88. Pesticide Fasts. Endocrine-Related Diseases & Pesticides. Available online: https://pesticidefacts.org/topics/pesticide-testing-and-regulation/endocrine-related-diseases-pesticides (accessed on 7 September 2024).

MDPI AG
Grosspeteranlage 5
4052 Basel
Switzerland
Tel.: +41 61 683 77 34

Biomedicines Editorial Office
E-mail: biomedicines@mdpi.com
www.mdpi.com/journal/biomedicines

MDPI

www.ingramcontent.com/pod-product-compliance
Lightning Source LLC
Chambersburg PA
CBHW041353210926
51850CB00013B/3065